The Scottish Café, birthplace of the Scottish Book problems, as it appeared in a postcard from the early 1970s. The Scottish Café is on the right, with the Café Roma on the left. According to Stanisław Ulam, this scene has changed little from the period preceding World War II

R. Daniel Mauldin

The Scottish Book

Mathematics from the Scottish Café

with Selected Problems from The New Scottish Book

Second Edition

 Birkhäuser

R. Daniel Mauldin
Department of Mathematics
University of North Texas
San Diego, CA, USA

ISBN 978-3-319-22896-9   ISBN 978-3-319-22897-6   (eBook)
DOI 10.1007/978-3-319-22897-6

Library of Congress Control Number: 2015947275

Mathematics Subject Classification (2010): 01A60.

Springer Cham Heidelberg New York Dordrecht London

Printed on acid-free paper

Springer International Publishing AG Switzerland is part of Springer Science+Business Media (www. springer.com)

Preface to the Second Edition

The Scottish Book, wrapped in the mists of the past, is a legend in the mathematics world. Its fascinating story and the legendary figures who formulated the problems in the book continue to hold our attention. It represents the best of café mathematics, an informal, free-wheeling style of mathematical conversation and interaction that seems almost lost today. For the solution to some of the problems, prizes were offered, ranging from the famous live goose, a bottle of whiskey of measure > 0, one kilo of bacon to one small beer.

One can imagine the atmosphere at the Scottish Café while the problems were being formulated. We should take to heart the playfulness and enjoyment that is on display, a good model for encouraging and stimulating mathematics at any age. One can also see that a few of the problems must have been stated after spending some time drinking tea or perhaps a brandy or two. In his book *Adventures of a Mathematician*, Ulam describes what a session (one lasting at least seventeen hours) at the café might be like and sketches some of the central figures. During the years I knew and worked with Ulam, he loved the café mode of discussion. Gian-Carlo Rota gives a detailed description of this in his article 'The Lost Café', in *From Cardinals to Chaos*, edited by N. G. Copper (Cambridge University Press, 1989). The entire school of mathematics in Lwów is wonderfully presented by Roman Duda in *Pearls From a Lost City* (American Mathematical Society, 2014), a translation of his 2008 Polish version. The tragedy that befell so many of the figures around the book has been traced many times, including the article by Joanna Diane Caytas, 'Survival of the Scottish Book: A Phoenix from the Holocaust of Polish Mathematics', available on the internet. There is even a collection of poems about the book by Susana H. Case, *The Scottish Café* (Slapering Hot press, 2002).

The problems and the ideas behind continue to effect mathematics today. In the 35 years since the first edition of the book, many more problems have been solved or partially solved. But even today, quite a few remain unsolved. In view of this, I decided to gather new commentaries and update some of the old ones. The appendices of this edition include a list of the unsolved and partially solved problems together with those that have no commentary, a list of unsolved prize problems, a list of problems posed by each author, and a list of problems by subject area. Besides correcting many errors in the first edition, for clarification a few changes in Ulam's translation of the original Scottish Book have been made.

In addition to some of the lectures given at the 1979 Scottish Book conference, this edition also includes a brief history of Wrocław's *New Scottish Book* and some selected problems from it.

This edition would not be possible without the generous contributions and suggestions of the commentators. I offer them my heartfelt gratitude. I want to thank Kirby Baker, Larry Lindsay, Bill Bernard, and Sue DeMeritt for their assistance in getting this project underway. I thank Al Hales, Joe Buhler, and Jan Mycielski for their counsel. I thank Allen Mann, Christopher Tominich, and Benjamin Levitt at Birkhäuser for their support during the preparation of this new edition. Finally, I thank Diana, my wife, for supporting me throughout this long project.

It is my sincere hope that this collection will bring to the reader, as it has to me, many hours of enjoyment and an image of what must have been a most wonderful place, The Scottish Café.

San Diego, CA, USA   R. Daniel Mauldin

Preface to the First Edition

Once while working on a problem, someone was kind enough to point out that my problem was in the "Scottish Book." I had not heard of the Scottish Book and certainly did not realize that this book had no connection with Scotland. But, since that introduction, I've become more and more aware of the magic of the mathematicians and the mathematics involved in its birth.

The Scottish Book offers a unique opportunity to communicate with the men (no women were on the scene, as I understand it) and ideas of a time and place, Lwów, Poland, which have had an enormous influence on the development of mathematics. The history of the Scottish Book as detailed in the following lectures by Ulam, Kac, and Zygmund provides amazing insights into the mathematical environment of Lwów before World War II.

There are many collections of problems, but this set has become world-renowned. Perhaps, a primary reason for this renown is that the problems are clearly and simply formulated, accessible to the general mathematical community, and yet strike at the heart of the concepts involved.

It is my pleasure and honor to edit this version of the Scottish Book, which includes a collection of some of the talks given at the "'Scottish Book Conference'" held at North Texas State University in May of 1979. The purpose of the conference was to examine the history, development, and influence of the Scottish Book. As John Oxtoby toasted at the conference, there was a "'condensation of Poles'" at this conference. Among them were some of the original contributors to the Scottish Book, Professors Ulam, Kac, and Zygmund. Their edited talks appear here, together with the talk given by one close to them in spirit and collaboration, Professor Paul Erdós. Also presented here is a talk by a member of a younger generation, Professor Andrzej Granas, in which one problem of Schauder is discussed with its many-faceted implications and connections. It should come as no surprise that the conference was held in Texas; the mathematical similarities between the Texas school and the Polish school have long been noted, beginning with the fact that the first American to publish in *Fundamenta Mathematicae* is R. L. Moore.

From a glance at the problems, one sees that they cover a wide range of mathematics. I think this simply reflects the wide interests of the unusual group, which assembled the collection. The problems are concentrated in the areas of summability theory, functional and real analysis, group theory, point set topology, measure theory, set theory, and probability. It is likewise easy to confirm that some of the contributors to the Book were, as R. H. Bing toasted, the "'leading lights'" in these fields.

I have attempted to obtain an appropriate commentary for each of the problems, although quite a few of the problems remain without comment. Some of these, as well as a number of problems with comments, remain unsolved to this day. For others I simply failed to get an appropriate expert comment (I would be grateful for contributions from readers of this edition).

Following a problem there may appear the word *Addendum*. This indicates a comment that was entered into the Scottish Book during the time when the problems were being collected in Lwów. Later commentaries, remarks, and solution to problems which are presented here for the first time follow the original addenda.

The problems and original addenda appear here essentially as they have in the two earlier English-language editions of the Scottish Book, both edited (and one produced) by Stanisław Ulam. The first, in 1957, was a mimeographed version of Ulam's own translation from the original languages in which the problems were inscribed in the Book (mostly Polish), which he distributed on personal request from his professional base at Los Alamos National Laboratory. By 1977, the volume of requests addressed to both Professor Ulam and the Los Alamos Laboratory's library made it only reasonable to prepare a somewhat more formal edition. This edition again presented only the translated problems and their contemporary addenda, and has been distributed by the Los Alamos laboratory since then. The recent reconcentration and expansion of interest in the Book, including the 1979 conference, had made a place for a new edition, including a collection of at least some of the work which has been stimulated by the Scottish Book problems in the years since they were first collected.

This project enjoyed the aid of several individuals and institutions, beginning with the encouragement of Stan Ulam and Gian-Carlo Rota. I sincerely thank all of the commentators for their generosity in providing the commentaries and suggestions. It is obvious that a major contributor to this edition is Jan Mycielski. His encouragement and constant flow of comments and references kept life in the project. Bill Beyer provided many significant comments on the formulation of the problems.

The Scottish Book conference, which was held in Denton in May 1979, focused our efforts. It was my hope that some of the spirit of that time and place would be recaptured at the conference. Perhaps, it was through the contributions of the speakers including R. D. Anderson and D. A. Martin, both of whom traced some of the most outstanding work in their fields back to the Scottish Book.

The National Science Foundation, through grant MCS-79-0971 and North Texas State University, provided funding for the conference. A number of commentaries were written under the auspices of a Faculty Research grant from North Texas State University. I sincerely thank Lynn Holick for the superb typing, and the people at Birkhäuser Boston for their help in bringing the project to fruition.

San Diego, CA, USA   R. Daniel Mauldin
1981

Preface to the Limited Los Alamos Edition of 1957

The enclosed collection of mathematical problems has its origin in a notebook, which was started in Lwów, in Poland in 1935. If I remember correctly, it was S. Banach who suggested keeping track of some of the problems occupying the group of mathematicians there. The mathematical life was very intense in Lwów. Some of us met practically every day, informally in small groups, at all times of the day to discuss problems of common interest, communicating to each other the latest work and results. Apart from the more official meetings of the local sections of the Mathematical Society (which took place Saturday evenings, almost every week!), there were frequent informal discussions mostly held in one of the coffee houses located near the University building - one of them a coffee house named "Roma," and the other "The Scottish Coffee House." This explains the name of the collection. A large notebook was purchased by Banach and deposited with the headwaiter of the Scottish Coffee House, who, upon demand, would bring it out of some secure hiding place, leave it at the table, and after the guests departed, return it to its secret location.

Many of the problems date from years before 1935. They were discussed a great deal among the persons whose names are included in the text, and then gradually inscribed into the "'book'" in ink. Most of the questions proposed were supposed to have had considerable attention devoted to them before an "'official'" inclusion into the "'book'" was considered. As the reader will see, this general rule could not guarantee against an occasional question to which the answer was quite simple or even trivial.

In several instances, the problems were solved, right on the spot or within a short time, and the answers were inscribed, perhaps some time after the first formulation of the problem under question.

As most readers will realize, the city of Lwów, and with it the "Scottish Book," was fated to have a very stormy history within a few years of the book's inception. A few weeks after the outbreak of World War II, the city was occupied by the Russians. From items at the end of this collection, it is seen that some Russian mathematicians must have visited the town; they left several problems (and prizes for their solutions). The last date figuring in the book is May 31, 1941. Item Number

193 contains a rather cryptic set of numerical results, signed by Steinhaus, dealing with the distribution of the number of matches in a box! After the start of war between Germany and Russia, the city was occupied by German troops that same summer and the inscriptions ceased.

The fate of the Scottish Book during the remaining years of war is not known to me. According to Steinhaus, this document was brought back to the city of Wrocław by Banach's son, now a physician in Poland. (Many of the surviving mathematicians from Lwów continued their work in Wrocłow. The tradition of the Scottish Book continues. Since 1945, new problems have been formulated and inscribed and a new volume is in progress.)

A general word of explanation may be in order here. I left Poland late in 1935 but, before the war, visited Lwów every summer in 1936, '37,'38, and '39. The last visit was during the summer preceding the outbreak of World War II, and I remember just a few days before I left Poland, around August 15, the conversation with Mazur on the likelihood of war. It seems that in general people were expecting another crisis like that of Munich in the preceding year, but were not prepared for the imminent world war. Mazur, in a discussion concerning such possibilities, suddenly said to me " A world war may break out. What shall we do with the Scottish Book and our joint unpublished papers? You are leaving for the United States shortly, and presumably will be safe. In case of a bombardment of the city, I shall put all the manuscripts and the Scottish Book into a case which I shall bury in the ground." We even decided upon a location of this secret hiding place; it was to be near the goal post of a football field outside the city. It is not known to me whether anything of the sort really happened. Apparently, the manuscript of the Scottish Book survived in good enough shape to have a typewritten copy made, which Professor Steinhaus sent to me last year (1956).

The existence of such a collection of problems was mentioned on several occasions, during the last 20 years, to mathematical friends in this country. I have received, since, many requests for copies of this document. It was in answer to such oral and written requests that the present translation was made. This spring in an article, "Can We Grow Geniuses in Science?", which appears in Harper's June 1957 issue, L. L. Whyte alluded to the existence of the Scottish Book. Apparently, the diffusion of this small mystery became somewhat widespread, and this provided another incentive for this translation.

Before deciding to make such an informal distribution, I consulted my teacher and friend (and senior member of the group of authors of the problems), Professor Steinhaus, about the propriety of circulating this collection. With his agreement, I have translated the original text (the original is mostly in Polish) in order to make it available through this private communication.

Even as an author or co-author of some of the problems, I have felt that the only practical and proper thing to do was to translate them verbatim. No explanations or reformulations of the problems have been made.

Many of the problems have since found their solution, some in the form of published papers (I know of some of my own problems, solutions to which were published in periodicals, among them Problem 17.1, Z. Zahorski, Fund. Math., Vol. 34, pp. 183–245 and Problem 77(a), R. H. Fox, Fund. Math., Vol. 34, pp. 278–287).

The work of following the literature in the several fields with which the problems deal would have been prohibitive for me. The time necessary for supplying the definitions or explanations of terms, all very well understood among mathematicians in Lwów, but perhaps not in current use now, would also be considerable. Some of the authors of the problems are no longer living, and since one could not treat uniformly all the material, I have decided to make no changes whatsoever.

Perhaps, some of the problems will still present an actual interest to mathematicians. At least the collection gives some picture of the interests of a compact mathematical group, an illustration of the mode of their work and thought; and reflects informal features of life in a very vital mathematical center. I should be grateful if the recipients of this collection were willing to point out errors, supply information about solutions to problems, or indicate developments contained in recent literature in topics connected with the subjects discussed in the problems.

It is with great pleasure that I express thanks to Miss Marie Odell for her help in editing the manuscript and to Dr. Milton Wing for looking over the translated manuscript.

Los Alamos, NM, USA   S. Ulam
May 1977

Preface to the Limited Los Alamos Edition of 1977 Monograph

Numerous requests for copies of this document, addressed to Los Alamos Scientific Laboratory library or to me, appear to make it worthwhile (after a lapse of some 20 yr) to reprint, with some corrections, this collection of problems.

This project was made possible through the interest and active help of Robert Krohn of the laboratory.

It is a pleasure to give special thanks to Dr. Bill Beyer for his perspicacious review of the changes and the revised version of some formulations. Thanks are due to Martha Lee Delanoy for editorial work.

Los Alamos, NM, USA   S. Ulam
May 1977

Contents

Part I   The Scottish Book Conference Lectures

1   An Anecdotal History of the Scottish Book, *S. Ulam* 3

2   A Personal History of the Scottish Book, *Mark Kac* 13

3   Steinhaus and the Development of Polish Mathematics, *A. Zygmund* .. 21

4   My Scottish Book 'Problems', *Paul Erdös* 27

5   KKM-Maps, *Andrzej Granas* 35
  5.1   Definition and examples 35
  5.2   Ky Fan fixed point theorem and the minimax inequality 39
  5.3   KKM-maps and variational inequalities 41
  5.4   KKM-maps and the theory of games.................................. 43
  5.5   Bibliographical and historical comments 45
  References 47

Part II   The Scottish Book Problems

6   Problems with Commentary 51

7   Appendices to the Problems 283
  7.1   Problem Subject Index 283
  7.2   Authors of the problems 285
  7.3   Problems of Note.. 287

Part III   A Brief History of Wrocław's New Scottish Book

8   Lwów of the West, *P. Biler, P. Krupski, G. Plebanek*
  and *W. A. Woyczyński* 291

Part IV   The New Scottish Book

9   Selected Problems 299

Part I
The Scottish Book Conference Lectures

Chapter 1
An Anecdotal History of the Scottish Book,
S. Ulam

For those readers who may not know, I should start by saying that the so-called Scottish Book is an informal collection of problems in mathematics. It was begun in Lwów, Poland—my home town—in 1935; how and why will be explained in due course. Most of the problems are due to a few local mathematicians, myself included. Actually, many of the earlier problems originated well before 1935—perhaps 6 or 7 years before—during the period when I was still a student. As a budding mathematician, I regularly attended all the seminars and lectures in my field of interest, and made friends with several of the older, established mathematicians. I was then able to take part in the informal discussions—generally among two or three of us at a time—which were a standard feature of mathematical life in pre-World War II Lwów. For several years I was invariably the youngest person in any such group; ultimately, Mark Kac made his appearance, and I lost my special position to him, my junior by some five years.

The story of the Scottish Book could also be called the "Tale of Two Coffee Houses," the Café Roma and, right next to it, the Café Szkocka, or Scottish Café. These two establishments are situated on a little square 100 or 200 yards from the University of Lwów. A few years ago my friend Mazur—one of the more prolific authors represented in the Scottish Book—sent me a post card which shows these two coffee houses as they were in the early 70s (and presumably still are). The postcard has been reproduced as the frontispiece of this edition of the Scottish Book. So far as I can tell, nothing has changed since the days before World War II.

For our story, the Café Roma was, in the beginning, the more important of the two coffee houses. It was there that the mathematicians first gathered after the weekly meetings of our local chapter of the Polish Mathematical Society. The meetings were usually held on Saturday in a seminar room at the University—hence close to the Cafés. The time could be either afternoon or evening. The usual program consisted of four or five ten-minute talks; half-hour talks were not very common, and hour-long talks were mercifully rare. There was of course some discussion at the seminar, but the really fruitful discussions took place at the Café Roma after the meeting was officially over.

© Springer International Publishing Switzerland 2015
R.D. Mauldin, *The Scottish Book*, DOI 10.1007/978-3-319-22897-6_1

Among the senior mathematicians who frequented the Café Roma, the most prominent was undoubtedly Banach. The other full professors or associate professors were Stożek, Ruziewicz, and Łomnicki. There were also younger lecturers and docents as well as a few students like myself. Kuratowski, who was a professor at the Polytechnic Institute, and Steinhaus, who was at the University, preferred a more elegant and genteel pastry shop. But Banach, Mazur and various visitors, including Sierpiński, patronized the Roma. There we sat discussing mathematics, lingering over a single cup of coffee or a glass of tea for three or four hours at a time—something one can still do in some Paris cafés.

Besides mathematics, there was chess. Auerbach was a very strong player. Frequently he would play a game or two with Stożek or Nikliborc while Banach watched and, of course, kibitzed. That is something I too love to do, although I know it can be extremely annoying to serious players. Many years later I read a story about an Englishman who habitually kibitzed in his club. When the players objected violently, he wrote a letter to the *Times* saying, "As a free Englishman, I believe I have the right to express my opinions freely, and evaluate the position for both players."

But above all, we mathematicians continued the discussions which had been aired earlier at the meetings of the Mathematical Society. The whole atmosphere, in Lwów especially, was one of enthusiastic collaborations; people were really interested in each other's problems. This was true in Warsaw too, where there was much collaboration among the topologists, set theoreticians and logicians. In Lwów, the interest was not only in set theory, but, owing to Steinhaus' and Banach's influence, also in functional analysis and several other fields.

It was Steinhaus who discovered Banach; in fact, he used to say that this was his greatest discovery. Steinhaus was a young professor in Kraków, a city about two hundred miles west of Lwów. One evening while walking in a park, he overheard two young men sitting on a bench discussing the Lebesgue integral. Lebesgue's integral was a rather new theory at the time (this was 1917). Steinhaus was intrigued and started talking to the two young men, one of whom was Banach. Steinhaus was greatly impressed, and he encouraged Banach to continue his studies. Banach, by the way, was a very eccentric person in his habits and personal life. He would not take any examinations at all, disliking them intensely. But he wrote so many original papers and proposed so many new ideas that he was granted a doctor's degree several years later without passing any of the regular exams. All this happened at the end of World War I around 1919.

Collaboration was of course not unknown in other mathematical centers. For example, the book by Felix Klein on the history of mathematics in the 19th century mentions that groups of mathematicians (small groups—pairs, triplets at most) discussed mathematical problems in Göttingen. This was not so prevalent in Paris.

At one time I thought it would be interesting to try to write a history of the development of mathematical collaboration. I have the impression that the profusion of joint papers is a rather recent trend. Not being a historian, I don't know the detailed course of the development of mathematics in Italy in the days of the Renaissance, but it is certainly true that even in antiquity mathematicians were writing letters to each other. There are letters of Archimedes containing mathematical problems and theorems. As for joint papers, I don't know their ancient history. There are of course some very famous 19th Century papers like Russell-Whitehead. Atiyah-Singer is an example of a more recent and celebrated mathematical paper, as are the Kac-Feynman formulae, and so on.

Today collaboration and the number of joint papers seem to be increasing. In fact, joint papers appear to constitute a sizable proportion of all current original mathematical work. There is almost as much as in physics, where, especially in experimental physics, there may be ten or twenty authors for a single paper. In theoretical physics too, one sees quite a few joint papers.

This seems to me a curious phenomenon both epistemologically and psychologically. Somehow, in some cases, collaboration is more fruitful than the efforts of a single individual. Certain single individuals still produce the main ideas, but it is interesting to compare this parallel work to work on computers. Without any doubt, the human brain, even in a single person, operates on many parallel channels simultaneously. This is not so on present day computers, which can only perform one operation at a time.

The germ of cooperation between elements exists in the brain of even very primitive animals, and is well-known in mammals. Certainly, creative activity in mathematics requires the putting together of very many elements. Suppose we had two or three brains working together in parallel on a subject; it is fascinating to speculate on what this might lead to. No one doubts that we shall witness the development of computers able to work in parallel; in fact, this development has already begun. Of course, it is dangerous to be too certain of what the future will bring. I cannot refrain from quoting a statement by Niels Bohr: "It is very difficult to predict; especially the future."

But to go back to the Polish school of Mathematics, to the cafés and to the Scottish Book—I should point out that the subjects studied partook of a certain novelty. Set theory itself was still rather new, and set theoretical topology was newer yet. The theory of functions of real variables and the idea of function spaces were to some extent fostered and developed in Poland, and in Lwów specifically.

Another point that should be made is that the definition of Banach spaces gave a very general framework and yet embraced many examples, each having, so to speak, a different flavor; these were sufficiently different to excite great interest. Generalizations about objects that are too similar to each other are less interesting. But where one can identify common properties of objects which appear quite different from each other, it is comparable to the living world where there exist so many species that are close but not alike. The richness depends on the combination of diversity and similarity.

Those who have followed the subject know that there are many different types of Banach spaces: the space of continuous functions, Hilbert space, not to mention the finite-dimensional Banach spaces with different Minkowski metrics, spaces of measurable functions, analytic functions, and so on, all having made their appearance implicitly in problems of mathematical analysis. And that class of spaces and transformations is of special interest mainly because the spaces deal with non-compact phenomena.

The usual approximation methods, the "epsilon" approaches, are almost by definition suitable for treating compactness in the classes of objects under discussion. Many problems of analysis however, are, so to say, very noncompact and yet somehow homogeneous and amenable to methods of a general analysis with limiting processes, which are encompassed by Banach's original definition. A similar definition was given independently by Norbert Wiener, but as he wrote in his autobiography, he somehow lost interest in the subject without developing a theory of such spaces. Some of the problems in the little collection gathered in the Scottish Book deal with function spaces.

I have not yet explained how this collection came about. Let us therefore go back to the Café Roma and Banach. He used to spend hours, even days there, especially towards the end of the month before the university salary was paid. One day he became irritated with the credit situation at the Roma and decided to move to the Szkocka next door, a mere twenty yards away. Stożek and some chemists and physicists continued to frequent the Roma, but the Scottish Café now became the meeting place of a smaller group of mathematicians, including Banach, Mazur, myself, and occasionally some others. It is owing to this that so many of the problems in this collection are entered in our names. There were of course visitors, my friend Schreier among others, but the regular habitués were just the three of us.

How did the book come about? One day Banach decided that because we talked about so very many things, we should write the ideas down whenever possible in order not to forget them. He bought a large and well-bounded notebook in which we started to enter problems. The first one bears the date July 17, 1935. This was while I was still living in Poland, before I received an invitation from von Neumann to visit him in Princeton. (It was during this visit to Princeton that the late G. D. Birkhoff, at a tea at von Neumann's, asked me whether I would come to Harvard to join the Society of Fellows there. I accepted of course, and consequently was able to remain in the United States.) During the summers I used to return to Poland to visit my family and my mathematical friends. These were the summers of 1936, '37, '38, and '39.

The notebook was kept at the Scottish Café by a waiter who knew the ritual—when Banach or Mazur came in it was sufficient to say, "The book please," and he would bring it with the cups of coffee.

As years passed, there were more and more entries by other Polish mathematicians, Borsuk for instance—a topologist friend of mine from Warsaw—and many others. The "Book" grew to become a collection of some 190 problems, of which by now, nearly fifty years later, about three-quarters have been solved. Some of the problems were entered without too much previous work or thought; a few were solved on the spot. All of this is noted in the book.

The document stayed in Poland. On my last prewar trip in the summer of 1939, Mazur, more realistic about the world situation than I (I thought we would only see more crises like that of Munich or Czechoslovakia), said he believed a great war was imminent. He said that our results, about countable groups among other subjects, some of which are unpublished to this day, should not be lost, so he proposed that when war came he would put the book in a little box and bury it where it could be found later, near the goal post of a certain soccer field. I don't know whether this is the way the Scottish Book was preserved or not, for when I saw Mazur a few years ago in Warsaw I forgot to ask him about it. At any rate, the Scottish Book survived the war and was in Banach's possession. When Banach died in 1945, his son Stephan Banach, Jr. (now a neurosurgeon in Warsaw) found it, and showed it to Steinhaus immediately after the war. Steinhaus then copied it verbatim by hand, and in 1956 sent this copy to me at Los Alamos. I translated it, and had some three hundred mimeographed copies of the translation made. I had to pay for this myself—Los Alamos is a government laboratory, and one cannot use taxpayers' money for such frivolous purposes. I mailed these copies to various universities both here and abroad, and also to a few friends. Since then, as the book became known in mathematical circles, people kept writing to Los Alamos for copies. There were so many requests over the years that the laboratory decided in 1977 to print another edition, under the supervision of W. A. Beyer. Photocopies of the Polish original have been preserved. If someone were interested in graphology or handwriting—the handwriting of Banach or Mazur, for instance—he could look at it. (Some is reproduced on pages 14 and 15.)

So much for the origin of the "Book." Many problems are still unsolved, and according to experts, have some value. I think it is fair to say that these problems did exert an influence on the development of some subjects in the areas of functional analysis, in the theory of infinite series, in real variable theory, in topology, in the theory of probability (including measure theory), in group theory, and so on. It was later in the game in Lwów that algebraic problems became of interest. Schreier and I, along with Mazur, began to discuss problems concerning groups, as well as various questions in the theory of Lie algebras. I remember that when I first learned about the latter at the age of twenty-two or twenty-three, they seemed too formal to me. Only later did I begin to appreciate their importance and applications. There were also some problems in geometry.

It was the variety of examples and a certain concreteness in these abstract ideas that made this whole subject, for me and perhaps for many other mathematicians, so vivid and alive. There are examples of spaces, examples of transformations, examples of functions, of sets. Recently, by the way, and quite by chance, I came upon the following phrase of Shakespeare's, in Henry VIII: "Things done without example, in their issue are to be feared." Is this an anti-"new-math" statement? I can certainly agree with the sentiment, even if, as I suspect, the word "example" was meant in quite a different way.

Central to the theme I am trying to develop is this class of examples which have something, but not too much, in common. Here we see almost a biological or genetic development, an evolutionary development of the objects which mathematics creates and which take on a life of their own. In the beginning, in the foundations of mathematics, you might say there are only sets, and next come spaces. In the next stage, where the sets are "animated," we have topology. Further development results in greater specificity, *ergo*, metric spaces. One could go on to mention certain algebras, and so forth. These, we might say, correspond to *nouns*. When we start operating on them, that is, when we consider transformations and functions, it is like introducing *verbs* into the language. It occurred to me long ago that many words in the common language can, in the mind of some young person of imagination, become germs of a mathematical theory. What is topology if not the study of an elaboration of the word "continuous" or "continuity"? There are many other words which could stimulate people to build theories, or at least "mini-theories."

I can give some examples of how my own interest was originally stimulated by problems of the Scottish Book type. For example, one important thread going through some of the problems is the idea of "approximate," or more properly "epsilon-approximability" by finite or generally simpler structures. Many problems of the book deal with properties of approximation, of reduction from the infinite to the finite. Of course it is the finite that interests physicists, but the idea of infinity, as all of us know, is useful because it puts in a more succinct way some properties of very large or very small numbers, just as the infinitesimal calculus is more concise and efficient than the calculus of finite differences. So the study of infinity per se and its relation to finite approximation is of great interest. I am speaking vaguely here, in general terms, but one can find many concrete examples in the Scottish Book. Speaking of concreteness, I should like to say that there may be a more tangible aspect to the ultra-set-theoretical investigation of very high cardinals, the incredibly large infinities which may be measurable. These big sets, cardinal numbers in our speculations, do throw as it were a shadow on the lower infinities. And indeed, there are more concrete or semi-concrete formulations or expressions of mathematical objects suggested by speculation on the existence of these superinfinities.

Speaking of "epsilons," I want to mention a number of little amusements I have indulged in over the years concerning what I call "epsilon stability," not just of equations and their solutions, but more generally of mathematical properties.

As an example of this "epsilon stability," consider the simple functional equation: $f(x+y) = f(x) + f(y)$, i.e., the equation defining the automorphism of the group of real numbers under addition. The "epsilonic" analogue of this equation is $|g(x+y) - g(x) - g(y)| < \varepsilon$. The question is then: Is the solution g necessarily near some solution $\bar{f}$ of the strictly linear equation? As D. Hyers and I showed, the answer is yes. In fact, $|g - \bar{f}| < \varepsilon$, with the same epsilon as above. This is not a very deep theorem. What about the more general case? Suppose I have a group for which I replace the group operation by one that is "close" to it in some appropriate sense. This of course requires a notion of distance in a group. The result of the replacement is an "almost endomorphism." Then we may ask: Is it of necessity "near" a strict endomorphism? The answer is not known in general, even for compact groups. Recently, D. Cenzer obtained an approximation result for some easy groups, e.g., the group of rotations on a circle.

In the same spirit, we may take the idea of a transformation which is an isometry, a transformation which preserves distances. What about transformations which do not exactly preserve distances but change them by very little, i.e., at most by a given $\varepsilon > 0$? Suppose I have a transformation of a Banach space or some space which transforms into itself, and where every distance is changed by less than a fixed ε. Is such a transformation "near" one which is a true isometry? Hyers and I proved, in a series of short papers, that this is true for Euclidean space, for Hilbert space, for the C space, and so forth. If you have such a transformation, it must then be within a fixed multiple of a strictly true isometry.

Recently I became more ambitious and looked at some other mathematical statements from this point of view. One could try to "epsilonize" in this sense theorems on projective geometry, on conics, and so on. More generally, take as an example some famous theorem like the theorem on functions with an algebraic addition. It is a well-known statement that the only functions which satisfy an algebraic addition theorem are, in addition to sine, cosine and elementary functions, the elliptic functions. One could ask (perhaps this question is not yet properly formulated): Is it true that a function which "almost" satisfies an algebraic addition theorem must be "almost" an elliptic function?

And in a similar vein: If we have a function which is differentiable, let us say five times, and its derivative vanishes and changes sign at a point, then any sufficiently differentiable function which is sufficiently close, in the sense of absolute value alone, must also have a vanishing fifth derivative at a nearby point. This is almost trivial to prove, though at first it seems false. Why is this true? Because the fifth derivative can be obtained by finite differences. This is all very nice and easy for functions of one variable. For functions of several variables the analog becomes interesting and not too well-known or established. The same is true, *mutatis mutandis*, for spaces of infinitely many dimensions, and is of possible interest to physicists as a general "stability" property.

Finally, I want to mention another class of problems which appears here and there in the Scottish Book. These problems are attempts to *characterize* certain spaces or certain transformations. For example, suppose one wants to characterize the Hilbert space among other Banach spaces by some properties of homogeneity or by the wealth of isometric transformations into itself which it allows. There is already a result in a finite number of dimensions due to Auerbach, Mazur, and myself on one way to characterize an ellipsoid.

We wrote a paper where we proved that a convex body, all of whose sections through a certain point are affine to each other, must be an ellipsoid. We did not prove it for all dimensions, only in three dimensions. This paper appeared before World War II in *Monatshefte für Mathematik*. It is merely another example of what I mean by a *characterization*. Recently this topic has been developed by many people, notably by Anderson in this country and Pełczyński in Poland.

There are other common threads going through the problems of the Scottish Book but it would not be true to say that the problems are all cast in a similar mold. Some are just momentary curiosities, spur-of-the-moment thoughts of the habitués of the Scottish Café, or of casual visitors such as von Neumann. After 1939, one notes a curious change: suddenly the contributors include Russian names, the names of well-known mathematicians like Sobolev. This was after the city was occupied by Russia, in September 1939.

Today more problem books are appearing. There are problem sections in the "Notices" and in the *American Mathematical Monthly*. Another little Scottish Book is being kept in Boulder, where I was a professor for some ten or twelve years; J. Mycielski is keeping track of it. There is another one currently kept in Wrocław, Poland—I don't know about Hungary. Erdős has written monumental papers containing selections or collections of problems in set theory and in number theory. One of them, written jointly with R. Graham, is not yet published, but I have seen the manuscript—it is a very interesting and exciting book.

Finally let me mention a few ideas which are not in the Scottish Book but which I remember from conversations and discussions with Mazur. As an example, Mazur and I discussed the possibility of establishing, at first only purely mathematical, but later physical objects, which could replicate or almost replicate themselves. This was a very sketchy and premature idea. Years later, as is well known, von Neumann discussed this question in some detail.

We also considered the purely theoretical (at the time) possibility of comprehensive computing machines. Neither of us had sufficient knowledge of electronics to even approximate the present schemata, but we discussed the concept on a purely abstract level.

We had some other very curious conversations. I specifically remember discussions among ourselves and with visitors about what is now known as nonlinear mathematics—truly a strange expression, for it is like saying "I will discuss nonelephant animals"—it was more specific than that. In fact Mazur and Orlicz had started a study of polynomial operations; their paper appeared in *Studia Mathematica*. Then we discussed iterations in one or more variables of transformations showing the sort of phenomena which very recently I and many other mathematicians have studied both experimentally and theoretically, and which seem to now present some interest even for physicists. But to go into this would take me too far afield.

From the original Scottish Book; the handwriting in the Mazur-Orlicz problem (bottom) is Mazur's.

From the original Scottish Book; the handwriting in the Banach-Ulam problem is Banach's.

Chapter 2
A Personal History of the Scottish Book,
Mark Kac

It is a special pleasure to be introduced by my old friend Erdős. The use of the adjective "old" is slightly depressing, and I would like to forget about it, but somehow Erdős will not let me do it.

I should like to begin my remarks by pointing out the remarkable thing that we celebrated the Scottish Book in Denton, Texas. It is remarkable not only because of the energy, dedication and interest of one man, namely Dan Mauldin, but it is also, for me at least, typically American. It represents the kind of combination of generosity and sentiment which runs through the whole history of this young civilization. I cannot think of any other country on the surface of the earth which would be interested in celebrating a somewhat obscure event which occurred in another country in what now seems like the dim past. And so on my own behalf, and I am sure also on behalf of all my former and present compatriots, I would like to express our thanks not only to Dan Mauldin but also to the spirit of America in Denton, Texas.

Before I come to Mathematics and to my connection (tenuous as it was) with the Scottish Book let me engage in a little of what Stan Ulam, quoting Disraeli, referred to as "anecdotage."

As you can see by perusing the Scottish Book, a significant number of problems were inscribed by distinguished foreign mathematicians who passed through Lwów. One of the most famous of these visitors and probably the most famous one, was Henri Lebesgue.

Lebesgue came to Lwów in May 1938 to receive an honorary doctorate from the University. At that time, since Stan Ulam, who was the Secretary of the Lwów Section of the Polish Mathematical Society, was away in the USA, I was substituting for him and was given the extremely pleasant job of showing Lebesgue around the city. I reminisced about this event in 1974 in Geneva, when the centennial of Lebesgue's birth was celebrated. My remarks were published in *L'Enseignment Mathématique* in French, and were later translated into Polish (not by me since my knowledge of my mother tongue is no longer sufficiently reliable). Today I give

© Springer International Publishing Switzerland 2015
R.D. Mauldin, *The Scottish Book*, DOI 10.1007/978-3-319-22897-6_2

you an abbreviated English version of these remarks. In fact I will tell you only two stories, one of which is directly connected with the Scottish Café, the birthplace and home of the Scottish Book.

At the time of his visit Lebesgue was no longer interested in anything but elementary mathematics; he refused to discuss measure, integrals, projection of Borel sets, or anything of that sort. He gave two lectures, both extremely beautiful, but entirely elementary: one on construction by ruler and compass, and the other on iterated radicals.[1]

As a footnote to the political atmosphere of those days it may be of interest to record the following. The Polish press, which was inept above and beyond the call of duty, confused Lebesgue with Hadamard. Hadamard was a known leftist. Lebesgue, on the other hand, was a man of rather conservative views, though by no means a reactionary. He was greeted upon arrival by a violent editorial against the leftist, communist French professor being honored by the Poles. The confusion was soon cleared up, but nobody bothered with a retraction. So you can see the press is the same the world over, and not much has changed in this respect over the years.

As I showed Lebesgue around the city he was extremely disappointed with me—he was very much interested in the churches, and wanted to know all about their history, and I was unable to provide him with much information on that subject. Lwów by the way was an extremely interesting city from the religious point of view, because it was, with the possible exception of Jerusalem, the See of all three lines of Catholicism. There were in fact three archbishops in Lwów, representing the Roman, Greek, and Armenian branches.

The Armenian Cathedral, one of the most beautiful churches in Europe, especially interested Lebesgue. To his chagrin I could not tell him anything about it, and I was equally disappointed by Lebesgue's refusal to discuss measure, integrals, and other mathematical topics. Still, we became reasonably friendly, and he merely pitied me as one doomed to some terrible fate for lack of interest in history.

That afternoon we had a 5 o'clock reception for Lebesgue in the Scottish Café. Fewer than 15 people attended, which goes to show how small the number of mathematicians was in those days. The waiter gave all of us menus, and not realizing that Lebesgue was not a Pole he gave him one too. Lebesgue looked at the menu for about 30 seconds with utmost seriousness and said, "Merci, je ne mange que des choses bien definies" (Thank you, I eat only well-defined things). At this moment I had an inspiration, and by changing a little a well-known phrase of Poincaré directed against Cantorism I said, "Ne mangez jamais que des objets susceptibles d'être définis par un nombre fini de mots" (Never eat things which cannot be defined in

---

[1]Professor Granas brought with him to the Texas conference a copy of *Summaries of the Proceedings* of the meetings of the Mathematical Society in Lwów. From these documents we can ascertain that one of the lectures took place on May 25, 1938, and was on iterated square roots. It also turned out that my recollection as reported in my Geneva talk was not entirely correct, but the errors are minor.

a finite number of words). "Ah," said Lebesgue, "you are familiar a little bit with Poincaré's philosophy," and I think that he forgave me at that moment my ignorance of the history of the Armenian Cathedral.

My second remark of an anecdotal nature has to do with the beautiful talk by Professor Martin, my former colleague at Rockefeller University, which was presented at the conference in Texas. It should be of interest, because it characterizes the way Steinhaus felt about mathematics, and especially about the axiom of determinacy. I am sure of this because I attended lectures by Steinhaus at both Rockefeller and the Courant Institute in the early sixties. I also had many occasions to speak to him about it.

I will now give Steinhaus' "proof" of the determinacy of the Ulam game.

We of course all remember the Ulam game, where Player One picks a zero or one and Player Two picks a zero or one, and one then constructs what Tony Martin called a decimal binary (which is an excellent name for what ordinary mortals call simply a binary). [Editor Note: The term binary decimal goes back to Turing's classic 1937 paper on computable numbers and Hardy and Wright's introduction to the theory of numbers.] If it falls into a set E Player One wins, and if it is not in E Player Two wins. The question is: Is there a winning strategy for either one of the players? Here is a "proof" that there is one:

Let me denote by x_1, x_2, ... the moves of Player One, and by y_1, y_2, ... the moves of Player Two. I will give in logical symbols, which I use very infrequently, the statement that Player One has a winning strategy:

$$\exists\, x_1(y_1)\ \exists\, x_2(y_2) \cdots \frac{x_1}{2} + \frac{y_1}{2^2} + \frac{x_2}{2^3} + \frac{y_2}{2^4} + \cdots \in E.$$

It says, "There is a first move of Player One such that for every first move of Player Two there is a second move of Player One, such that for every second move of Player Two, etc., the fraction $\frac{x_1}{2} + \frac{y_1}{2^2} + \cdots$ belongs to E." This is merely a transcription in logical symbols of the statement that there is a strategy for Player One.

Now suppose there is no such strategy; then you put the symbol $\sim$ in front of the string of quantifiers in the formula above and use the DeMorgan rule, obtaining

$$(x_1)\ \exists\, y_1(x_2)\ \exists\, y_2 \cdots \frac{x_1}{2} + \frac{y_1}{2^2} + \frac{x_2}{2^3} + \frac{y_2}{2^4} + \cdots \notin E.$$

Now if you translate this into human language, it means that Player Two has a winning strategy. So if Player One doesn't have a strategy, Player Two has a strategy, and, consequently, the axiom of determinacy in this case merely allows one to use DeMorgan's law for an infinite number of quantifiers. Now of course you can see where the difficulty comes in. It is that difficulty which plagues the whole beastly subject, and it is, namely, where you ask, "How does one know whether something does or does not belong to set E?" It is here, of course, that we get into all the difficulties, and Steinhaus merely felt—and I have enormous sympathy for it—that his axiom had a chance to distinguish those sets E that are worthy to be called sets from those that are not.

Axioms like the axiom of choice allow us—give us a legal license—to create certain objects and then call them sets. Steinhaus thought that his axiom would be of the kind that would distinguish between constructible and nonconstructible sets.

This little argument reminds me—and now I am only almost serious; up to this point I was dead serious—of an imperfect analogy with what happens in quantum mechanics where certain statements, although they sound perfectly all right, are not allowable. For instance, when you say, "The amount of energy in a radiation field in a subvolume," then it sounds like a perfectly well-defined thing. But if you really follow the dicta of quantum mechanics, you have to express it in terms of a Hermitian operator—every physical quantity has to be represented by a Hermitian operator—and it turns out that it is not unique. In fact, how to interpret this may very well depend on the method of measurement. You have something of the sort here—nothing is really defined until you come to grips with saying, "How do you know whether a number constructed by an infinite number of operations does or does not belong to a set?"

Now, one final observation in connection with other people's involvement in the Scottish Book Conference, namely with Professor Zygmund's, who referred in his talk to one of the greatest Polish discoveries, the category method. As a matter of fact, this discovery is so well known that one does not even recognize what a remarkable discovery it was. It was remarkable because it showed that sometimes it is easier to prove that *most* objects have a certain property than to exhibit a particular example.

Professor Zygmund asked about the rearrangement of the Fourier series in connection with the question of convergence, and bemoaned the fact, which many of us bemoan, that there is no decent, sensible measure in the set of all permutations. However, if one goes back to the Polish invention of the method of category, then of course the set of all permutations can be easily metrized by the Frechet trick. Consequently, the concept of sets of first and second category is perfectly well defined. There is in fact a book by Professor Oxtoby, who attended the conference (and I even ascertained from him that it was published in 1971 by Springer-Verlag), called *Measure and Category*. The message of the book is that whenever both can be defined and whenever the measure is reasonable, then second category and measure one, other than in very exceptional situations, are the same. One can rephrase Professor Zygmund's question to ask whether the set of all permutations of Fourier series which lead to divergence is of second category. A very simple case—similar abut much simpler—was considered by my colleague at the time, and still a good friend, Professor Ralph Palmer Agnew of Cornell, in response to a question posed during a conversation we had many years ago. If you take a conditionally convergent series of real numbers, then of course we know that it can be rearranged so as to make it converge to any prescribed number, and it can also be rearranged into a divergent series. Now it is easy to prove, and in fact Agnew proved it (it was published around 1940 in the *Bulletin of the American Mathematical Society*) that the set of permutations which lead to divergent rearrangements is indeed of second category. You might say that everything bad which one might expect to happen is going to happen in a plentiful sort of way.

Now to some of the more personal things. I am not really, in a certain sense, a product, or at least not a typical product, of the Polish school. When I came to Lwów as a student in October 1931, I did not know any of the great masters; my first contact was with the late Marceli Stark, a remarkable man and a tremendously well-educated mathematician who died recently and to whose memory I would like to pay tribute. I was very concretely minded, and I still am—in fact even more so. Yet I felt a little bit that I also ought to do these abstract things, and Steinhaus, whom I met a little later, said, "You shouldn't; you must earn the right to generalize." I have not yet earned that right.

I became interested in probability theory in a way that I am not even going to tell you in detail, because I can't give you a full autobiography. Some day I am going to get even with Stan Ulam and write my own adventures, which, however, are not nearly as exciting as his.

It was through Steinhaus that I became interested in probability theory, and, with the exception of one problem (I put altogether four problems into the Scottish Book—numbers 126, 161, 177, and 178) and I really do not know why I put it in; it is not even properly stated—these problems deal directly or indirectly with probability theory. The first one is a minor technicality, which Hinčin proved in response to a letter.

The problem that I cannot for the life of me remember how and why I thought of it, is the problem of characterizing continuous functions, $\phi(x, y)$, such that if A and B are real symmetric matrices, then ϕ is positive definite (Problem no. 177). Now, because of noncommutativity, $\phi(A, B)$ is not properly defined. But that is easily remedied if ϕ is a polynomial in two variables—one simply replaces ϕ by a symmetrized polynomial, in which case it makes perfect sense, and the question can still be asked. Whether it is of any interest I have no idea. I do not have any recollection as to why it interested me at the time, and I probably should have appealed to Dan Mauldin to put this problem in a footnote because there is no particular reason to bother the next generation with this one—unless in the meantime I remember what it was I really wanted.

The first, as I have already told you, was a minor technical problem, but the fourth (Problem no. 178) has a certain degree of interest, and I may as well say what it is. It is unsolved not because it is necessarily difficult, but because nobody has tried. I am not going to give any prizes for it. It might, however, be of some interest to those of you who are analytically-minded.

There is a well-known theorem of Cramér that if a product of two characteristic functions $\phi_1(\xi)$, $\phi_2(\xi)$ is $\exp(-\xi^2/2)$ then both ϕ_1 and ϕ_2 must themselves be Gaussian, i.e.,

$$\phi_1 = \exp(-\alpha_1 \xi^2 + \beta_1 \xi)$$

$$\phi_2 = \exp(-\alpha_2 \xi^2 + \beta_2 \xi)$$

with $\alpha_1 + \alpha_2 = 1/2$ and $\beta_1 + \beta_2 = 0$. (In Problem 178 the theorem is slightly misstated.)

In probabilistic terms, if a sum of two *independent* random variable is Gaussian, then the random variables themselves must be Gaussian. Similar theorems hold for stable distributions. My Problem 178 raised the question of whether other distributions can be similarly characterized. One must, of course, get away from the product since the product is intimately tied to addition of random variables and therefore to stable distributions, and I hit upon

$$\left(\frac{1}{x} + \frac{1}{y} - 1\right)^{-1}$$

as a candidate for the characterization of the class of characteristic functions

$$\frac{1}{1 + \alpha\xi^2}, \ \alpha > 0.$$

The problem is closely related to the following problem which is perhaps of greater general interest:

What are the functions $F(x,y)$ of two variables such that $F(\phi_1(\xi), \phi_2(\xi))$ is a characteristic function of a probability distribution whenever ϕ_1 and ϕ_2 are?

I strongly suspect that F must be a function of the product xy, i.e., $F(x,y) \equiv G(x,y)$ with G satisfying some additional conditions, but I have no idea how to go about proving it.

The only one of my four problems which was destined to have a future was Problem 161. There is not much point in going into details since an interested reader can consult my 1949 address, "Probability Methods in some problems of analysis and number theory" (*Bull. Am. Math. Soc. 55*, (1949), 390–408). Echoes of this problem are still reverberating, as witness a recent paper by I. Berkes, "A Central Limit Theorem for Trigonometric Series with Small Gaps," (*Z. für Wahrsch.*, 47 (1979), 157–161), but the original source will only become known with the publication of the Scottish Book. As it is, not even my 1949 address is cited, which is some kind of a price one must pay for pioneering.

Problem 161 bears the date June 10, 1937, which was five days after I repeated the ancient oath, "*Spondeo ac polliceor ...*" and was awarded the degree of Doctor of Philosophy of the John Casimir University in Lwów. Actually, not knowing Latin, I got into my head that in *spondeo* the accent is on the second syllable and not, as is correct, on the first. Steinhaus, who was my sponsor *(promotor)* and who was a stickler for proper usage of all languages, used to make me practice the correct pronunciation before the actual ceremony. When the moment arrived for me to reply to a Latin oath read with pomp (though not with pomposity) by the *Rector Magnificus* I forgot all the practice and put an emphatic accent on the wrong (second) syllable. Steinhaus cringed and so did my father, who knew Latin and who journeyed to Lwów to witness the occasion.

Returning to the Scottish Book, I would like to point out that although the problems in it range over most of the principal branches of Mathematics, one branch is conspicuously absent, and that is Number Theory. The reason is simple, and it is that Number Theory was not in vogue in Poland at the time. Sierpiński in his younger years (and also toward the end of his life) did important and interesting things in Number Theory, and the Warsaw school did produce two "mutants": A. Walfisz (who left Poland for the Soviet Union and was Professor at Tiblisi) and S. Lubelski. There was even a serious journal, *Acta Arithmetica* (which continues to this day), devoted to Number Theory, but this beautiful and important area was far from the forefront of mathematical preoccupation in Poland before World War II.

I cannot remember at all how I came to think about number theoretic problems in connection with Probability Theory, but I do remember making what appeared to me then to be a great discovery (it wasn't).

If $\phi(n)$ is the familiar Euler function, one has

$$\frac{\phi(n)}{n} = \prod_{p|n} \left(1 - \frac{1}{p}\right)$$

which can be written in the form

$$\frac{\phi(n)}{n} = \prod_{p} \left(1 - \frac{\rho_p(n)}{p}\right)$$

where

$$\rho_p(n) = \begin{cases} 1, & p \mid n, \\ 0, & p \nmid n. \end{cases}$$

This of course was a well-known elementary fact, but the method also yielded at once

$$M\left\{\left(\frac{\phi(n)}{n}\right)^\ell\right\} = \prod_{p} M\left\{\left(1 - \frac{\rho_p(n)}{p}\right)^\ell\right\}$$

$$= \prod_{p}\left[1 - \frac{1}{p} + \frac{1}{p}\left(1 - \frac{1}{p}\right)^\ell\right],$$

for all ℓ such that the infinite product converges, and hence one had a handle on the distribution of $\phi(n)/n$.

When late in November 1938 I left for the United States, the boat (M.S. *Pilsudski*, sunk in the early days of World War II) stopped for about six hours in Copenhagen, which gave me a chance to meet Professor Børge Jessen. I communicated my number theoretic discovery to him only to learn that the same result had been obtained and already published by I.J. Schoenberg. The probabilistic nature of the result was, however, somewhat hidden in Schoenberg's proof, and I had the advantage (because of my deep involvement with the normal distribution in unexpected contexts, as illustrated by Problem 161) of being—so to speak—on the ground floor. It was therefore a small step to suspect that the number of prime divisors $v(n)$ of n given by the formula

$$v(n) = \sum \rho_p(n)$$

should behave like a sum of independent random variables and hence be normally distributed after subtracting an appropriate mean ($\log \log n$) and scaling down by an appropriate standard deviation ($\sqrt{\log \log n}$). But here my ignorance of Number Theory proved an impediment. The number of terms in the sum $\sum \rho_p(n)$ depends on n, preventing a straightforward application of the Central Limit Theorem. I struggled unsuccessfully with the problem until I stated my difficulties during a lecture in March 1939 in Princeton. Fortunately Erdős was in the audience and he perked up at the mention of Number Theory. He made me repeat my problem, and before the lecture was over he had a proof. Thus did the Normal Distribution enter Number Theory and thus was born its probabilistic branch. While stretching a bit the historical truth I hereby assign the role of godmother of this branch to the Scottish Book.

Chapter 3
Steinhaus and the Development of Polish Mathematics, *A. Zygmund*

The origin and history of the Scottish Book is described by Professor Ulam in his own lecture and I could not add much here.

The book is a product of one of the mathematical schools in Poland, that of Lwów, while I myself, born and educated in Warsaw, belonged to what was then known, both in Poland and abroad, as the Warsaw mathematical school. There was a close collaboration between individuals of both schools, and though my personal contact with Lwów was rather loose, I was very much interested in the work going on there, and it had considerable influence on my own work.

In what follows I shall give a few facts about the development of Polish Mathematics, limiting myself to those which have some pertinence to the Scottish Book.

The Polish mathematical school of the period 1919–1939 was an interesting phenomenon, first because of its achievements, and secondly because of the place and circumstances in which it arose. One might say that before 1919 there had been Polish mathematicians but there was no Polish mathematical school. The rapid growth of Polish Mathematics after 1919 was partly spontaneous, helped by the recent freeing of the country from foreign occupation, and partly a result of thoughtful planning.

The development of Polish Mathematics was in the first place due to Janiszewski, Mazurkiewicz, and Sierpiński in Warsaw and to Banach and Steinhaus in Lwów. The role of Janiszewski here was particularly significant and unique. Born in 1888, he died in 1920 and so did not live to see the fruition of his ideas, but he was the chief planner of the Polish school. A talented mathematician (topologist) himself, he realized the difficulties of organizing good mathematical research in a country without a strong and continuous mathematical tradition. His idea was that the surest and quickest way to success would be first through concentration on a particular mathematical discipline which would be the main source of interest and of problems for a larger group of mathematicians, and secondly, through starting a mathematical

© Springer International Publishing Switzerland 2015
R.D. Mauldin, *The Scottish Book*, DOI 10.1007/978-3-319-22897-6_3

publication specializing in this selected branch of mathematics. Once a strong point was established, gradual extension of interest to other fields of mathematics was expected.

At that time the theory of Sets, Topology, and Real Variables were attracting a number of Polish mathematicians. It was natural to make a starting point here and in 1920 the first volume of the publication *Fundamenta Mathematicae* appeared in Warsaw. It was a success from the start. It gave an outlet to Polish mathematical production and attracted foreign papers. Before September 1939, thirty-two volumes of *Fundamenta* had been published.

Let me pass to another Polish mathematical school, that of Lwów. (After 1945 the city of Lwów was no longer within the boundaries of Poland.) When we think of Lwów mathematics, two names usually come to our minds. One of them is Stefan Banach; the other is Hugo Steinhaus, Banach's teacher and later collaborator. While the importance of Banach's mathematical work is widely recognized, and the name is essentially an adjective in Mathematics, few people outside Poland appreciate the importance of Steinhaus' influence on Polish Mathematics. Without Steinhaus, Banach as we know him probably would not have existed, and Polish Mathematics would have had a different character. It is for this reason that I would like to devote most of the time at my disposal to the role played by Steinhaus in the development of Polish Mathematics. It is my personal feeling that despite generally high respect for his work, Steinhaus' role is not sufficiently appreciated here. In what follows, I would like to indicate some of the achievements of Steinhaus and his collaborators.

Born in 1888, Steinhaus studied in Germany and Paris before the outbreak of the first World War. When he returned to Poland, just before the war, he was appointed first a docent and then a professor at the University of Lwów. That was the beginning of his impact on Polish mathematics, for he brought from abroad, to what was a rather provincial mathematical milieu, not only new ideas but also personal contacts with outstanding foreign mathematicians, which were very beneficial to Polish mathematics and contributed very much to its development. Let me illustrate this by one story.

While in Germany, Steinhaus had become a personal friend of Otto Toeplitz, a German mathematician who was also partly of Polish origin. Under Steinhaus' influence, Toeplitz published a short paper in a relatively little-known Polish mathematical periodical *Prace Matematyczno-Fizyczne*, which mostly published Polish papers. The title of the paper was "Über lineare Mittelbildungen"; it appeared in Volume 22 (1911) of the *Prace* and is essentially a paper about the method of condensation of singularities. I was curious that a German mathematician should want to publish such a paper in a very little-known Polish journal. But looking back, one may say that this was an important step in the development of modern functional analysis. Let me be more specific.

The main result of Toeplitz was as follows. Given a matrix $\{a_{mn}\}$ $(m, n = 0, 1, \ldots)$ of real or complex numbers, we may associate with every numerical sequence $\{s_n\}$, $n = 0, 1, \ldots$ a transformed sequence

$$t_m = \sum_n \{a_{mn}s_n\} \qquad (m = 0, 1, \ldots).$$

The problem was to find necessary and sufficient conditions for the matrix $\{a_{mn}\}$ to have the property that every convergent sequence $\{s_n\}$ is transformed into a convergent sequence $\{t_n\}$. The problem is, in today's perspective, very elementary and Toeplitz in his paper (that was in 1911) gave such necessary and sufficient conditions. One of those conditions, the basic one, is very familiar by now. It is

$$\sum_n |a_{mn}| \leq \text{constant}, \qquad \text{for all } m.$$

Toeplitz proved both the necessity and the sufficiency of his conditions. Sufficiency alone was proved independently and at about the same time by the American mathematician Silverman, and the theorem itself is occasionally quoted as the Toeplitz-Silverman theorem.

Obviously, each t_m is a linear operation defined in the space of sequences $\{s_n\}$, and for mathematicians in Lwów interested in functional analysis Toeplitz' result raised a question of abstract generalizations. In 1928, Banach and Steinhaus sent a paper to *Fundamenta* giving one such generalization. The main result of the paper was as follows: Let $\{u_m(x)\}$ be a sequence of bounded linear operations defined in a normed linear space E, and let M_{u_m} be the norm of the operation u_m. If $\sup_m \|u_m(x)\|$ is finite for every point x belonging to a set F of the second category in E (in particular, if it is finite for every $x \in E$), then the sequence M_{u_m} is bounded. In other words, there is a constant M such that

$$\{|u_m(x)| \leq M|x| \text{ for } x \in E \text{ and } m = 1, 2, \ldots\}.$$

Of course, the result of Toeplitz is a consequence of this general theorem. Sierpiński, the editor of *Fundamenta*, gave a paper to Saks, his former pupil, for refereeing, and I remember that Saks showed me the manuscript and pointed out that the argument could be much simplified by replacing the rather cumbersome method of condensation of singularities by the application of the notion of sets of second category. For example, if the functionals are not uniformly bounded, then by merely considering sets of first and second category one can prove, without computation, the existence of a point at which all the functionals are unbounded. The paper appeared in a revised form in *Fundamenta* in 1928, and marks an important point in the development of functional analysis through the application of sets of first and second category. It is perhaps regrettable that the paper, rewritten by Saks, nowhere mentions the fact that it was he who introduced the new method, and the authorship of the method remains unknown except to very few people.

Let me mention another result (due to Steinhaus himself) which had considerable influence upon my own work. The story begins with a theorem of Hurwitz which gives the following. Suppose we have a power series $\sum c_n z^n$ of radius of convergence equal to 1. The function then must have at least one singularity on the circle of convergence. Hurwitz proved that if we select a suitable sequence $\{\pm 1\}$, the series $\sum(\pm c_n)z^n$ is nowhere continuable across the circle $|z| = 1$. In this connection, Steinhaus proved the following: if instead of ± 1 we introduce a Gaussian random variable, then what happened in the case of Hurwitz for a particular sequence of signs becomes true, in the new situation, with probability 1. In other words, if we introduce Gaussian random variable into the coefficients of a power series of finite radius, then with probability 1 this series becomes nowhere continuable, and what was initially an individual situation, which occurred due to a special selection of the values of the random variable, tends to be a general phenomenon. This result of Steinhaus was the beginning of a certain development, randomization of series, which plays a distinctive role in the theory of functions, both of real and complex variable. Steinhaus used a special method and for this reason he had to use Gaussian random variables, but it turns out that this is merely a special case of a much more general theorem. For example, under the assumptions of the theorem of Hurwitz almost all series $\sum(\pm c_n)z^n$ are nowhere continuable.

It was Steinhaus' idea to introduce methods of probability into construction of functions with required properties. Here the property was noncontinuability, but there are many similar situations. Let me describe one which is elementary but very useful.

Let $\phi_0(t)$ be a function of period 1 which is equal to 1 for $0 \leq t < 1/2$ and to -1 for $1/2 \leq t < 1$. Let $\phi_n(t) = \phi_0(2^n t)$, $n = 0, 1, \ldots$. The $\phi_n(t)$—called *Rademacher functions*—form an orthonormal system on $0 \leq t \leq 1$ and are known to possess the following properties: For any sequence $\{c_n\}$, $n = 0, 1, 2, \ldots$ of real or complex numbers, if $\sum |c_n|^2 < \infty$, then the series $\sum c_n \phi_n$ converges almost everywhere and its sum is in L^p on the interval $0 \leq t \leq 1$, no matter how large p is (it is even exponentially integrable). If, on the contrary, $\sum |c_n|^2 = \infty$, then $\sum c_n \phi_n$ not only diverges almost everywhere but is almost everywhere nonsummable by any linear method of summability. Consider now any trigonometric series

$$(1/2)a_0 + \sum_{n=1}^{\infty} (a_n \cos nx + b_n \sin nx)$$

with, say, real coefficients. Then using very elementary methods, one can show that if $\sum(a_n^2 + b_n^2) < \infty$, then almost all series $S_t = \sum \phi_n(t)(a_n \cos nx + b_n \sin nx)$ converge almost everywhere and are in the class L_p for every finite p, and if $\sum(a_n^2 + b_n^2) = \infty$ then almost all S_t are nonsummable by any linear method of summability, and in particular, are not Fourier series. The situation is rather typical and the method when applied to various series leads to examples illustrating various points of the theory of functional developments. It is an important method in the theory of Fourier series.

Let me mention in this connection one problem to which I do not know the answer but which intrigues me. There is a celebrated theorem of Carleson which says that the Fourier series of a function in L^2 converges almost everywhere. On the other hand, Kolmogorov and Zahorski showed that there is an L^2 function whose Fourier series suitably rearranged diverges almost everywhere. Thus if we do not fix the order of terms in a Fourier series, we may have convergence almost everywhere as well as divergence almost everywhere. The question naturally arises which situation occurs "more frequently." The question may be a little foolish and may have no obvious answer since we have no measure in the space of rearrangements of natural numbers. Still, it is of a certain interest since in analysis we have situations when a sequence of functions has no "natural" ordering (take, for example, a general orthogonal system).

The school of Lwów is technically no longer in existence and its organ *Studia Mathematica*, begun in 1932, is now being published in Warsaw. But the influence of the work of its founders and their pupils continues and grows in various Polish mathematical centers. The names of Banach, Steinhaus, Schauder, Kaczmarz, Auerbach, Ulam, Mazur, Orlicz, Nikliborc, Schreier, Ruziewicz, Kac, and others symbolize the achievements of this school.

Chapter 4
My Scottish Book 'Problems', *Paul Erdös*

I shall discuss several problems which are connected with the Scottish Book. Let me start with a problem which F. Bagemihl and I solved. Everybody knows Riemann's theorem: A nonabsolutely convergent series of real numbers has the property that for any preassigned number a the series can be reordered to converge to a. Bagemihl and I proved that for the Cesàro sum of the series there are three possibilities: On reordering the domain of convergence is just one point or it is the whole line, or it is an arithmetic progression (this last possibility does not occur in Riemann's theorem). Our paper appeared in *Acta Mathematica* in 1954; later we found that it is Problem 28 of the Scottish Book, due to Mazur. Some interesting questions remain. First of all one could investigate what happens with other summability methods, for example with C_k, the kth Cesàro mean, and more complicated summability methods. Lorentz and Zeller proved that for an arbitrary analytic set there is a matrix summability method such that by reordering one can get that analytic set. But this still leaves the problem of what happens if one uses a decent summability method like C_k or Abel or some other fixed scheme. Bagemihl and I did not investigate what happens for complex series under reordering and Cesàro summability. There are interesting possibilities here. For example, there is a very pretty theorem by Steinitz: for a reordered complex series, there are three possibilities for the convergent sums; they constitute (a) a single point, (b) a flat, or (c) the whole complex plane. The analogue for n-dimensional vectors also holds. I believe it was Banach who raised the question as to whether this theorem can be generalized to function spaces, including, of course, Hilbert space. This was answered in the negative by Marcinkiewicz or Mazur.

The Scottish Book's Problem 8, due to Mazur, is a very nice question. There is a classical theorem which states that the Cauchy product of two convergent series U and V need not be convergent, but the product series is always C_1-summable to the sum UV. Mazur asks the converse: Is every series summable by the first mean representable as the Cauchy product of two convergent series? I tried to do this but I couldn't, and it should be looked at by somebody who knows more about it than I.

© Springer International Publishing Switzerland 2015

R.D. Mauldin, *The Scottish Book*, DOI 10.1007/978-3-319-22897-6_4

Problems 22 (Ulam-Schreier) and 99 (Ulam) are as follows:
(Problem 22) Is every set z of real numbers a Borel set with respect to sets G which are additive groups of real numbers?
(Problem 99) Can every set in the plane be gotten by Borel operations on squares?

Ulam and I settled many of these questions long ago, but we never got around to publishing our results. These results were rediscovered and published by the Indian mathematician B.V. Rao; when Rao sent me a preprint I urged him to publish. Naturally, I did not tell him that Ulam and I had already done the work. Eventually he found out, however, and asked me why I hadn't said anything about it. I replied that this was the one respect in which I did not want to imitate Gauss, who had the nasty habit of "putting down" younger mathematicians by telling them he had long ago obtained their supposedly new results.

I have some remarks on Problem 88. This problem is a curious question about infinite series, due to Mazur:
Consider a sequence of numbers $\{a_n\}_{n=1}^\infty$ with the following property: if $x_1, x_2, \ldots$ is a bounded sequence, then

$$\left| \sum_{i=1}^\infty a_i x_i \right| + \left| \sum_{i=1}^\infty a_{i+1} x_i \right| + \left| \sum_{i=1}^\infty a_{i+2} x_i \right| + \cdots$$

converges. Is it true that $\sum_{n=1}^\infty n|a_n|$ converges?

I have no idea why it should be true and I haven't been able to settle it. [Editor's Note: The problem has been solved; see Commentary.]

Now let me talk about some of the problems which don't seem to be very difficult but still may be of some interest even now, after many years. One of these is a very pretty conjecture by Borsuk which says that if one has a set of diameter 1 in n-dimensional space it can be decomposed into $n+1$ sets of diameter < 1. This is trivial on the line, easy in the plane, difficult in 3-space, and unsolved higher (I suspect that it is false for sufficiently high dimension). There is another difference between two and three dimensions. For two dimensions one knows the extremal solution to be an equilateral triangle. The decomposition is as follows: Construct the circumscribed circle about the equilateral triangle, and draw three radii so that they divide the circle into three equal areas.

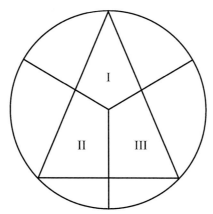

This induces the desired decomposition of the triangle—the three sets are congruent and have diameter about .88, and that is the extreme situation.

Now in three dimensions nobody knows what the extreme result is. The three-dimensional case was done by Eggleston and then independently and much more simply by Grünbaum and Heppes.

There is a very simple theorem of Steinhaus which says that the difference set of a set of positive measure (say on the line) contains an interval. This is an almost trivial theorem by our present standards. It follows instantly from the Lebesgue density theorem, and therefore by this method one obtains the following theorem: Any set of positive measure on the line, or, more generally, in k-dimensional space, contains all finite sets in the space to within a similarity transformation. The proof is almost immediate because by the Lebesgue density theorem there is an interval or a sphere in which the density is as close to 1 as one wishes, and therefore it follows that set will contain a set which is similar to any finite set. A related question is Problem 146 (Ulam), to which the answer is negative. For a set of positive measure (on the line, say) one can find an interval in which the density is $> 1 - \varepsilon$. Problem 146 asks: Can one determine how fast the density will tend to 1 as a function of the length of the interval? It is easy to see that the answer is negative—no general statement can be made as to how fast the density will tend to 1.

I have the following problem; perhaps it is easy, but it has remained unsolved for so long that I should offer \$100. Consider a sequence of positive reals $\{x_n\}_{n=1}^{\infty}$, with $x_n \to 0$. Does there exist a set of positive measure which does not contain a set similar to this sequence? If the answer is yes, it would show that this simple extension of Steinhaus' theorem does not hold for infinite sets. In this case one could ask for the minimum of the measure of a set which has this property. I don't think the problem is difficult, but perhaps it is not quite trivial.

An amusing consequence of Steinhaus' theorem is the following: A set of infinite measure in the plane contains the vertices of some triangle with a preassigned area. If one wants to find a triangle of area 1 whose vertices are required to lie in the set, it is easy to do it. The same is true of a set E in the plane which has a line which E intersects in a set of positive measure, and which has points arbitrarily far from the line, i.e., it contains triangles of any area. This is an immediate consequence of Steinhaus' theorem; I want to pose a slightly different problem.

Is it true that there is an absolute constant c so that a set with planar measure $> c$ contains three points which form a triangle with area 1? I don't know what the answer is. The extremal case might take the form: Choose a circle so that the inscribed equilateral triangle has area 1, and take the interior of that circle. Then this set will not contain a triangle with area 1 and the corresponding area may be the minimum value of c. This surely should be disprovable if it is false.

Incidentally, there is an interesting question of Ulam and myself, the proof of which is lost. We had the following result: Take an ideal in the integers and take the Boolean algebra modulo that ideal—for example, one can take the Boolean algebra of subsets of integers modulo the finite sets. In other words, two subsets are distinct if they differ by an infinite set. First of all we wanted to prove that there are 2^c nonisomorphic ideals; we didn't completely do this, but it has been done in the meantime.

The thing which is lost is the following: Consider the ideal of the finite sets, the ideal of the sets of density zero, and the ideal of the sets of logarithmic density zero. Now it is clear that if a sequence has density 0, then it has logarithmic density 0, and it is clear that the converse is false. (The integers between $n!$ and $2(n!)$ clearly don't have density 0 and clearly do have logarithmic density 0. By clearly, I mean a good freshman should be able to do it, although it's not completely trivial.) We proved easily that the algebra modulo finite sets is not isomorphic to the other two because the Boolean algebra modulo finite sets has no upper bound and the other two have. Now we allegedly proved that the Boolean algebra modulo the sequences of density 0 and logarithmic density 0 are not isomorphic. When I first visited Ulam in 1943 or 1944 in Madison we had the proof, then six months later we had forgotten the proof, and had to reconstruct it, so it seems that the proof should have been correct. Now the proof is gone and nobody can prove it. This problem should be settled; perhaps I should offer a hundred dollars for a proof (or a disproof) that these two Boolean algebras are not isomorphic. If it is trivial I well deserve to have to pay the hundred dollars.

I also want to mention a very nice problem of Tarski which should be settled: squaring the circle. Can a square and a circle of the same area be decomposed into a finite number of congruent parts? This is a very beautiful problem, and rather well known. If it were my problem I would offer $1000 for it—a very nice question, possibly very difficult. Really one has no obvious method of attack. In higher dimensions this is no longer true. As everybody knows, this is the famous Banach-Tarski paradox, the basic idea of which really goes back to Hausdorff in his 1914 book. [Ed. Note: This problem was solved by M. Laczkovich in his beautiful paper, Equidecomposability and discrepancy: a solution to Tarksi's circle squaring problem, *J. Reine Angew. Math.* 404 (1990), 77–117. His solution has led to many research directions today (2015).]

Now let me say a few things about the Cauchy equation $f(x+y) = f(x) + f(y)$. Assume $f(x+y) - f(x)$ is a continuous function of x for every h. I conjectured that $f(x) = g(x) + h(x)$, where $g(x)$ is continuous and $h(x)$ is a Hamel function.

I didn't know how to prove it, so I did the next best thing—I guessed who would be able to prove it. I wrote to de Bruijn, and he proved my conjecture. The paper appeared in the *Niewu Archief vorr Wiskunde* about 28 years ago. I also conjectured that if $f(x+h) - f(x)$ is a measurable function of x, for every h, then $f(x)$ can be decomposed into three parts: $g(x)$ measurable, $h(x)$ Hamel, and $r(x)$, where $r(x+h) - r(x)$ is zero almost everywhere. This conjecture has recently been proved by Laczkovich, a young Hungarian mathematician. [Ed. Note. See M. Laczkovich, Functions with measurable differences, *Acta Math. Acad. Sci. Hungar.* 35 (1980), 217–235. Laczkovich wrote a survey about all of this: The difference property. In: *Paul Erdős and his Mathematics* (editors: G. Halász, L. Lovász, M. Simonovits and V. T. Sós), Springer, 2002, Vol. I, 363–410.] The nicest problem here is due to Kemperman; if it were my problem I would offer $500 for it. This is the problem of Kemperman: If for every x and positive h, $2f(x) \leq f(x+h) + f(x+2h)$, then $f(x)$ is monotonic. Now at first sight this seems harmless—it looks as though anyone could prove it or find a counterexample, but nobody has succeeded. It is rather easy to show that if $f(x)$ is measurable and satisfies the above condition property it must be monotonic. That is a simple exercise, but one can define such a function which is not monotonic on the rational numbers, so one has to use more than just a countable subset, and this is all I know about it. Nobody else has made any progress on this problem—the question remains open. I think it is very surprising that this should be so difficult. [Ed. Note. This problem was solved by M. Laczkovich, *On Kemperman's inequality* $2f(x) \leq f(x+h) + f(x+2h)$. *Colloq. Math.* 49, (1984), 109–115, also see his paper: *A generalization of Kemperman's functional inequality*, $2f(x) \leq f(x+h) + f(x+2h)$. *General Inequalities 3* Oberwolfach, 1981, Proceedings, edited by E. F. Beckenbach; Birkháuser, (1983), 281–293. As Laczkovich points out these papers deal with the following general unsolved problem. Suppose f is a function defined on the real line satisfying the inequality

$$a_1 f(x+b_1 h) + \cdots + a_n f(x+b_n h) \geq 0$$

for every real x and nonnegative h. For which values a_i, b_i does this condition imply that f is monotonic?]

Let me talk about a different kind of problem which still bears some relation to the problems discussed above. Kakutani and I proved in 1942 the following the theorem: $c = \aleph_1 \iff$ the real line (continuum) can be decomposed as the union of $\aleph_0$ Hamel bases. In other words, if $c = \aleph_1$, then one can decompose the real line into countably many rationally independent sets, and conversely. This is not very hard to prove; it appeared in in *Bulletin of the AMS* in 1943. Now I ask the following question: Can an n-dimensional Euclidean space be decomposed into countably many sets so that each set has the property that the distances are all distinct? The relation of this to the previous theorem is that if one decomposes the real line into $\aleph_0$ Hamel bases and if one takes a Hamel basis, then all distances will be distinct between two points of the Hamel basis, because of the rational independence. In a Hamel basis, with elements a, the numbers $a_1 - a_2$ are all distinct. So for the real line the theorem of Kakutani and myself settles it, but for the plane there are already great difficulties. Nevertheless R.O. Davies, the English mathematician, succeeded in carrying out the proof. Recently I had a letter from K. Kunen in which he says that he has proved my conjecture for n-dimensional space, and that the proof is very complicated. [See K. Kunen, Partitioning Euclidean space. *Math. Proc. Cambridge Philos. Soc.* 102 (1987), 379–383.] So it seems this problem is settled now. For every n, n-dimensional space can be decomposed into the union of countably many sets S_k where each S_k has the property that any four points give six distinct distances; in other words, all the distances are defined. The continuum hypothesis must be used; without the continuum hypothesis, it cannot possibly be true. It is already wrong for the line. If $c > \aleph_1$ and one decomposes (theorem of Hajnal and myself) the real line into $\aleph_0$ sets, then there are always four points which determine only four distances, i.e., at least one of the sets contains four points which determine only four distances. It will contain, namely, the following configuration: ●● ●●. This is not hard to prove, using a partition theorem due to Hajnal and myself. If each point of a set of size $\aleph_1$ is connected to every point of a set of size $\aleph_2$ and the edges of the resulting graph are colored with $\aleph_0$ colors, then there is a monochromatic circuit of length four in the graph. From this one can obtain the above configuration. Now in Hilbert space the situation is completely different. Hajnal and I have an easy example in Hilbert space of a set of power c where all triangles are isosceles, so in this case one can't even find three points where all distances are distinct. I asked, and Posa settled, the following question: Is there a set of power c in Hilbert space so that every subset of power c contains an n-dimensional regular simplex? Every subset of power c contains an equilateral triangle or regular simplex and this is true even for infinite dimensional simplices if the continuum hypothesis is assumed, so Hilbert space behaves completely differently even in this simple case. Now it frequently happens in problems of this sort that the infinite dimensional case is easier to settle than the finite dimensional analogues. This moved Ulam and me to paraphrase a well-known maxim of the American armed forces in World War II: "The difficult we do immediately, the impossible takes a little longer," viz: "The infinite we do immediately, the finite takes a little longer."

There is a beautiful theorem of Sierpiński. I remember how surprised I was when I first saw it. If $c = \aleph_1$ the plane can be decomposed into two sets S_1 and S_2 so that every vertical line meets S_1 in a countable set and every horizontal line meets S_2 in a countable set. It is a very simple theorem by present standards but it was very startling then. I made the following generalization, which is also very simple. Split the lines into two classes. Then if $c = \aleph_1$ one can decompose the plane into two sets so that S_1 meets every line of the first class in a countable set and S_2 meets every line of the second class in a countable set. Thus one is not restricted to the vertical and horizontal lines. The proof is almost trivial but rather standard, and there are various generalizations. R.O. Davies has investigated and settled almost all the problems here.

Now there is a very pretty three-dimensional generalization of Sierpiński which goes as follows: If $c = \aleph_1$, three-dimensional space can be decomposed into three sets, E_1, E_2, E_3, so that every line parallel to the x-axis meets E_1 in a finite set, every line parallel to the y-axis meets E_2 in a finite set, and every line parallel to the z-axis meets E_3 in a finite set. It is a very pretty theorem. This is necessary and sufficient for $c = \aleph_1$. After the war Sierpiński returned to elementary number theory, which was his first love. He did his first work in number theory and his last work in number theory: in between he did set theory and real functions. This, however, was one of the few really new things which he did after the war. Actually when I first lectured in Poland in 1956 on the partition calculus of Rado and myself, certain things were called Sierpińskizations. But Sierpiński was not very interested in that—he really was very much more interested by that time in number theory. At any rate, this is a very pretty theorem. Kuratowski has various generalizations, and Hajnal and I raised the following problem which was one of the few problems in partition calculus which was unsolved until very recently. This is the problem (I offered \$50 for it): Choose three sets of power $\aleph_1$: A, B, and C. Take the triples (x, y, z), $x \in A$, $y \in B$, $z \in C$, and decompose them in an arbitrary way into two classes. Then is it true that there is a set A_1 of size $\aleph_0$ in A, a set B_1 of size $\aleph_0$ in B, and a set C_1 of size $\aleph_0$ in C so that all triples from A_1, B_1, and C_1 are in the same class? That was one question which remained unsolved. During the last meeting on logic and set theory in Cambridge, Prikry and Mills thought of this negatively, and they disproved it. I think the paper will appear soon.

I will close with some comments on number theory. In 1947 there was a meeting in Oslo, and MacLane handed me a paper to referee, by a young man called Mills, the son of the older Mills. He proved the following theorem: There is a real number $c > 1$ so that $\left[c^{3n} \right]$ is a prime, for every n. I was very excited just because he had written down an arbitrarily large prime which seems far better than humanity deserves in this case. Well, I was a little disappointed when I looked at the paper, which was very nice and the first of its kind, but in a way it was cheating—it has nothing to do with primes. All one needs to know about the primes is that there is a prime between two consecutive cubes and then one constructs the c by a descending sequence of integers which are products of primes. So actually one doesn't get a single new prime. It is a nice remark but it is useless for the theory of primes. The existence of a polynomial of many variables which whenever positive is also a prime is of no use to number theory. It tells nothing about primes and a similar result holds for any recursively enumerable set.

Chapter 5
KKM-Maps, *Andrzej Granas*

5.1 Definition and examples

Let E be a real vector space and $X \subset E$ be an arbitrary subset. A set-valued function $G : X \to 2^E$ is called a Knaster-Kuratowski-Mazurkiewicz map or simply a KKM-map[1] if

$$\operatorname{conv}\{x_1, \dots, x_s\} \subset \bigcup_{i=1}^{s} G(x_i)$$

for each finite subset $\{x_1, \dots, x_s\} \subset X$.

We now give some examples of KKM-maps.

(i) *Variational problems.* Let C be a convex subset of E and $\phi : C \to \mathbb{R}$ be a convex functional[2] combination $\sum \lambda_i x_i$ in C; if $\phi : C \to \mathbb{R}$ is convex, then the sets $\{y \in C \mid \phi(y) < \lambda\}$ and $\{y \in C \mid \phi(y) \le \lambda\}$ are convex for each $\lambda \in \mathbb{R}$. For each $x \in C$, let

$$G(x) = \{y \in C \mid \phi(y) \le \phi(x)\}.$$

We show that $G : C \to 2^C$ is a KKM-map. For a contradiction let $y_0 = \sum \lambda_i x_i$ be convex combination in C such that $y_0 \notin \bigcup_{i=1}^{n} G(x_i)$. Then $\phi(x_i) < \phi(y_0)$ for $i = 1, 2, \dots, n$ and this means that each x_i lies in $\{x \mid \phi(x) < \phi(y_0)\}$; since this set is convex we have a contradiction $\phi(y_0) = \phi(\sum \lambda_i x_i) < \phi(y_0)$.

(ii) *Best approximation.* (a) Let $E = (E, \|\cdot\|)$ be a normed linear space, $C \subset E$ be a convex set and $f : C \to E$ be a map. For each $x \in C$ let

$$G(x) = \{y \in C \mid \|fy - y\| \le \|fy - x\|\}.$$

---

[1] 2^E stands for the set of subsets of E and $\operatorname{conv}(A)$ for the convex hull of $A \subset E$.
[2] Recall that a real-valued functional ϕ on C is convex if $\phi(\sum \lambda_i x_i)$ for any convex

© Springer International Publishing Switzerland 2015
R.D. Mauldin, *The Scottish Book*, DOI 10.1007/978-3-319-22897-6_5

We show that $G : C \to 2^C$ is a KKM-map. Indeed, let $y_0 = \sum \lambda_i x_i$ be a convex combination in C. If $y_0 \notin \bigcup_{i=1}^n G(x_i)$, we would have $\|fy_0 - y_0\| > \|fy_0 - x_i\|$ for each $i = 1, 2, \ldots, n$, i.e., that x_i lies in the open ball $\{x \in E \mid \|fy_0 - x\| < \|fy_0 - y_0\|\}$. Since this ball is convex it contains $y_0 \in \text{conv}\{x_1, \ldots, x_n\}$ and we have a contradiction: $\|fy_0 - y_0\| < \|fy_0 - y_0\|$.

   (b) Let E be a vector space, $C \subset E$ be convex, p be a seminorm on E and let $f : C \to E$ be any map. For each $x \in C$ let

$$G(x) = \{y \in C \mid p(fy - y) \le p(fy - x)\}.$$

The same argument as in the previous example shows that $G : C \to 2^C$ is a KKM-map.

(iii) *Variational inequalities.* Let $E = (H, (\ ,\ ))$ be a pre-Hilbert space, C a convex subset of H and $f : C \to H$ any map. For each $x \in C$, let

$$G(x) = \{y \in C \mid (fy, y - x) \le 0\}.$$

We show that $G : C \to 2^E$ is a KKM-map. Indeed, let $y_0 \in \text{conv}\{x_1, \ldots, x_n\}$. If $y_0 \notin \bigcup_{i=1}^n G(x_i)$, we would have $(fy_0, y_0 - x_i) > 0$ for each $i = 1, 2, \ldots, n$, i.e., that each x_i lies in the set $\{x \in E \mid (fy_0, y_0) > (fy_0, x)\}$. Since this set is convex it also contains $y_0 = \sum \lambda_i x_i$ and we have a contradiction: $(fy_0, y_0) < (fy_0, y_0)$.

The principle of KKM-maps

The following fundamental result represents a version of the well-known Knaster-Kuratowski-Mazurkiewicz theorem [19], which was used in their simple proof of Brouwer's fixed point theorem:

Theorem 5.1. *Let E be a vector space, X an arbitrary subset of E, and $G : X \to 2^E$ a KKM-map such that each $G(x)$ is finitely closed[3]. Then the family $\{G(x) \mid x \in X\}$ of sets has the finite intersection property.*

Proof. We argue by contradiction, so assume that $\bigcap_1^n G(x_i) = \emptyset$. Working in the finite-dimensional subspace L spanned by $\{x_1, \ldots, x_n\}$, let d be the Euclidean metric in L and $C = \text{conv}\{x_1, \ldots, x_n\} \subset L$. Note that because each $L \cap G(x_i)$ is closed in L, and since $\bigcap_1^n L \cap G(x_i) = \emptyset$ by hypothesis, the function $\phi : C \to \mathbb{R}$ given by $x \mapsto \sum_1^n d(x, L \cap G(x_i))$ does not vanish. We now define a continuous map $f : C \to C$ by setting

$$f(x) = \frac{1}{\phi(x)} \sum_{i=1}^n d(x, L \cap G(x_i)) \cdot x_i.$$

---

[3]A subset $A \subset E$ is finitely closed if its intersection with each finite dimensional linear subspace $L \subset E$ is closed in the Euclidean topology of L.

By Brouwer's fixed point theorem, f would have a fixed point $x_0 \in C$. Letting $I = \{ i \mid d(x_0, L \cap G(x_i)) \neq 0 \}$, the fixed point x_0 cannot belong to $\bigcup \{ G(x_i) \mid i \in I \}$; however,

$$x_0 = f(x_0) \in \text{conv}\{ x_i \mid i \in I \} \subset \{ G(x_i) \mid i \in I \}$$

and, with this contradiction, the proof is complete.

As an immediate corollary we obtain:

Theorem 5.2. *(Ky Fan [7]). Let E be a topological vector space[4], $X \subset E$ an arbitrary subset, and $G : X \to 2^E$ a KKM-map. If all the sets $G(x)$ are closed in E, and if one is compact, then $\bigcap \{ G(x) \mid x \in X \} \neq \emptyset$.*

We now observe that the conclusion $\bigcap G(x) \neq \emptyset$ can be reached in another way which avoids placing any compactness restriction on the sets $G(x)$; it involves using an auxiliary family of sets and a suitable topology on E:

Theorem 5.3. *Let E be a vector space, X an arbitrary subset of E, and $G : X \to 2^E$ a KKM-map. Assume there is a set-valued map $\Gamma : X \to 2^E$ such that $G(x) \subset \Gamma(x)$ for each $x \in X$, and for which $\bigcap_{x \in X} G(x) \neq \emptyset \implies \bigcap_{x \in X} \Gamma(x) \neq \emptyset$.*

Because of (1.1) the proof is obvious.

Simple Applications

We give now some simple applications of KKM-maps.

Theorem 5.4. *(Mazur-Schauder [22]) Let E be a reflexive Banach space and C a closed convex set in E. Let ϕ be a lower-semicontinuous[5] convex and coercive (i.e., $|\phi(x)| \to \infty$ as $\|x\| \to \infty$) functional on C. If ϕ is bounded from below, then at some $x_0 \in C$ the functional ϕ attains its minimum.*

Proof. Let $d = \inf\{ \phi(x) \mid x \in C \}$; because ϕ is coercive, we can find a number $\rho > 0$ such that $K = B(0, \rho) \cap C \neq \emptyset$ and $\phi(x) > d + 1$ for all $x \in C \backslash K$. It is enough now to show that there is a point $x_0 \in K$ such that $\phi(x_0) \leq \phi(x)$ for all $x \in K$. For each $x \in K$, let $G(x) = \{ y \in K \mid \phi(y) \leq \phi(x) \}$; since $d = \inf \phi(x)$, the theorem will be proved by showing $\bigcap G(x) \neq \emptyset$. Since $G : K \to 2^E$ is a KKM-map (cf. example (i)), the conclusion is obtained by observing that in the weak topology of E each $G(x)$ (being closed and convex) is compact.

---

[4]All topological spaces are assumed to be Hausdorff.

[5]A map $f : X \to \mathbb{R}$ on a topological space is lower semicontinuous (l.s.c.) if $\{ x \in X \mid f(x) > \lambda \}$ is open for each $\lambda \in \mathbb{R}$; it is upper semicontinuous (u.s.c.) if $\{ x \in X \mid f(x) < \lambda \}$ is open for each $\lambda \in \mathbb{R}$.

The next result generalizes one of the forms of the Schauder fixed point theorem; it follows that the principle of KKM-maps is in fact equivalent to the Brouwer fixed point theorem.

Theorem 5.5. *(Ky Fan [10]) Let C be a compact convex set in a normed space E and let $f : C \to E$ be continuous. Assume further that, for each $x \in C$ with $x \neq f(x)$ the line segment $[x, f(x)]$ contains at least two points of C. Then f has a fixed point.*

Proof. Define $G : C \to 2^E$ by

$$G(x) = \{ y \in C \mid \|y - f(y)\| \leq \|x - f(y)\| \}.$$

We know (cf. example (ii)) that G is a KKM-map. Because f is continuous, the sets $G(x)$ are closed in C, and therefore compact. Consequently, we find a point y_0 such that $y_0 \in \bigcap_{x \in C} G(x)$ and hence $\|y_0 - f(y_0)\| \leq \|x - f(y_0)\|$ for all $x \in C$. We show that y_0 is a fixed point: the segment $[y_0, f(y_0)]$ must contain a point of C other than y_0, say $x = t y_0 + (1 - t) f(y_0)$ for some $0 \leq t < 1$; then $\|y_0 - f(y_0)\| leqt \|y_0 - f(y_0)\|$ and, since $t < 1$, we must have $y_0 - f(y_0) = 0$.

Theorem of Tychonoff and two of its generalizations

Theorem 5.6. *(Tychonoff [31]) Let C be a compact convex set in a locally convex topological space E. Then every continuous $f : C \to C$ has a fixed point.*

Proof. Let $\{p_i\}_{i \in I}$ be the family of all continuous seminorms in E. For each $i \in I$ set

$$A_i = \{ y \in C \mid p_i(y - f(y)) = 0 \}.$$

A point $y_0 \in C$ is a fixed point for f if and only if $y_0 \in \bigcap_{i \in I} A_i$. By compactness of C we need show only that $\bigcap_{j \in J} A_j \neq \emptyset$ for each finite subset $J \subset I$. Define $G : C \to 2^E$ by

$$G(x) = \{ y \in C \mid \sum_{j \in J} p_j(y - f(y)) \leq \sum_{j \in J} p_j(x - f(y)) \}.$$

It is easy to verify that G is a KKM-map; consequently, there is a point $y_0 \in C$ such that

$$\sum_{j \in J} p_j(y_0 - f(y_0)) \leq \sum_{j \in J} p_j(x - f(y_0))$$

for all $x \in C$. This clearly implies that $p_j(y_0 - f(y_0)) = 0$ for $j \in J$ and thus $y_0 \in \bigcap_{j \in J} A_j$.

The Tychonoff fixed point theorem (1.6) is a special case of the following result of Ky Fan [10] which extends Theorem (1.5) to locally convex spaces:

Theorem 5.7. *Let C be a compact convex set in a locally convex topological vector space E and let f : C → E be continuous. Assume that, for each x ∈ C with x ≠ f(x), the line segment [x, f(x)] contains at least two points of C. Then f has a fixed point.*

Proof.   Assume $f(x) \neq x$ for all $x \in C$. Then for some continuous seminorm p we would have $\inf_{y \in C} p[f(y) - y] > 0$. Define $G : C \to 2^C$ by $G(x) = \{ y \in C \mid p(fy - y) \leq p(fy - x) \}$. Since G is a compact valued KKM-map (cf. example (ii)b), we get a point $y_0 \in C$ such that

$$0 < p(fy_0 - y_0) \leq p(fy_0 - x) \quad \text{for all} \quad x \in C.$$

Now the same simple argument as in (1.5) gives a contradiction $p(fy_0 - y_0) < p(fy_0 - y_0)$. The proof is completed.

As an immediate application of (1.7) we derive a fixed point theorem for inward and outward maps in the sense of B. Halpern. Let C be a convex subset of a vector space E; for each $x \in C$, let

$$I_C(x) = \{ y \in C \mid \text{there exists } y_0 \in C \text{ and } \lambda > 0 \text{ such that } y = x + \lambda(y_0 - x) \}$$

and

$$O_C(x) = \{ y \in C \mid \text{there exists } y_0 \in C \text{ and } \lambda > 0 \text{ such that } y = x - \lambda(y_0 - x) \}.$$

A map $f : C \to E$ is said to be inward (resp. outward) if $f(x) \in I_C(x)$ (resp. $f(x) \in O_C(x)$) for each $x \in C$.

Theorem 5.8. *Let C be a convex compact subset of a locally convex topological vector space E. Then every continuous inward (resp. every continuous outward) map f : C → E has a fixed point.*

Proof.   The case of an inward map follows directly from (1.7); if f is outward, then $g : C \to E$ given by $x \mapsto 2x - f(x)$ is inward with the same set of fixed points as f and the conclusion follows.

5.2   Ky Fan fixed point theorem and the minimax inequality

The following result is an important application of the KKM-map principle:

Theorem 5.9. *(Ky Fan [10]) Let C be a nonempty compact convex set in a linear topological space E and let $A : C \to 2^C$ be a set-valued map such that*

(i) $A^{-1}y$[6] is open for each $y \in C$;
(ii)   Ax is convex nonempty for each $x \in C$. Then there is a $w \in C$ such that $w \in Aw$.

---

[6]If $A : X \to 2^Y$, then $A^{-1} : Y \to 2^X$ is defined by the condition $x \in A^{-1} \iff y \in Ax$.

Proof. Define $G : C \to 2^C$ by $y \mapsto C \backslash A^{-1} y$; each $G(y)$ is a nonempty set closed in C, therefore compact. We observe that $C = \bigcup \{ A^{-1} y \mid y \in C \}$:

Given any $x_0 \in C$ choose a y_0 in the nonempty set $A x_0$; then $x_0 \in A^{-1} y_0$. Thus $\bigcap \{ G(y) \mid y \in C \} = \emptyset$ and G cannot be a KKM-map. Therefore for some convex combination $w = \sum_{i=1}^s \lambda_i y_i \notin \bigcup_{i=1}^s G(y_i)$ and hence $w \in C \backslash \bigcup_{i=1}^s G(y_i) = \bigcap_{i=1}^s A^{-1} y_i$. Thus $w \in A^{-1} y$ for each $i = 1, 2, \ldots, s$ and therefore $y_i \in A w$ for each $i = 1, 2, \ldots, s$. Since $A w$ is convex we get $w = \sum \lambda_i y_i \in A w$ and the proof is completed.

Using Theorem (2.1) we shall now derive two other generalizations of the Tychonoff Theorem. We first establish the following.

Theorem 5.10. *Let C be a nonempty compact convex subset of a linear topological space E, V an open convex nbd of 0 and let $f : C \to E$ be a continuous map such that $f(C) \subset C + V$. Then there is an $x_0 \in C$ satisfying $f(x_0) \in x_0 + V$.*

Proof. Define $A : C \to 2^C$ by

$$A x = \{ y \in C \mid f x - y \in V \}.$$

Each $A x$ is convex and each $A^{-1} y$ is open. Supposing $A x_0 = \emptyset$ for some x_0, we get $f(x_0) \notin C + V$ contrary to $f(C) \subset C + V$. Thus by the Ky Fan fixed point theorem we get $x_0 \in A x_0$ for some $x_0 \in C$, i.e., $f(x_0) \in x_0 + V$ and the proof is complete.

Theorem 5.11. *(Schauder-Tychonoff) Let C be a convex subset of a locally convex linear topological space E and let $f : C \to C$ be a continuous compact map (i.e., $f(C)$ is relatively compact in C). Then f has a fixed point.*

Proof. Let V be a convex symmetric nbd of 0. Because E is locally convex it is enough to show that f has a V-fixed point, i.e., a point x_0 such that $f(x_0) \in x_0 + V$. Let $\{ x_i + V \}_{i=1}^k$ be a finite covering of the compact set $f(\overline{C})$ and let $K = \text{conv} \{ x_1, \ldots, x_k \}$. Since $f(K) \subset K + V$ there is by Lemma (2.2) a point $x_0 \in K \cap C$ such that $f(x_0) \in x_0 + V$ and the proof is completed.

Theorem 5.12. *(Ky Fan-Iokhvidov) Let C and K be two convex compact subsets of a locally convex space E and let $f : C \to E$ be a continuous map such that $f(C) \subset C + K$. Then there is a point $x_0 \in C$ such that $f(x_0) \in x_0 + K$.*

Minimax inequality

The following result due to Ky Fan [11] represents an analytic formulation of the Ky Fan fixed point theorem and at the same time generalizes the Mazur-Schauder theorem (2.4):

Theorem 5.13. *(Minimax inequality) Let C be a compact convex set in a topological vector space. Let $f : C \times C \to \mathbb{R}$ be a real-valued function such that:*

(i) $y \mapsto f(x, y)$ is l.s.c. on C for each $x \in C$;

(ii) $x \mapsto f(x,y)$ is quasi- concave[7] on C for each $y \in C$. Then $\min_{y \in C} \sup_{x \in C} f(x,y) \leq \sup_{x \in C} f(x,x)$.

Proof. Note that $y \mapsto \sup f(x,y)$ is l.s.c. and hence its minimum $\min_{y \in C} \sup_{x \in C} f(x,y)$ on the compact set C exists. Let $\mu = \sup_{x \in C} f(x,x)$; clearly, we may assume that $\mu < \infty$. Define $G : C \to 2^E$ by

$$G(x) = \{ y \in C \,|\, f(x,y) \leq \mu \}.$$

As in Example (i), it can be easily verified that G is a KKM-map; furthermore, each $G(x)$ is compact because $y \mapsto f(x,y)$ is l.s.c. By Theorem (1.2) we infer that $\bigcap_{x \in C} G(x) \neq \emptyset$ and hence there is a $y_0 \in C$ such that $f(x,y_0) \leq \mu$ for all $x \in C$; this clearly implies the assertion of the theorem and thus the proof is completed.

Among numerous applications of the Ky Fan minimax inequality we mention the following fundamental existence theorem in potential theory:

Theorem 5.14. *Let X be a compact space and $G : X \times X \to \mathbb{R}^+$ a continuous function such that $G(x,x) > 0$ for all $x \in X$. Then there exists a positive Radon measure μ on X such that*

$$\int G(x,y) d\mu(y) \geq 1$$

for all $x \in X$, and

$$\int G(x,y) d\mu(y) = 1$$

for x in the support of μ.

For a proof, we refer to Ky Fan [11].

5.3 KKM-maps and variational inequalities

KKM-maps can be used to get some of the basic facts in the theory of variational inequalities.

Let $(H,(\,,\,))$ be a Hilbert space and C be any subset of H. We recall that a map $f : C \to H$ is called monotone[8] on C if $(f(x) - f(y), x - y) \geq 0$ for all $x, y \in C$. We say that $f : C \to H$ is hemi-continuous if $f|_{L \cap C}$ is continuous for each one-dimensional flat $L \subset H$.

---

[7]Recall that a real-valued function ϕ defined on a convex set C is *quasi-concave* if the set $\{x \in C \,|\, f(x) > r\}$ is convex for each $r \in \mathbb{R}$; ϕ is *quasiconvex* if $-\phi$ is quasi-concave.

[8]If $\phi : H \to \mathbb{R}$ is Gateaux differentiable, then $\phi' : H \to H$ is monotone if and only if ϕ is convex; thus, the notion of a monotone operator arises naturally in the classical context of the calculus of variations.

Theorem 5.15. *(Hartman-Stampacchia [15]) Let H be a Hilbert space, C a closed bounded convex subset of H, and $f : C \to H$ monotone and hemi-continuous. Then there exists a $y_0 \in C$ such that $(f(y_0), y_0 - x) \leq 0$ for all $x \in C$.*

Proof.   For each $x \in C$, let

$$G(x) = \{ y \in C \mid (f(y), y - x) \leq 0 \};$$

the theorem will be proved by showing $\bigcap \{ G(x) \mid x \in C \} \neq \emptyset$.

We know (cf. example (iii)) that $G : C \to 2^H$ is a KKM-map. Consider now the map $\Gamma : C \to 2^H$ given by

$$\Gamma(x) = \{ y \in C \mid (f(x), y - x) \leq 0 \};$$

we show that Γ satisfies the requirements of (2.4):

(i) $G(x) \subset \Gamma(x)$ for each $x \in C$. For, let $y \in G(x)$, so that $0 \geq (f(y), y - x)$. By monotonicity of $f : C \to H$ we have $(f(y) - f(x), y - x) \geq 0$ so $0 \geq (f(x), y - x)$ and $y \in \Gamma(x)$.

(ii)   Because of (i), it is enough to show $\bigcap \{ \Gamma(x) \mid x \in C \} \subset \bigcap \{ G(x) \mid x \in C \}$. Assume $y_0 \in \bigcap \Gamma(x)$. Choose any $x \in C$ and let $z_t = tx + (1-t)y_0 = y_0 - t(y_0 - x)$; because C is convex, we have $z_t \in C$ for each $0 \leq t \leq 1$. Since $y_0 \in \Gamma(z_t)$ for each $t \in [0, 1]$, we find that $(f(z_t), y_0 - z_t) \leq 0$ for all $t \in [0, 1]$. This says that $t(f(z_t), y_0 - x) \leq 0$ for all $t \in [0, 1]$ and, in particular, that $(f(z_t), y_0 - x) \leq 0$ for $0 < t \leq 1$. Now let $t \to 0$; the continuity of f on the ray joining y_0 and x gives $f(z_t) \to f(y_0)$ and therefore that $(f(y_0), y_0 - x) \leq 0$. Thus, $y_0 \in G(x)$ for each $x \in C$ and $\bigcap \Gamma(x) = \bigcap G(x)$.

(iii)   We now equip H with the weak topology. Then each $\Gamma(x)$, being the intersection of the closed half-space $\{ y \in H \mid (f(x), y) \leq (f(x), x) \}$ with C, is closed convex and bounded and therefore weakly compact.

Thus, all the requirements in (1.4) are satisfied; therefore, $\bigcap \{ G(x) \mid x \in C \} \neq \emptyset$ and, as we have observed, the proof is complete.

Corollary 5.1. *(Browder-Goedhe-Kirk) Let C be a closed bounded convex subset of H and $F : C \to C$ a nonexpansive map, i.e., $\|Fx - Fy\| \leq \|x - y\|$ for all $x, y \in C$. Then F has a fixed point.*

Proof.   Putting $f(x) = x - F(x)$ for $x \in C$, we verify by simple calculation that $f : C \to H$ is a continuous monotone map; so by theorem (3.1) there is a $y_0 \in C$ such that $(y_0 - Fy_0, y_0) = (fy_0, y_0 - x) \leq 0$ for all $x \in C$. By taking in the above inequality $x = F(y_0)$ we get $y_0 = Fy_0$, and the proof is complete.

Corollary 5.2. *(Nikodym [26]) Let $C \subset H$ be a closed bounded convex set. Then for each $x_0 \in H$ there is a unique $y_0 \in C$ with $\|x_0 - y_0\| = \inf\{ \|x_0 - x\| \mid x \in C \}$.*

Proof.   Uniqueness being evident, let $f : C \to H$ be given by $y \mapsto y - x_0$; clearly, f is continuous and monotone. By (3.1) there is $y_0 \in C$ with $(y_0 - x_0, y_0 - x) \leq 0$ for all $x \in C$; this being equivalent to $\|x_0 - y_0\| = \inf_C \|x_0 - x\|$, the assertion of the theorem follows.

5.4   KKM-maps and the theory of games

The notion of a KKM-map can be used to establish general geometric results which have many applications in the theory of games.

The Coincidence Theorem and the Minimax Principle

Theorem 5.16.   *(Ky Fan) Let $X \subset E$ and $Y \subset F$ be nonempty compact convex sets in the linear topological spaces E and F. Let $A, B : X \to 2^Y$ be two set-valued maps such that*

(i) *Ax is open and Bx is a nonempty convex set for each $x \in X$;*
(ii) *$B^{-1}y$ is open and $A^{-1}y$ is a nonempty convex set for each $y \in Y$. Then there is an $x_0 \in X$ such that $Ax_0 \cap Bx_0 \neq \emptyset$.*

Proof.   Let $Z = X \times Y$ and define $G : X \times Y \to 2^{E \times F}$ by $(x, y) \mapsto Z - (B^{-1}y \times Ax)$; each $G(x, y)$ is a nonempty set closed in $X \times Y$, therefore compact. As in the proof of Theorem (2.1) one verifies easily that G cannot be a KKM-map. Therefore there are elements $z_1, \ldots, z_n$ in Z such that $\text{conv}(z_1, \ldots, z_n)$ is not contained in $\bigcup_1^n G(z_i)$, so that some convex combination $w = \sum_1^n \lambda_i z_i \notin \bigcup_1^n G(z_i)$. Because Z is convex, the point w belongs to Z, so $w \in Z - \bigcup_1^n G(z_i) = \bigcap_1^n B^{-1}(y_i) \times Ax_i$. Writing $w = (\sum \lambda_i x_i, \sum \lambda_i y_i)$ we have $\sum_1^n \lambda_i x_i \in B^{-1}(y_i)$ for each $i = 1, \ldots, n$ and $\sum_1^n \lambda_i y_i \in Ax_i$ for each $i = 1, \ldots, n$. The first inclusion shows each $y_i \in B(\sum_1^n \lambda_i x_i)$ and therefore that $\sum \lambda_i y_i \in B(\sum \lambda_i x_i)$. The second inclusion shows each $x_i \in A^{-1}(\sum \lambda_i y_i)$, therefore $\sum \lambda_i x_i \in A^{-1}(\sum \lambda_i y_i)$, and consequently $\sum \lambda_i y_i \in A(\sum \lambda_i x_i)$. Thus, $A(\sum \lambda_i x_i) \cap B(\sum \lambda_i x_i) \neq \emptyset$, and the proof is complete.

We give an immediate application to game theory by establishing a general version of the von Newmann minimax principle due to M. Sion [29].

Theorem 5.17.   *(Minimax principle) Let X and Y be two nonempty compact convex sets in the linear topological spaces E and F. Let $f : X \times Y \to \mathbb{R}$ satisfy*

(i) *$y \mapsto f(x, y)$ is lsc and quasi-convex for each fixed $x \in X$;*
(ii) *$x \mapsto f(x, y)$ is usc and quasi-concave for each fixed $y \in Y$. Then $\max_x \min_y f(x, y) = \min_y \max_x f(x, y)$.*

Proof. Because of upper semicontinuity, $\max_x f(x,y)$ exists for each y and is a lower semicontinuous function of y, so $\min_y \max_x f(x,y)$ exists; similarly, $\max_x \min_y f(x,y)$ exists. Since $f(x,y) \leq \max_x f(x,y)$ we have $\min_y f(x,y) \leq \min_y \max_x f(x,y)$; therefore,

$$\max_x \min_y f(x,y) \leq \min_y \max_x f(x,y).$$

We shall show that inequality cannot hold. Assume it did; then there would be some r with $\max_x \min_y f(x,y) < r < \min_y \max_x f(x,y)$. Define $A, B : X \to 2^Y$ by $Ax = \{y \mid f(x,y) > r\}$ and $Bx = \{y \mid f(x,y) < r\}$. We verify that these set-valued maps satisfy the conditions of the coincidence theorem: Each Ax is open by lower semicontinuity of $y \mapsto f(x,y)$, each Bx is convex by the quasi-convexity of $y \mapsto f(x,y)$, and is nonempty because $\max_x \min_y f(x,y) < r$. Since $A^{-1}y = \{x \mid f(x,y) > r\}$ and $B^{-1}y = \{x \mid f(x,y) < r\}$, we find in the same way that each $A^{-1}y$ is nonempty and convex and each $B^{-1}y$ is open. Then by the coincidence theorem, there would be some (x_0, y_0) with $y_0 \in A(x_0) \cap B(x_0)$, which gives the contradiction $r < f(x_0, y_0) < r$. Thus, the inequality cannot hold, and the proof is complete.

The Intersection Theorem and the Nash Equilibrium Theorem

Given a cartesian product $X = \prod_{i=1}^n X_i$ of topological spaces, let $X^j = \prod_{i \neq j} X_i$ and let $p_i : X \to X_i$, $p^i : X \to X^i$ denote their projections; write $p_i(c) = x_i$ and $p^i(x) = x^i$. Given $x, y \in X$ we let

$$(y_i, x^i) = (x_1, \dots, x_{i-1}, y_i, x_{i+1}, \dots, x_n).$$

The following geometrical result of Ky Fan [8] generalizes the coincidence theorem:

Theorem 5.18. *Let $X_1, X_2, \dots, X_n$ be nonempty compact convex sets in linear topological spaces and let $A_1, A_2, \dots, A_n$ be n subsets of X such that*

(i) for each $x \in X$ and each $i = 1, 2, \dots, n$,

$$A_i(x) = \{y \in X \mid (y_i, x^i) \in A_i\}$$

is convex and nonempty;
(ii) for each $y \in X$ and each $i = 1, 2, \dots, n$,

$$A^i(y) = \{x \in X \mid (y_i, x^i) \in A_i\}$$

is open. Then $\bigcap_{i=1}^n A_i \neq \emptyset$.

Proof. As in (4.1) define $G : X \to 2^X$ by $y \mapsto X \setminus \bigcap_{i=1}^n A^i(y)$; one verifies that G is not a KKM-map and if a convex combination $w = \sum \lambda_i x_i \notin \bigcup G(x_i)$, then $w \in \bigcap_{i=1}^n A_i$.

As an immediate corollary:

Theorem 5.19. *(Nash equilibrium theorem [24])*
Let $X_1, X_2, \ldots, X_n$ be nonempty compact convex sets each in a topological vector space. Let $f_1, f_2, \ldots, f_n$ be n real-valued continuous functions defined on $X = \prod_{i=1}^{n} X_i$ such that for each $y \in X$ and each $i = 1, 2, \ldots, n$ the function $x_i \mapsto f_i(x_i, y^i)$ is quasi-concave on X_i. Then there is a point $y_0 \in X$ such that $f_i(y_0) = \max_{x_i} f_i(x_i y_0^i)$.

Proof. We briefly indicate the proof. Given $\varepsilon > 0$, define for each $i = 1, 2, \ldots, n$,

$$A_i^{\varepsilon} = \{\, y \in X \mid f_i(y) > \max_{x_i \in X_i} f_i(x_i, y^i) - \varepsilon \,\}.$$

One verifies easily that the conditions of (4.3) are satisfied and hence $\bigcap_{i=1}^{n} A_i^{\varepsilon} \neq \emptyset$. Then by a compactness argument one gets a point $y_0 \in X$ such that $y_0 \in \bigcap_{i=1}^{n} A_i^{\varepsilon}$ for each $\varepsilon > 0$, and this point y_0 satisfies the assertion of the theorem.

The coincidence theorem of Ky Fan also has applications in areas other than the theory of games. Among such applications we will mention the following result which extends the Tychonoff fixed point theorem to an important class of nonlocally convex spaces:

Theorem 5.20. *(Ky Fan [8]) Let E be a linear topological space with sufficiently many continuous linear functionals[9], and let C be a convex and compact subset of E. Then every continuous map $f : C \to C$ has a fixed point.*

5.5 Bibliographical and historical comments

1. In the special case when X is the set of vertices of a simplex in $\mathbb{R}^n$, Theorem (1.1) was discovered by Knaster-Kuratowski-Mazurkiewicz [19]; their method of proof was based on Sperner's Lemma. The abbreviation KKM stands for Knaster-Kuratowski-Mazurkiewicz. The principle of KKM-maps (Theorem (1.1)), established in a somewhat different form by Ky Fan [7], represents an infinite-dimensional analog of the Knaster-Kuratowski-Mazurkiewicz theorem; its formulation and the proof are taken from Dugandji-Granas [5]. Ky Fan demonstrated the importance of the principle of KKM-maps by giving numerous applications to various fields.

   Theorem (1.4) of Mazur-Schauder [22] (and an earlier Theorem (3.3) of Nikodym [26]) initiated the abstract approach to problems in calculus of variations. Mazur and Schauder gave applications of Theorem (1.4) to a number of concrete problems in calculus of variations; these results, however, were never published (cf. Scottish Book Problem 105).

---

[9]For example, the Hardy spaces H^p ($0 < p < 1$) have this property but are nonlocally convex.

Theorem (1.6) of Tychonoff [31] gives a positive answer to the second part of
Problem 54. The proof of (1.6) given here is due to Ky Fan [7]. Theorems (1.7)
and (1.8) were established by Ky Fan [10] and Browder [3] respectively.

2. Theorem (2.1) was established (in a different form) by Ky Fan [7]; the formulation of (2.1) given here is due to Browder [3], who obtained it from the Brouwer
   theorem.

   Theorem (2.3), due (in a slightly less general form) to Hukuhara [16] (cf. also
   an earlier result by Mazur [21]), gives a positive answer to the third part of
   Problem 54; the proof of (2.3) given here is due to Lassonde [20]. Theorem (2.4),
   established by Ky Fan, generalizes an earlier result of Iokhvidov [17].
   The fact that the minimax inequality of Ky Fan is equivalent to Theorem (2.1)
   is proved in Ky Fan [11]. For other applications of the Ky Fan fixed point
   theorem (2.1) (or of the minimax inequality) the reader is referred to Ky Fan
   [11], Lassonde [20], and Browder [4].

3. Variational inequalities (the systematic study of which began around 1965)
   have in recent years assumed increasing importance in many applied problems (cf. the survey by Stampacchia [30] for an introductory account and
   further references). The proof of Theorem (3.1) is from Dugundji-Granas [5].
   The same type of proof works for semimonotone operators in the sense of
   F. Browder (cf. Lassonde [20]). For more general results, see also Brezis-
   Nirenberg-Stampacchia 479 [2], Lassonde [20], and Mosco [23].

4. The Coincidence Theorem (4.1) is a special case of the intersection theorem (4.3)
   proved by Ky Fan in [9]. Theorem (4.2), established by Sion [29] evolved from
   several earlier results; in the special case when $X \subset \mathbb{R}^n$, $Y \subset \mathbb{R}^k$ are simplexes and
   f is bilinear, Theorem (4.2) was discovered by von Neumann [25], who deduced
   it from the Brouwer theorem. The direct proof of (4.2) is a modification of an
   earlier proof by Ky Fan [8] and is taken from Dugundji-Granas [6]. The proof of
   the Nash equilibrium theorem is due to Ky Fan [9]. For more general results and
   further references, see Browder [3], Lassonde [20], and Ky Fan [12].

   In connection with Theorem (4.5) we remark that the first part of Problem 54
   remains unanswered; it is not known whether a compact convex subset of an
   F-space has the fixed point property. Theorem (4.5) represents the best-known
   partial answer to this question. For other fixed point results in nonlocally convex
   spaces, see Klee [18], Granas [14], and also Riedrich [27], Granas [13], where
   further references will be found.

   Problem 54 was an inspiration for numerous later investigations both in fixed
   point theory and in nonlinear functional analysis. The literature is too extensive
   to be summarized here and we refer to Dugundji-Granas [6] and Granas [13] for
   bibliographies on the topics in fixed point theory related to this problem.

References

1. S. Banach, Die Theorie der Operationen und ihre Bedeutung für die Analysis, Proc. Int. Congress in Oslo (1936), 261–268.
2. H. Brezis, L. Nirenberg, G. Stampacchia, A remark on Ky Fan's minimax principle, *Boll. Unione Mat. Ital.* (4) 6 (1972), 293–300.
3. F.E. Browder, The fixed point theory of multi-valued mappings in topological vector spaces, *Math. Ann.* 177 (1968), 283–301.
4. F.E. Browder, On a sharpened form of the Schauder fixed-point theorem, *Proc. Natl. Acad. Sci. USA*, 74 (1977) 4749–4751.
5. J. Dugundji, A. Granas, KKM-maps and variational inequalities, *Ann. Scu. Norm. di Pisa*, V (1978), 69–682.
6. J. Dugundji, A. Granas, *Fixed point theory, Vol. I*, Monografie Matematyczne 61, Warszawa, 1982. Springer Monographs in Mathematics, Springer-Verlag, New York, 1982.
7. K. Fan, A generalization of Tychonoff's fixed point theorem, *Math. Ann.* 142 (1961), 305–310.
8. K. Fan, Sur un theoreme minimax, *C.R. Acad. Sci. Paris*, Groupe 1, 259 (1962), 3925–3928.
9. K. Fan, Applications of a theorem concerning sets with convex sections, *Math. Ann.* 163 (1966), 189–203.
10. K. Fan, Extensions of two fixed point theorems of F.E. Browder, *Math. Z.* 112 (1969), 234–240.
11. K. Fan, A minimax inequality and applications, in O. Shisha, ed., *Inequalities III*, Academic Press, New York and London, 1972, 103–113.
12. K. Fan, Fixed point and related theorems for non-compact convex sets. Game theory and related topics (Proc. Sem, Bonn and Hagen, 1978), pp. 151–156, North Holland, New York, 1979.
13. A. Granas, *Points fixes pour les applications compactes: Espaces de Lefschetz et la théorie de l'indice*. Cours professe pendant la session du Seminaire de Mathematiques Superieures a l'Université de Montreal (1973), Presses de l'Université de Montreal, 1980.
14. A. Granas, Sur la methode de continuité de Poincare, *C.R. Acad. Sci. Paris* 200 (1976), 983–985.
15. P. Hartman, G. Stampacchia, On some non-linear elliptic differential equations, *Acta Math.* 115 (1966), 271–310.
16. M. Hukuhara, Sur l'existence des points invariants d'une transformation dans l'espace fonctionnel, *Jap. J. Math.* 20 (1950), 1–4.
17. I.S. Iokhvidov, On a lemma of Ky Fan generalizing the fixed-point principle of A.N. Tikhonov (Russian), *Dokl. Akad. Nauk SSSR* 159 (1964), 501–504. English transl.: *Soviet Math. Dokl.* 5 (1964), 1523–1526.
18. V. Klee, Leray-Schauder theory without local convexity, *Math. Ann.* 141 (1960), 286–296.
19. B. Knaster, C. Kuratowski, S. Mazurkiewicz, Ein Beweis des Fixpunktsatzes für n-dimensionale Simplexe, *Fund. Math.* 14 (1929), 132–137.
20. M. Lassonde, Multiapplications KKM en analyse non lineaire, These Ph.D. Math., Univ. de Montreal (1978).
21. S. Mazur, Un theoreme sur les points invariants, *Ann. Soc. Polon. Math.* XVII (1938), 110.
22. S. Mazur, J. Schauder, Über ein Prinzip in der Variationsrechnung, *Proc. Int. Congress Math.*, Oslo (1936), 65.
23. U. Mosco, Implicit variational problems and quasi variational inequalities, in Lecture Notes in Mathematics, 543: *Non-linear operators and the calculus of variations*, Bruxelles (1975), Springer-Verlag, Berlin-Heidelberg-New York (1976), 83–156.
24. J. Nash, Non-cooperative games, *Ann. of Math.* 54 (1951), 286–295.
25. J.V. Neumann, *Ueber ein okonomisches Gleichungssystem und eine Verallgemeinerung des Brouwerschen Fixpunktsatzes*, Ergebnisse eines Mathematischen Kolloquiums 8 (1935), 73–83.
26. O. Nikodym, Sur le principe de minimum dans le probleme de Dirichlet, *Ann. Soc. Polon. Math.* X (1931), 120–121.

27. T. Riedrich, *Vorlesungen über nichtlineare Operator-gleichungen*, Teubner-Texte, Teubner-Verlag, Leipzig (1975).
28. J. Schauder, Einige Anwendungen der Topologie der Funktionalraume, *Recueil mathematique* I (1936), 747–753.
29. M. Sion, On general minimax theorems, *Pacific J. Math.* 8 (1958), 171–176.
30. G. Stampacchia, *Variational inequalities*, Summer School on theory and applications of monotone operators, Venice, (1968).
31. A. Tychonoff, Ein Fixpunktsatz, *Math. Ann.* 111 (1935), 767–776.

Front Row l. to r.: Stan Ulam, Mark Kac, Paul Erdos, A. Zygmund. Back Row l. to r.: Dan Mauldin, Tony Martin, A. Granas, R. D. Anderson.

Part II
The Scottish Book Problems

Chapter 6
Problems with Commentary

Stefan Banach's Last Picture, at the End of the War in Poland. Sent to Stan Ulam by his son.

PROBLEM 1: BANACH
July 17, 1935

(a) When can a metric space [possibly of type (B)] be so metrized that it will become complete and compact, and so that all the sequences converging originally should also converge in the new metric?
(b) Can, for example, the space c_0 be so metrized?

Commentary

A space of type (B) is the terminology from Banach's monograph, Théorie des Opérations Linéaires, Warszawa, 1932, for a Banach space.

© Springer International Publishing Switzerland 2015
R.D. Mauldin, *The Scottish Book*, DOI 10.1007/978-3-319-22897-6_6

There is probably no satisfactory answer to part (a). This can be seen as follows. First, notice that if X is a metric space, then X admits a new metric under which X is compact and such that all sequences which converge in the original metric should converge in the new metric if and only if there is a continuous one-to-one map f of X (with the original metric) onto a compact metric space. Next, notice that if g is a function from $[0,1]$ into $[0,1]$ and X is the graph of g in the unit square provided with the usual Euclidean metric, then the projection of X into the first axis is a continuous one-to-one map of X onto $[0,1]$. Since there are 2^c maps of $[0,1]$ into $[0,1]$, a majority of the spaces X so obtained are very strange.

However, there are restricted cases of this general problem which seem to be unresolved. For example:

Let X be a complete separable metric space. Are there some simple conditions such that there is a continuous one-to-one map of X onto a compact metric space?

For example, if X is a locally compact separable metric space, then there is such a map. The space N^N or equivalently the space of all irrational numbers has this property [4]. Also if $X = \prod_{n=1}^{\infty} X_n$, where each X_n has this property, then X also has this mapping property. In particular, $\mathbb{R}^\omega$ has this property. Since it is now known that every infinite-dimensional Banach space is homeomorphic to $\mathbb{R}^\omega$ [1, 2] the answer to part (b) is yes. This was first proved in a different way by Klee [3].

1. R. D. Anderson and R. H. Bing, A complete elementary proof that Hilbert space is homeomorphic to the countable infinite product of lines, *Bull. Amer. Math. Soc.* 74 (1968).
2. M. I. Kadeč, On homeomorphisms of certain Banach spaces, *Dokl. Akad. Nauk SSSR* 92 (1955), 465–468.
3. V. Klee, On a problem of Banach, *Coll. Math.* 5 (1957), 280–285.
4. W. Sierpiński, Sur les images continues et biunivoques de l'ensemble de tous les nombres irrationels, *Mathematica* 1 (1929), 18–21.

R. Daniel Mauldin

PROBLEM 2: BANACH, ULAM

(a) Can one define, in every compact metric space E, a measure (finitely additive) so that Borel sets which are congruent should have equal measure?

(b) Suppose $E = E_1 + E_2 + \ldots + E_n$, and $E_1 \cong E_2 \cong \cdots \cong E_n$ and $\{E_i\}$ are disjoint; then we write $E_i = \frac{1}{n}E$. Can it occur that $\frac{1}{n}E = \frac{1}{m}E$, $n \neq m$, if we assume that $\frac{1}{n}E$ are Borel sets and E is compact?

Commentary

Two Borel sets A and B are congruent means there is an isometry of A onto B.

By a result of Tarski ([3], Theorem 16.12) problems (a) and (b) are equivalent to each other.

The answer is yes if the space E is also supposed to be countable [4].

It is known that for every compact metric space there exists a Borel measure such that congruent open sets have equal measures (see [1, 2]). It follows that if two Borel sets are congruent by an isometry which extends to some open sets then those Borel sets have equal measures.

1. J. Mycielski, Remarks on invariant measures in metric spaces, *Coll. Math.* 32 (1974), 105–112.
2. _____, A conjecture of Ulam on the invariance of measure in Hilbert's cube, *Studia Math.* 60 (1977), 1–10.
3. A. Tarski, Cardinal Algebras, *Oxford University Press*, 1949.
4. R. O. Davies and A. J. Ostaszewski, Denumerable compact metric spaces admit isometry-invariant finitely additive measures, *Mathematika* 26 (1979), 184–186.

Jan Mycielski

Second Edition Commentary

Problem 2 remains open for uncountable compact metric spaces. This problem has given rise to several interesting related results and questions concerning the existence of countably additive, metrically invariant σ-finite measures on locally compact metric spaces.

We say a metric space (X, d) is *locally homogeneous* if for any two points $x, y \in X$ there is some $\varepsilon > 0$ and an isometry ϕ of the open ball $B(x, \varepsilon)$ onto $B(y, \varepsilon)$ which sends x to y. We say that a measure (finitely additive or countably additive) on the Borel subsets of X is *metrically invariant* provided whenever A and B are congruent Borel sets, then $\mu(A) = \mu(B)$. We say μ is *open-invariant* provided $\mu(A) = \mu(B)$, whenever A and B are congruent open sets. Bandt and Babaki in [1] proved the following.

Theorem 1. *Let (X, d) be locally compact and locally homogeneous and let A be a compact subset of X with non-empty interior. Then there is a unique metrically invariant Borel measure μ on X with $\mu(A) = 1$.*

This class includes all locally compact groups with a left-invariant metric.

We note the following example of Bandt and Baraki demonstrating why one cannot expect countable additivity without local homogeneity. Let $X = \cup_{n=1}^{\infty} C_n$, where C_n is the n-dimensional cube $[0, 1/n]^n$ and let d be the Euclidean metric on each C_n and let $d(x, y) = \max\{1/n, 1/m\}$, for $x \in C_n, y \in C_m, n \neq m$. For each n, Lebesgue measure λ_n on C_n induces an open-invariant measure λ'_n on X and positive linear combinations of these are open-invariant. However, there is no σ-finite metrically invariant measure, μ. For otherwise, for some n, $\mu(C_n) > 0$. Then a subcube of C_n with edge length $1/(n+1)$ will have positive measure. Thus, C_{n+1} would *impossibly* contain uncountably many disjoint isometric copies with the same measure.

Let me mention that Bandt and Baraki also incorporated this example into hyperspaces answering a question of McMullen:

Theorem 2. *Let $n > 1$ and let $\mathscr{K}(R^n)$ be the space of all compact convex subsets of R^n with d_H, the Hausdorff metric. Then there is no nontrivial σ-finite Borel measure on $\mathscr{K}(R^n)$ which is invariant even under all isometries from the whole space into itself. (On the other hand, there is an open-invariant measure.)*

Ulam posed the following specific problem concerning metrically invariant measures.

Open Problem [Ulam]: Let μ be Lebesgue product measure on $[0,1]^\infty$. Is μ metrically invariant with respect to metrics of the type
$d(x,y) = (\sum a_n^2 \cdot (x_n - y_n)^2)^{1/2}$, where $a = (a_n) \in \ell_2$?
Mycielski in his referenced 1977 paper showed that invariant open sets A and B have the same measure under these metrics but whether arbitrary invariant Borel sets have the same measure remains open. There has been some progress. J. Fickett in [2] proved the following.

Theorem 3. *Lebesgue measure is metrically invariant on $[0,1]^\infty$, the Hilbert cube, provided the sequence a_n is very rapidly decreasing:*
$$\lim_n \frac{a_n^{1/2^n}}{a_{n-1}} = 0.$$

1. Christoph Bandt and Gebreselassie Baraki, Metrically invariant measures on locally homogeneous spaces and hyperspaces. Pacific J. Math. 121 (1986), no. 1, 13-28.
2. James W. Fickett, Approximate isometries on bounded sets with an application to measure theory. Studia Math. 72 (1982), no. 1, 37-46.

                                                                    R. Daniel Mauldin

PROBLEM 3: BANACH, ULAM
Theorem. It is proved very simply that a compact set cannot be congruent to a proper subset of itself.

Commentary

A stronger theorem has been proved by A. Lindenbaum [1], namely a set in a compact metric space which is both an F_σ and a G_δ cannot be congruent to a proper subset of itself. For sets which are only F_σ or only G_δ this is false as the set $\{e^{in} : n = 1, 2, \dots\}$ in the unit circle and its complement show.

1. A. Lindenbaum, Contributions à l'étude de l'espace métrique I, *Fund. Math.* 8 (1926), 209–222.

                                                                    Jan Mycielski

PROBLEM 4: SCHREIER
Theorem. If $\{\xi_n\}$ is a bounded sequence, summable by the first mean to ξ, then almost every subsequence of it is also summable by the first mean to ξ.

Commentary

Problem 4 probably arose in the context of the "normal numbers" discoveries of Borel and others, dealing with sequences $\{t_n\}$ of 0's and 1's; if $t = \sum_1^\infty t_n 2^{-n}$, then Lebesgue measure on $[0,1]$ enables one to speak of a random sequence $\{t_n\}$, and to prove theorems about the density of 1's in such a sequence. [*Rend. Circ. Matem. Palermo* 27 (1909), 247–271] Input also came from the foundations of probability theory; if the probability of 1 is p, and $\{t_n\}$ is the sequence of observations, and this is $(C, 1)$ summable to p, then one expects that a random choice of a subsequence from $\{t_n\}$ would have the same property. The result cited to J. Schreier in Problem 4 is a generalization of this: if $\{x_n\}$ is bounded and $(C, 1)$ summable to L, then so is almost every subsequence. (See Birnbaum and Schreier, *Studia Math.* 4 (1933), 85–89; Birnbaum & Zuckerman, *Amer. J. Math.* 62 (1940), 787–791.)

In 1943, independently motivated, Buck and Pollard proved the following assertions: (1) if $\{x_n\}$ is divergent, so are almost all its subsequences; (2) if every subsequence of $\{x_n\}$ is $(C, 1)$ summable, $\{x_n\}$ is convergent; (3) if $\{x_n\}$ is not $(C, 1)$ summable, neither are almost all its subsequences; (4) there is a sequence $\{x_n\}$ that is $(C, 1)$ summable but such that almost all its subsequences fail to be $(C, 1)$ summable; (5) if $\{x_n\}$ is summable to L and $\sum_1^\infty (x_n/n)^2 < \infty$, then so are almost all its subsequences. (See *Bull. Amer. Math. Soc.* 49 (1943), 924–931.) "Almost all" was interpreted in terms of the standard mapping between selection sequences of 0's and 1's, and $[0,1]$. Of these five assertions, (2) was a special case of an earlier result of Buck [*MR* 5, 117] showing that a sequence $\{x_n\}$ must be convergent if every subsequence of it is summable by any fixed regular matrix method. (See also Agnew [*MR*6, 46] and Buck [*MR* 18, 478].) In subsequent papers, assertion (1) was extended to generalized sequences and cluster sets by Buck [*MR* 5, 235] and Day [*MR* 5, 236] and a best possible strengthening of assertion (5) was obtained by Tsuchikura [*MR* 12, 820] who proved that the condition $\sum_1^\infty (x_n/n)^2 < \infty$ could be replaced by $\sum_1^\infty x_R^2 = o(n^2/\log\log n)$. Techniques used involved properties of the Rademacher functions $R_n(t)$, and the strong law of large numbers.

R.C. Buck

Madison, Wisconsin

February, 1979

PROBLEM 5: MAZUR

Definition. A sequence $\{\xi_n\}$ is *asymptotically convergent to* ξ if there exists a subsequence of density 1 convergent to ξ.

Theorem. (Mazur) In the domain of all sequences this notion is not equivalent to any Toeplitz method.

How is it in the domain of bounded sequences?

Addendum. We have the following theorems:

(1) If a method (a_{ik}) sums all the asymptotically convergent sequences, the $a_{ik} = 0$ for $k > k_0$, $i = 1, 2, \ldots$ and there exist finite $\lim a_{ik}$ for $k = 1, \ldots, k_0$, such that the method sums all the sequences.

(2) If a method (a_{ik}) sums all the convergent sequences and every bounded sequence summable by the sequence is asymptotically convergent, then there exists a sequence of increasing integers $\{k_n\}$ with density 1, such that for every bounded sequence $\{\xi_n\}$ summable by this method, the sequence $\{\xi_{k_n}\}$ is convergent.

From (1) it follows that there does not exist a permanent method summing all the asymptotically convergent sequences; from (2) it follows that a permanent method summing all bounded asymptotically convergent sequences must also sum other bounded sequences.

Mazur
July 22, 1935

Commentary

In 1980, matrix summability has lost the interest it had in the '30s. However, this problem, which remains unsolved, dealt with a concept that perhaps ought to receive more attention. A sequence (of numbers, functions, operators, etc.) may diverge and yet have a subsequence of density $d > 0$ that converges; if $d = 1$, Mazur called the sequence "asymptotically convergent." Samples: (a) A measurable transformation T is mixing if $\lim m(T^{-n}A \cap B) = m(A)m(B)$ for each pair of measurable sets A and B; T is called *weakly* mixing if the sequence is asymptotically convergent [Halmos, *Lectures on Ergodic Theory,* Chelsea Publ. 1956]. (b) Let $f(x)$ be an entire function of exponential type $c < \pi$ with

$$\int_1^\infty \frac{\log |f(x)f(-x)|}{x^2} dx < \infty.$$

Then the sequence $s_n = n^{-1} \log |f(n)|$ has a subsequence of positive density that converges to zero [Levinson, Gap and Density Theorems, *Amer. Math. Soc. Colloq.* 1940].

It has been noted that asymptotic convergence is closely related to Cesàro summability. Thus, a bounded sequence $\{x_n\}$ that is $(C, 1)$ summable to one of its outer limit points must be asymptotically convergent to it; this is also related to what is called strong $(C, 1)$ summability. There are enough results to suggest that a more structured theory lies in the background. A key result is the *additivity theorem* for generalized asymptotic density, regarded as a finitely additive measure (See [1, 2, 4, 5]).

1. R. C. Buck, The measure theoretic approach to density, *Am. J. Math.* 68 (1946), 560–580.
2. _____, Generalized asymptotic density, *Am. J. Math.* 75 (1953), 335–346.
3. _____, Some Remarks on Tauberian Conditions, *Quart. J. Math. Oxford Ser.* (2) 6 (1955), 128–131.
4. _____, Convergence theorems for finitely additive integrals, *J. Indiana Math. Soc.* (N.S.) 23 (1959), 1–9.
5. A. R. Freedman, On the additivity theorem for *n*-dimensional asymptotic density, *Pacific J.Math.* 49 (1973), 357–363.
6. C. T. Rajagopal, Some theorems on convergence in density, *Publ. Math. Debrecen* 5 (1957), 77–92.
7. Hans Rohrbach, Bodo Volkmann, Zur Theorie der asymptotische Dichte, *J. Reine Angew. Math.* 192 (1953), 102–112.
8. _____, Verallgemeinerte asymptotische Dichten, *J. Reine Angew. Math.* 194 (1955), 195–209.

R. C. Buck

PROBLEM 6: MAZUR, ORLICZ

Prize: Bottle of wine, S. Mazur

Is a matrix, finite in each row and invertible (in a one-to-one way), equivalent to a normal matrix?

Second Edition Commentary

The terminology used in the formulation of Problem 6 can be found in S. Mazur's paper [1]. The paper gives a functional analysis approach to summability theory. The subject was quite popular at the beginning of the 20th century and then mostly abandoned. In particular, nowadays the meaning of "normal matrix" is very different from that in Problem 6. Probably for this reason the problem escaped an attention of readers of the Scottish Book. An answer to Problem 6 is in negative. To present it we recall some definitions from [1] and introduce some new ones. So:

c is the Banach space of all convergent sequences of real numbers with the norm $||x||_\infty = \sup_{n \in N} |\alpha_n|$ for $x = (\alpha_n)_{n \in N} \in \mathbf{c}$.

s is the space of all sequences of real numbers equipped with the topology of coordinate convergence. It is an **F** space or Fréchet space. For $x = (\alpha_n)_{n \in N} \in \mathbf{s}$ we write $x(n)$ for α_n.

$T = (\theta_{n,m})_{n,m \in N}$ is said to be a *Toeplitz matrix* if there are only finite nonzero entries in each row. (Note that the present meaning of Toeplitz matrix is different).

For a Toeplitz matrix $T = (\theta_{n,m})_{n,m \in N}$ and $x = (\alpha_n)_{n \in N}$ the sequence $T(x)$ is defined by $T(x)(n) = \sum_{m=1}^{\infty} \theta_{n,m} \alpha_m$. This formula gives a $1 - 1$ correspondence between Toeplitz matrices and continuous linear operators in **s**. The matrix product of

Teoplitz matrices is well defined, it is a Toeplitz matrix, and it corresponds to the composition of the corresponding operators in **s**. If a Toeplitz matrix T corresponds to a $1-1$ and "onto" operator in **s**, then there is unique inverse matrix T^{-1} (it has to be a Toeplitz matrix then).

Toeplitz matrices T, R are said to be *equivalent* if $\{x \in \mathbf{s} : T(x) \in \mathbf{c}\} = \{x \in \mathbf{s} : R(x) \in \mathbf{c}\}$

A matrix $D = (\delta_{n,m})_{n,m \in N}$ is said to be *normal* if $\delta_{n,m} = 0$ and $\delta_{n,n} \neq 0$ for all $n < m$.

A normal matrix D is a Toeplitz matrix, it has an inverse matrix which is normal and in which the entries on the diagonal are the inverses of corresponding diagonal entries in D. Problem 6 can be reformulated as follows. Is it true that for each invertible Toeplitz matrix T there exits a normal matrix D such that the $T^{-1}(\mathbf{c}) = D^{-1}(\mathbf{c})$. The last equality means that the matrix TD^{-1} which is a continuous and $1-1$ linear map of **s** onto itself transforms **c** onto itself. By the Closed Graph Theorem in **F** spaces it follows that the matrix TD^{-1} corresponds to a bounded linear invertible operator L in **c**.

A matrix $T = (\theta_{n,m})_{n,m \in N}$ is called *semi-normal* if there exists a sequence of nonsingular square matrices $A_k = (\alpha_{n,m}^k)_{1 \leq n,m \leq l_k}$ such that for the sequence $s_0 = 0, s_k = l_1 + \ldots + l_k$ we have $\theta_{n,m} = \alpha_{n-s_{k-1}, m-s_{k-1}}^k$ for each $s_{k-1} < n, m \leq s_k$ and $\theta_{n,m} = 0$ if $n \leq s_k < m$ for some k. Then (l_k) is called a *rank sequence* and (A_k) a *diagonal sequence of matrices* of T.

It is clear that each semi-normal matrix is a Toeplitz matrix. An easy inspection shows that if T, R are semi-normal matrices with the same rank sequence (l_k) and the diagonal sequences of matrices (A_k) of T and (B_k) of R then the matrix TR is semi-normal with rank sequence (l_k) and diagonal sequence of matrices $(A_k B_k)$. Each semi-normal matrix T with diagonal sequence (A_k) is invertible and the inverse matrix T^{-1} is semi-normal which has the same rank sequence as T and (A_k^{-1}) is the diagonal sequence of matrices of T^{-1}. Each normal matrix T is semi-normal. For such matrix each sequence (l_k) of positive integers is a rank sequence of T and the corresponding diagonal matrix sequence $A_k = (\alpha_{m,n}^k)_{1 \leq n,m \leq l_k}$ is defined as in the definition of semi-normal matrix, i.e. $\alpha_{m,n}^k = \theta_{n+s_{k-1}, m+s_{k-1}}$. We will construct a semi-normal matrix T which is not equivalent to a normal matrix. For each $k \in N$ let C_k be a matrix with 2^k rows and k columns with entries equal to ± 1 and such that no two rows are identical. Then the columns of C_k are orthogonal vectors in R^{2^k}. Let A_k be any square nonsingular matrix of rank 2^k in which the last k columns form the matrix C_k. Such a matrix exists in view of linear independence of the columns from C_k. Let $T = (\theta_{n,m})_{n,m \in N}$ be a semi-normal matrix such that (A_k) is its diagonal sequence. Then $(l_k) = (2^k)$ is the rank sequence of T and $s_k = 2^{k+1} - 2$ for $k \in N$.

Assume that T is equivalent to a normal matrix D. Then as explained earlier the normal matrix $B = D^{-1} = (\beta_{n,m})_{n,m \in N}$ is such that TB is a matrix which defines a bounded invertible linear operator L in **c**. Let us fix an integer $r > 3(\|L\|\|L^{-1}\|)^2$.

Finally let e_m be the sequence such that $e_m(n) = 0$ for $n \neq m$ and $e_m(n) = 1$. For $m \in (s_r - r, s_r]$ and the sequence $L(e_m)$ let f_m denote the sequence such that $L(e_m)(n) = f_m(n)$ for $m \leq s_r$ and $f_m(n) = 0$ for $n > s_r$. The first s_{r-1} coordinates of the both sequences are equal to 0 as well. For these reasons and since L^{-1} corresponds to a semi-normal matrix with the rank sequence (2^k) we obtain that the first s_r coordinates of $L^{-1}(f_m)$ and of $L^{-1}(L(e_m))$ are the same. Thus $||L^{-1}|| \, ||f_m||_\infty \geq |L^{-1}(f_m)(m)| = |L^{-1}L(e_m)(m)| = 1$. But $||f_m||_\infty = \sup_{s_{r-1} < n \leq s_r} |\sum_{m \leq j \leq s_r} \theta_{n,j} \beta_{j,m}| = \sum_{m \leq j \leq s_r} |\beta_{j,m}|$. The last equality follows because there exists $n \in (s_{r-1}, s_r]$ such that $\theta_{n,j} = \mathrm{sign}(\beta_{j,m})$ and $\theta_{n,j} = \pm 1$ for all $s_r - r < j < s_r$. Hence

$$\sum_{s_r - r < m \leq s_r} \sum_{m \leq j \leq s_r} |\beta_{j,m}| \geq r/||L^{-1}||.$$

Because TB corresponds to a bounded operator L in $\mathbf{c}$ the sum of absolute values of entries in each row does not exceed $||L||$. Therefore for each $n \in (s_{r-1}, s_r]$ we obtain the inequality $||L|| \geq \sum_{s_r - r < m \leq s_r} |\sum_{m \leq j \leq s_r} \theta_{n,j} \beta_{j,m}|$. Now averaging over $n \in (s_{r-1}, s_r]$ and taking into account the Khinchin inequality:

$2^{-r} \sum_{s_{r-1} < n \leq s_r} |\sum_{s_r - r \leq j \leq s_r} \theta_{n,j} \gamma_j| \geq (1/\sqrt{3})(\sum_{s_r - r \leq j \leq s_r} \gamma_j^2)^{\frac{1}{2}} \geq \frac{1}{\sqrt{3r}} \sum_{s_r - r \leq j \leq s_r} |\gamma_j|$

which holds for each sequence $(\gamma_j)_{s_r - r < j \leq s_r}$ we arrive at

$$||L|| \geq 2^{-r} \sum_{s_{r-1} < n \leq s_r} \sum_{s_r - r < m \leq s_r} |\sum_{m \leq j \leq s_r} \theta_{n,j} \beta_{j,m}| =$$

$$\sum_{s_r - r < m \leq s_r} 2^{-r} \sum_{s_{r-1} < n \leq s_r} |\sum_{m \leq j \leq s_r} \theta_{n,j} \beta_{j,m}| \geq \frac{1}{\sqrt{3r}} \sum_{s_r - r < m \leq s_r} \sum_{m \leq j \leq s_r} |\beta_{j,m}|.$$

This inequality together with the preceding one gives $r \leq 3(||L|| \, ||L^{-1}||)^2$ which yields a contradiction with the initial choice of r.

1. S. Mazur, *Eine Anwendung der Theorie der Operationen bei der Untersuchung der Toeplitzschen Limitierungsverfahren*, Studia Math. **2** (1930).40–50.

Stanisław Kwapień

PROBLEM 7: MAZUR, BANACH

Are two convex infinite-dimensional subsets of a Banach space [of type (B)] always homeomorphic?

Commentary

Keller [7] proved that all infinite-dimensional compact convex subsets of Hilbert space are homeomorphic with the cube $[0,1]^{\aleph_0}$, and this is easily extended to

Fréchet spaces. Klee [9] showed that if K is a locally compact closed convex subset of a Banach space then there are cardinal numbers m and n with $0 \le m \le \aleph_0$ and $0 \le n < \aleph_0$ such that K is homeomorphic with either $[0,1]^m \times (-\infty,\infty)^n$ or $[0,1]^m \times [0,\infty)$. The various possibilities indicated are topologically distinct.

Klee [8] showed that Hilbert space is homeomorphic with all of its closed convex bodies. Extending this and results of Stoker [10] and Corson and Klee [5], Bessaga and Klee [2] showed that if K is a closed convex body in an arbitrary topological linear space E, then K is homeomorphic with a closed halfspace in E or with the product (for some finite $m \ge 0$) of $[0,1]^m$ by a closed linear subspace of codimension m in E. From this, another result of Bessaga and Klee [3], and theorems of Kadeč [6] and Anderson [1] on topological equivalence of Fréchet spaces, it follows that every infinite-dimensional Fréchet space is homeomorphic with all of its closed bodies. Most of the above material appears in the book of Bessaga and Pełczyński [4].

1. R. D. Anderson, Hilbert space is homeomorphic to the countable infinite product of lines, *Bull. Amer. Math. Soc.* 72 (1966), 515–519.
2. C. Bessaga and V. Klee, Two topological properties of topological linear spaces, *Israel J. Math.* 2 (1064), 211–220.
3. _____, Every non-normable Fréchet space is homeomorphic with all of its closed convex bodies, *Math. Ann.* 163 (1966), 161–166.
4. C. Bessaga and A. Pełczyński, *Selected Topics in Infinite-Dimensional Topology, Monografie Matematyczne,* Vol. 58, Polish Scientific Publishers, Warsaw, 1975.
5. H. Corson and V. Klee, Topological classification of convex sets, *Convexity* (V. Klee, ed.), *Amer. Math. Soc. Proc. Symp. Pure Math.* 7 (1963), 37–51.
6. M.I. Kadeč, A proof of the topological equivalence of all separable infinite-dimensional Banach spaces (Russian), *Funktsional. Anal. i Prilozhen.* 1 (1967), 61–70.
7. O.H. Keller, Die Homeomorphie der kompakten konvexen Mengen in Hilbertschen Raum, *Math. Ann.* 105 (1931), 748–758.
8. V. Klee, Convex bodies and periodic homeomorphisms in Hilbert space, *Trans. Amer. Math. Soc.* 74 (1953), 10–43.
9. _____, Some topological properties of convex sets, *Trans. Amer. Math. Soc.* 78 (1955), 30–45.
10. J.J. Stoker, Unbounded convex point sets, *Amer. J. Math.* 62 (1940), 165–179.

V. Klee

PROBLEM 8: MAZUR

Prize: Five small beers, S. Mazur

(a) Is every series summable by the first mean representable as Cauchy product of two converging series? Or else, equivalently,

(b) Can one find for each convergent sequence $\{z_n\}$ two convergent sequences $\{x_n\}$, $\{y_n\}$ such that

$$z_n = \frac{x_1 y_n + x_2 y_{n-1} + \cdots + x_n y_1}{n}.$$

Second Edition Commentary

The problem reflects Mazur's interests in the summability theory. It was solved negatively in 1984, three years after Mazur had died. The solution was obtained independently by P.P. B. Eggermont, Y.J. Leung, [1] and by S. Kwapień, A. Pełczyński [2]. The problem is closely related to Problem 88. Both of them can be easily reduced to problems on Hankel matrices. In 1982 V.V. Peller, [3], obtained deep results on these matrices. The same author in [4] showed how to derive solutions to Problems 8 and 88 from a theorem in [3] in an easy way. Beside the paper [4] contains results on quantitative aspects of Problems 8 and 88 which strengthen theorems and answer questions from [5].

Another solution to Problem 8 and its extension to some other classes of sequences were obtained by C. Lennard and D. Redelet, [6].

1. P.P.B. Eggermont and Y.J. Leung, *On a factorization problem for convergent sequences and on Hankel forms in bounded sequences*, Proc. Amer. Math. Soc. **96** (1986),269–274.
2. S. Kwapień and A. Pełczyński, *On two problems of S.Mazur from The Scottish Book*, Lecture at the Colloquim dedicated to the memory of Stanisław Mazur, Warsaw Univeristy, (1985) (unpublished).
3. V.V. Peller, *Estimates of functions of power bounded operators on Hilbert space*, J. Operator Theory **7** (1982),341–371.
4. V.V. Peller, *On S.Mazur's Problems 8 and 88 from the Scottish Book*, Studia Math. **180** (2007),191–200.
5. A. Pełczyński and F. Sukochev, *Some remarks on Toeplitz mutltipliers and Hankel matrices*, Studia Math. **175** (2006),175–204
6. C. Lennard and D. Redelet, *The Mazur product map on Hardy type sequence spaces*, J. Math. Anal. Appl. **350** (2009),384–392.

Stanisław Kwapień

PROBLEM 9: MAZUR, ORLICZ

Theorem (Ulam). If E is a class of sets, each finite, each of which contains at most n elements, and such that every $n + 1$ of these sets have a common element, then there exists an element common to all sets of E.

Remark

The following is a slightly stronger statement of this theorem: If E is a class of sets which contain a set with less than m elements and every m sets of E have an element in common, then there exists an element common to all sets of E.

The proof is immediate.

J. Mycielski

PROBLEM 10: BANACH, MAZUR

Let H be an arbitrary abstract set and E the set of all real-valued functions defined on H. The sequence $x_n(t) \to x(t)$ (such that $t \in H$, $x_n, x \in E$) if $\lim x_n(t) = x(t)$ for each $t \in H$.

Theorem. Each linear functional $f(x)$ defined in E is of the form

$$f(x) = \sum_{i=1}^{n} \alpha_i x(t_i),$$

where α_i and t_i do not depend on x.

Commentary

Apparently, Banach and Mazur never published a proof of this theorem. An argument is given below. Note that the statement (in the terminology of that time) that f is a linear functional includes the condition that f is continuous; i.e., if x_n converges pointwise to x, then $f(x_n)$ converges to $f(x)$.

Let B denote the space of bounded real-valued functions on H with the uniform norm. Notice that f is a continuous linear functional on B, since if $\|x_n\| \to 0$, then $f(x_n) \to 0$. So there is a finitely additive set function μ defined on all subsets of H so that if $x \in B$, then

$$f(x) = \int_H x d\mu.$$

Let $\{A_n\}_{n=1}^{\infty}$ be a sequence of pairwise disjoint subsets of H. If $f(\chi_{A_n}) = c_n \neq 0$ for infinitely many n, then $f(x_n) = 1$ for infinitely many n, where $x_n \to 0$ pointwise. This contradiction establishes that μ is countably additive and that there do not exist infinitely many pairwise disjoint sets with nonzero measure. From this it follows that there are finitely many points $t_1, \ldots, t_n$ in H and numbers $\alpha_1, \ldots, \alpha_n$ so that

$$f(x) = \sum_{i=1}^{n} \alpha_i x(t_i)$$

for all $x \in B$.

Finally, if x is an unbounded real-valued function defined on H, then x is the pointwise limit of the functions x_n, where $x_n(t) = n$ if $|x(t)| \geq n$ and $x_n(t) = x(t)$ otherwise. Thus,

$$f(x) = \sum_{i=1}^{n} \alpha_i x(t_i)$$

for all $x \in E$. Clearly, every linear functional (in today's terminology) defined on E is of the given form if and only if H is finite.

R. Daniel Mauldin

PROBLEM 10.1: MAZUR, AUERBACH, ULAM, BANACH

Theorem. *If* $\{K_n\}_{n=1}^{\infty}$ *is a* sequence of convex bodies, each of diameter $\leq a$ and the sum of their volumes is $\leq b$, then there exists a cube with the diameter $c = f(a,b)$ such that one can put all the given bodies in it disjointly.

Corollary. One kilogram of potatoes can be put into a finite sack.

Determine the function $c = f(a,b)$.

Commentary

Known as the "sack of potatoes" theorem, the first published proof is due to Kosiński [1]. There it is established that in k-dimensional Euclidean space the bodies can be put in a rectangular parallelepiped with edges $3a, 3a, \ldots, 3a, (a + k!b/a^{k-1})$. An exact computation of the function $f(a,b)$ is not given, but clearly $f(a,b) \leq \sqrt{k} \max\{3a, a + k!b/a^{k-1}\}$. For $k \geq 3$ Moon and Moser [3] give an improvement of Kosinski's main lemma; it follows that the bodies can be put in a rectangular parallelepiped with edges $2a, 2a, \ldots, 2a, 2(a + k!b/a^{k-1})$, and a similar estimate for $f(a,b)$ can be derived. Several related questions are also investigated in [3] and [2].

1. A. Kosiński, A proof of an Auerbach-Banach-Mazur-Ulam theorem on convex bodies, *Coll. Math.* 4 (1957), 216–218.
2. A. Meir and L. Moser, On packing of squares and cubes, *J. Combin. Theory* 5 (1968), 126–134.
3. J.W. Moon and L. Moser, Some packing and covering theorems, *Coll. Math.* 17 (1967), 103–110.

Branko Grünbaum

PROBLEM 11: BANACH, ULAM

Assume that there is a measure defined in the space of all integers. This measure is finitely additive and any single point has measure zero. Let us extend this measure to product spaces over the set of integers (finite of infinite products) in such a way that the measure of a subproduct equals the numerical product of the measures of its projections.

(a) Is the set of all sequences convergent to infinity measurable?

(b) Is the set of all pairs (x,y) where x,y are relatively prime measurable?

(c) **Theorem** (Schreier).   The set of all pairs (x,y) where $x < y$ is nonmeasurable.

Remark. A set is not measurable if a measure can be defined in it at least two different ways and still satisfy the conditions above.

Solution

We show that the answer to (a) is no and that the answer to (b) depends on the measure in question.

Let X be a set, $\mathscr{B}$ an algebra of subsets of X, μ a real-valued, finitely additive, nonnegative function defined on $\mathscr{B}$. The proof of the following theorem may be found in [1].

Theorem. If $E \in \mathscr{P}(X) - \mathscr{B}$, then there is a finitely additive extension, $\hat{\mu}$, of μ to the algebra $\mathscr{A}$, generated by the sets in $\mathscr{B}$ and E. Moreover, any two extensions of μ take the same value on E if and only if

$$\sup\{\mu(B) : B \in \mathscr{B} \text{ and } B \subseteq E\}$$

$$= \inf\{\mu(B) : B \in \mathscr{B} \text{ and } B \supseteq E\}.$$

Let us remark that if the preceding equality holds, then there is a unique extension of μ to $\mathscr{A}$. In particular, if $\inf\{\mu(B) : B \in \mathscr{B} \text{ and } B \supseteq E\} = 0$, then there is a unique extension of μ to $\mathscr{A}$.

Let us show that the answer to (a) is no. Let $C = \{\langle x_n \rangle \in N^N : \lim_{n \to \infty} x_n = +\infty\}$. Let μ be a finitely additive probability measure defined on all subsets of N which gives measure zero to singletons. Let $\mathscr{M}$ be the algebra of subsets of N^N generated by $\mathscr{D}$, all sets of the form $A_1 \times A_2 \times \cdots \times A_n \times \cdots$ where for each i, $A_i \subseteq N$ and for all but finitely many i, $A_i = N$. Let m be the unique finitely additive measure defined on $\mathscr{M}$ such that

$$m(A_1 \times A_2 \times \cdots) = \prod_{i=1}^{\infty} \mu(A_i).$$

Suppose $B \in \mathscr{M}$, $B \subseteq C$ and $B \neq \emptyset$. There is some k such that B is of the form $T \times N \times N \times \cdots$ where T is a subset of N^k and $T \neq \emptyset$. Let $(x_1, \ldots, x_k) \in T$. Clearly, this finite sequence can be extended to an infinite sequence $\langle x_n \rangle$ which does not converge to infinity. This contradicts the fact that $T \subseteq C$. Thus,

$$\sup\{m(B) : B \in \mathscr{M} \text{ and } B \subseteq C\} = 0.$$

Similarly, if $T \in \mathscr{M}$ and $T \supseteq C$, then $T = N^N$. It follows from the theorem quoted above that C is not measurable.

Let us show that the answer to (b) depends on what probability measure μ is under consideration. Let $\mathscr{W}$ be the algebra of subsets of $N \times N$ generated by all sets of the form $A \times B$, where $A, B \subseteq n$ and let $R = \{(x,y) : x \text{ and } y \text{ are relatively prime}\}$. Let us note that a set K is in $\mathscr{W}$ if and only if K can be expressed as

$$K = \bigcup_{i=1}^{n} A_i \times B_i, \qquad (+)$$

where the sets A_i are pairwise disjoint.

Let μ be a 0-1 valued measure defined on all subsets of N such that μ gives measure zero to singletons and $\mu(Q_2) = 1$, where

$$Q_2 = \{2^n : n \in N\}.$$

Notice that $R \subseteq Q_2 \times (N - Q_2) \cup (N - Q_2) \times N$. From this it follows that

$$\begin{aligned}
0 = & \quad \sup\{\mu \times \mu(K) : K \in \mathscr{W} \text{ and } K \subseteq R\} \\
= & \quad \inf\{\mu \times \mu(K) : K \in \mathscr{W} \text{ and } K \supseteq R\}.
\end{aligned}$$

Thus, R is measurable with respect to the product measure $\mu \times \mu$.

Let v be a 0-1 valued measure defined on all subsets of N such that v gives measure zero to singletons and $v(P) = 1$, where P is the set of primes.

Suppose $A \times B \subseteq R$ and $v \times v(A \times B) = v(A)v(B) > 0$. Since v is 0-1 valued, there is some $z > 1$ with $z \in A \cap B$. But then $(z, z) \in B$. This contradiction shows that $\sup\{v \times v(K) : K \in \mathscr{W} \text{ and } K \subseteq R\} = 0$.

Suppose $K \in \mathscr{W}$ and $K \supseteq R$. Consider an expression for K of the form $(+)$. Since $A_q \cup \cdots \cup A_n \supseteq N - \{1\}$, there is some i so that $v(A_i) = 1$. Let g be a prime, $g \in A_i$. Since $\{g\} \times P - \{g\} \subseteq K$, $P - \{g\} \subseteq B_i$. Thus, $v \times v(K) \geq 1$. It follows from the quoted theorem that R is not measurable with respect to $v \times v$.

The theorem stated by Schreier can be proven by similar methods.

This problem naturally leads to the following problem. For each n, $n = 2, 3, 4, \ldots$, let $\mathscr{U}_n$ be the algebra of universally measurable subsets of N^n. In other words, for each finitely additive probability measure μ defined on all subsets of N which vanishes on singletons, let $\mathscr{M}_n(\mu)$ be the algebra of all subsets on N^n which are measurable with respect to μ^n, and let $\mathscr{U}_n$ be the intersection of all such families. Stan Williams has proved the following theorem.

Theorem. A subset E of N^n, for $n = 2, 3, \ldots$ is universally measurable if and only if there is a set B in the algebra generated by product sets and a finite subset F of N such that $B \subset E$ and

$$E - B \subset \bigcup\{\pi_i^{-1}(F) : 1 \leq i \leq n\},$$

where π_i is the projection of N^n into the ith coordinate.

In particular, if $n = 2$, E is universally measurable if and only if E is in the algebra generated by product sets.

Added 2014. Williams also characterized the universally measurable sets for infinite product spaces.

1. G. Birkhoff, *Lattice Theory*, Amer. Math. Soc. Coll. Publ. 25, revised edition, New York, 1967.
2. S. C. Williams, Universally measurable sets of finitely additive product measures, *Illinois J. Math.* 33 (1989), 451–463.

R. Daniel Mauldin

PROBLEM 12: BANACH

A surface S is homeomorphic to the surface of a sphere and it has:

(a) a tangent plane everywhere
(b) a continuously varying tangent plane.

Is S equivalent to the surface of a geometric sphere? (That is to say, does there exist a homeomorphism of the whole space which transforms the given surface S into the surface of the sphere?)

Solution

We use the definition of a tangent plane that is stated in Problem 156: A plane $T(q)$ in Euclidean three-dimensional space E^3 is *tangent* to a topological 2-sphere S at the point $q \in S$ if for every $\varepsilon > 0$, there exists a round ball B with center q such that any straight line joining q to a point of $S \cap B - \{q\}$ makes an angle of size less than ε with the plane $T(q)$. We say that a topological sphere S has a *continuous family of tangent planes* if S has a unique tangent plane at each of its points and, for any sequence $\{q_i\}$ of points of S converging to a point q, the sequence $\{T(q_i)\}$ converges to $T(q)$.

Using current terminology, we can state Problem 12 as follows:

Is a 2-sphere S tame in E^3 if S has a continuous family of tangent planes? We use E^3 to denote Euclidean 3-dimensional space. A topological 2-sphere S is defined to be *tame* in E^3 if there is a homeomorphism of E^3 onto itself that carries S onto the graph of $x^2 + y^2 + z^2 = 1$; otherwise, S is *wild* in E^3. The existence of wild spheres in E^3 became known in the early 1920s with Alexander's description of a "horned sphere" [1] and Antoine's construction of a wild Cantor set in E^3 [2, 3]. References for numerous other examples of wild spheres can be found in surveys of work on embeddings of surfaces in E^3 [5, 6]. It is our purpose in this note to describe a wild sphere that has a continuous family of tangent planes.

We notice, with Example 1 below, that the definition stated above for a tangent plane does not imply that there is a unique tangent plane at each point. However, we do not permit this situation in our definition of a continuous family of tangent planes.

Example 1. Let S denote the topological 2-sphere that is obtained by revolving the graph of $|x|^{1/2} + |z|^{1/2} = 1$ about the z-axis. Any vertical plane that contains the point $(0, 0, 1)$ would, under the definition given above for a tangent plane, be tangent to S at the point $(0, 0, 1)$. In this example, any normal line to S at $(0, 0, 1)$ fails to pierce S at $(0, 0, 1)$. (We say that a line L pierces S at a point $q \in L \cap S$ if there exist two points p and r of L such that q is between p and r on L and the open intervals (p, q) and (q, r) of L are in different components of the complement of S.)

We describe, with Example 2, a 2-sphere S which has a continuous family of tangent planes such that there are points of S where the normal line does not

Figure 12.1

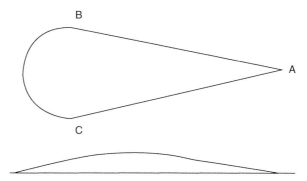

Figure 12.2

pierce S. While this sphere is tame in E^3, we use the example in our description, with Example 3, of a wild sphere that has a continuous family of tangent planes. Then we state a theorem about spheres that are pierced by their normal lines.

Example 2. Let D_1 denote a disk in the xy-plane as indicated in Figure 12.1. (The boundary of D_1 is the union of three arcs AB, AC, and BC, where AB and AC are closed intervals of two intersecting lines and BC is an arc of a circle that is tangent to both lines.)

We let $S = D_1 \cup D_2$, where D_2 is another disk, indicated as follows, which has the same boundary as D_1. Let P be any vertical plane that intersects the interior of D_1. The intersection of P with D_2 is required to be a curve of the type indicated in Figure 12.2. (Symmetry and similarity are not required.) This can be done so that S has a continuous family of tangent planes.

Example 3. We describe a sphere S' that is wildly embedded in E^3 and has a continuous family of tangent planes. To do this, we follow a procedure described by Fox and Artin [9] to "entangle" S in the vicinity of the point A so that a wild sphere S' is obtained. Let D_1' be a disk that is embedded in E^3 as indicated in Figure 12.3. (This is similar to the wild embedding of an arc described in Example 1.2 of the paper by Fox and Artin.)

The wild disk D_1' is in the disk D_1 of Example 2 except near the overcrossings (or undercrossings) where D_1' is raised slightly above the plane of D_1 to avoid self-intersections. Now we place another disk D_2' over D_1' in a manner analogous to the way D_2 was placed over D_1 in Example 2. We do this so that D_1' and D_2' have the same boundary and $D_1' \cup D_2'$ is a 2-sphere S' in the closure of the bounded component of the complement of the sphere S of Example 2. This can be done so that the wild sphere S' has a continuous family of tangent planes. For any vertical plane P that intersects the interior of S', each component of $P \cap$ Int S' would be an open disk with a boundary as indicated in Figure 12.2, except that the adjustments near the overcrossings might not permit the lower edge to be straight. If we picture the disk D_1' as a roadway approaching the point A, the heights of the overcrossings (or

Figure 12.3

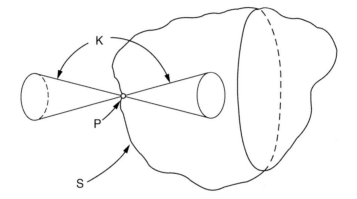

Figure 12.4

undercrossings) should approach zero and the slope of the roadway should approach zero. Also, the ratio of the height of D_2' above D_1' to the width of the road should approach zero as the road approaches A. We rely on the work of Fox and Artin in the paper cited above to know that the topological sphere S' is wild in E^3.

Theorem. If the topological 2-sphere S in E^3 has a tangent plane at each point and, for each $p \in S$, the line normal to S at p pierces S at p, then S is tame in E^3.

Proof. It follows from the definition of a tangent plane for a sphere S that each point $p \in S$ is the vertex of a solid double cone K that does not intersect S except at p. Since the normal line to S at p pierces S at p, it follows that the two components of $K - p$ are in different components of $E^3 - S$. (See Figure 12.4.) Cannon [7] and Bothe [4] have independently shown that the existence of such a double cone at each point of S implies that S is tame in E^3.

Remarks. We can construct a wild sphere like Alexander's horned sphere [1] so that it has a continuous family of tangent planes. We would first construct a tame sphere S such that the disk $x^2 + y^2 \leq 1$ in the xy-plane is in S and each vertical section of the interior of S would be like Figure 12.2. Then a wild sphere S' similar to the one described by Alexander [1] would be constructed in S together with the bounded component of its complement. The Cantor set of points where S' is locally wild would be a subset of the circle $x^2 + y^2 = 1$ in the xy-plane. As in Example 3, we would need to exercise some control on cross-sections, slopes, heights, and ratios as we approach a point of $x^2 + y^2 = 1$ along the horns.

A 2-sphere S would be tame in E^3 if for each point q of S there is plane that contains q and does not intersect $S - q$. We can readily show that the interior of such a sphere is convex, and this implies that the sphere is tame. A more general, and much more difficult theorem has very recently been proved by Daverman and Loveland [8]. They have shown that a 2-sphere S is tame in E^3 if there is a $\delta > 0$ such that, for each $p \in S$, there is a round ball B of diameter δ such that $p \in B$ and B does not intersect the interior of S.

1. J. W. Alexander, An example of a simply connected surface bounding a region which is not simply connected, *Proc. Nat. Acad. Sci. U.S.A.* 10 (1924), 8–10.
2. _____, Remarks on a point set constructed by Antoine, *Proc. Nat. Acad. Sci. U.S.A.* 10 (1924), 10–12.
3. L. Antoine, Sur L'homeomorphisme de Deux Figures et de leurs Voisinages, *J. Math. Pures Appl.* 4 (1921), 221–325.
4. H. G. Bothe, Differenzeirbare Flächen sind zahm, *Math. Nachr.* 43 (1970), 161–180.
5. C. E. Burgess, Embeddings of surfaces in Euclidean three-space, *Bull. Amer. Math. Soc.* 81 (1975), 797–818.
6. C. E. Burgess and J. W. Cannon, Embeddings of surfaces in E^3, *Rocky Mountain J. Math.* 1 (1971), 259–344.
7. J. W. Cannon, $\star$-Taming sets for crumpled cubes II: Horizontal sections in closed sets, *Trans. Amer. Math. Soc.* 161 (1971), 441–446.
8. R.J. Daverman and L.D. Loveland, Any 2-sphere in E^3 with uniform interior tangent balls is flat, *Canadian J. Math.* 33 (1981), 150–167.
9. R.H. Fox and E. Artin, Some wild cells and spheres in three-dimensional space, *Ann. of Math.* (2) 49 (1948), 979–990.

C.E. Burgess

University of Utah

PROBLEM 13: ULAM

Let E be the class of all subsets of the set of integers. Two subsets $K_1, K_2 \in E$ are called equivalent or $K_1 \equiv K_2$ if $K_1 - K_2$ and $K_2 - K_1$ are at most finite sets. There is given a function $F(K)$ defined for all $K \in E$; its range is contained in E and

$$F(K_1 + K_2) \equiv F(K_1) + F(K_2)$$

$$F(\text{compl. } K) \equiv \text{ compl. } F(K).$$

Question: Does there exist a function $f(x)$ (x and $f(x)$ natural integers) such that $f(K) \equiv F(K)$?

Commentary

The answer is no. As is well known today there are 2^c ultrafilters over the integers E. For each ultrafilter U, let $F_U(X)$ be defined by $F_U(X) = E$ if $X \in U$, $F_U(X) = \emptyset$, otherwise. Each ultrafilter defines a distinct homomorphism F. Also the corresponding functions f would have to be different, but there are only $2^{\aleph_0}$ such maps f.

Richard Laver

PROBLEM 14: SCHAUDER, MAZUR

Let $f(x_1, \ldots, x_n)$ be a function defined in the cube K_n. Let us suppose that f possesses almost everywhere all the partial derivatives up to the rth order and the derivatives up to the order $r-1$ are absolutely continuous on almost every straight line parallel to any axis. All the partial derivatives (up to the order r) $\in L^p, p > 1$.

Does there exist a sequence of polynomials $\{w_i\}$ which converges in the mean in the pth power to f and in all partial derivatives up to the order r?

For $r = 1$ this was settled positively by the authors. An analogous problem exists for domains other than K_n.

PROBLEM 15: SCHAUDER

Let $f(x_1, \ldots, x_n)$ be a function defined in K_n, i.e., in the n-dimensional cube. Does there exist for every n some $p_n \geq 2$ such that if $f \in L^{p_n}$ then there exists a function $u(x_1, \ldots, x_n)$ continuous on K_n:

(a) vanishing on the boundary of K_n,
(b) possessing first derivatives on almost every line parallel to the axes and absolutely continuous,
(c) possessing almost everywhere second partial derivatives ($\in L^{p_n}$) and satisfying the equation : $\Delta u = f$.

The author proved that for $n = 2, 3$; $p_n = 2$. Mazur observed that for $n = 4$, p_n cannot be equal to 2. For which n does there exist a $p_n > 2$?

Solution

To present our solution to this problem, let $Q = (-\frac{1}{2}, +\frac{1}{2})^n, \bar{Q} = [-\frac{1}{2}, +\frac{1}{2}]^n$ be the open and closed unit cubes in $\mathbb{R}^n$, respectively. Let ∂Q be the boundary of Q. Fix $p > n$.

<u>Theorem</u>: Given $f \in L^p(Q)$ there exists a continuous function u on $\bar{Q}$ with the following properties.

- $u = 0$ on ∂Q.
- For each $x = (x_1, \cdots, x_n) \in Q$ and $j = 1, \cdots, n$, the partial derivative $u_j(x) := \lim_{h \to 0}[u(x_1, \cdots, x_{j-1}, x_j + h, x_{j+1}, \cdots, x_n) - u(x_1, \cdots, x_n)]/h$ exists.

- For each $j, i = 1, \cdots, n$, and for almost every $(x_1, \cdots, x_{i-1}, x_{i+1}, \cdots, x_n) \in (-\frac{1}{2}, \frac{1}{2})^{n-1}$, the function $F : x_i \mapsto u_j(x_1, \cdots, x_n)$ is absolutely continuous.
- For each j, i, and for almost every $x \in Q$, the limit $u_{ij}(x) := \lim_{h \to 0} [u_j(x_1, \cdots, x_{i-1}, x_i + h, x_{i+1}, \cdots, x_n) - u_j(x_1, \cdots, x_n)]/h$ exists.
- For almost every $x \in Q$, we have $\sum_{i=1}^{n} u_{ii}(x) = f(x)$.

<u>Remarks</u>: This answers Schauder's problem. The methods used to prove it were out of reach when the problem was proposed, but are now standard.

Notation and Definitions

- We let $e_j = (0, 0, \cdots, 0, 1, 0, \cdots, 0)$ denote the jth unit vector in $\mathbb{R}^n$.
- We write c, C, C', etc. to denote constants depending only on n, p. These symbols may denote different constants in different occurrences.
- For $0 < \alpha < 1$, we write $||F||_{Lip(\alpha)} = \sup_{x \neq y} \frac{|F(x) - F(y)|}{|x - y|^\alpha}$. Note that $||F||_{lip(\alpha)} = 0$ if F is constant.
- A *rectangular box* in $\mathbb{R}^n$ is a Cartesian product $I = (a_1, b_1) \times \cdots \times (a_n, b_n)$ of open intervals (a_j, b_j) with $a_j < b_j$.
  The *volume* of $I = (a_1, b_1) \times \cdots \times (a_n, b_n)$, written as $|I|$, is defined as the product $(b_1 - a_1) \cdot (b_2 - a_2) \cdot \cdots \cdot (b_n - a_n)$.
- We write $L^p_{comp}(\mathbb{R}^n)$ to denote the space of all functions $F \in L^p(\mathbb{R}^n)$ that vanish a.e. outside a compact set.
- We write $C_0^\infty(\mathbb{R}^n)$ to denote the space of C^∞ functions of compact support on $\mathbb{R}^n$.
- The adjective "smooth" means C^∞ unless we say otherwise.
- We write $B(x, r)$ to denote the open ball about x with radius r in $\mathbb{R}^n$.
- If $\Omega \subset \mathbb{R}^n$, then $\bar{\Omega}$ denotes the closure of Ω.
- The expression $f * g$ denotes the convolution, $f * g(x) = \int_{\mathbb{R}^n} f(y) g(x - y) dy$.

Background

We start with basic properties of the Newtonian potential. References to the literature and/or sketch of proofs may be found at the end of this article.

Let $T\varphi(x) = -c_n \int_{\mathbb{R}^n} \frac{\varphi(x-y)}{|y|^{n-2}} dy$

(or $T\varphi(x) = -\text{const} \int_{\mathbb{R}^2} (\ln|y|) \varphi(x-y) dy$ if $n = 2$) for $\varphi \in C_0^\infty(\mathbb{R}^n)$. Then $T\varphi$ is smooth. Define

$T_j \varphi := \frac{\partial}{\partial x_j}(T\varphi)$ and $T_{ij}\varphi = \frac{\partial^2}{\partial x_i \partial x_j}(T\varphi)$ for $\varphi \in C_0^\infty$.

(A) For some $\alpha \in (0,1)$ depending only on n, p, we have the following result. (the Sobolev inequality).

T and T_j extend to operators from $L^p_{comp}(\mathbb{R}^n)$ to $\mathrm{Lip}(\alpha)$; we continue to denote these extensions by T, T_j, respectively.

We have the inequalities:

- $\|Tf\|_{Lip(\alpha)} \leq C(R)\|f\|_{L^p(\mathbb{R}^n)}$ if $\mathrm{supp}\, f \subset B(0, R)$; here $C(R)$ depends only on n, p, R.
- $\|T_j f\|_{Lip(\alpha)} \leq C\|f\|_{L^p(\mathbb{R}^n)}$.

(B) Moreover, T_{ij} extends to a bounded operator from $L^p(\mathbb{R}^n)$ to $L^p(\mathbb{R}^n)$. We continue to denote this extension by T_{ij}. We have the inequality (Calderón-Zygmund): $\|T_{ij}f\|_{L^p(\mathbb{R}^n)} \leq C\|f\|_{L^p(\mathbb{R}^n)}$

(C) If $f \in L^p(\mathbb{R}^n)$ and $f = 0$ a.e. in a neighborhood of x_0, then $Tf, T_i f, T_{ij} f$ are smooth in a neighborhood of x_0.

(D) (Dirichlet Problem)

- Let f be a continuous function on ∂Q. Then there exists a continuous function u on $\bar{Q}$, such that $u = f$ on ∂Q and u is harmonic in Q.
- Let $\Omega \subset \mathbb{R}^n$ be a bounded open set with a smooth boundary $\partial \Omega$, and let f be a continuous function on $\partial \Omega$. Then there exists one and only one continuous function u on $\bar{\Omega}$ such that $u = f$ on $\partial \Omega$ and u is harmonic in Ω. Moreover, if f is smooth in a neighborhood of some given $x_0 \in \partial \Omega$, then $u|_{\bar{\Omega} \cap B(x_0, \delta)}$ extends to a smooth function on $B(x_0, \delta)$ for some small $\delta > 0$.

(E) (Strong Maximal Function)

For $f \in L^p(\mathbb{R}^n)$ and $x \in \mathbb{R}^n$, define
$Mf(x) = \sup\{\frac{1}{|I|} \int_I |f(y)|dy \colon I \text{ is a rectangular box containing } x\}$
$= \sup\{\frac{1}{|I|} \int_I |f(x-y)|dy \colon I \text{ is a rectangular box containing } 0\}$.
Mf is called the "strong maximal function" of f.
Then we have the inequality:
$\|Mf\|_{L^p(\mathbb{R}^n)} \leq C\|f\|_{L^p(\mathbb{R}^n)}$ for all $f \in L^p(\mathbb{R}^n)$.

A computation with the strong max function

We fix a smooth function $\theta(t)$ supported in $[-\frac{1}{2}, \frac{1}{2}]$, with $\int_{-\infty}^{\infty} \theta(t)dt = 1$. For $\delta > 0$ and $x = (x_1, \cdots, x_n) \in \mathbb{R}^n$, define $\theta_\delta(x_1, \cdots, x_n) = \delta^{-n}\theta(\frac{x_1}{\delta}) \cdots \theta(\frac{x_n}{\delta})$.

Then $\int_{\mathbb{R}^n} \theta_\delta(x)dx = 1$, and
$|\theta_\delta(x_1, \cdots, x_n)| \leq \frac{C}{\delta^n} \prod_{j=1}^n \mathbb{I}_{[-\frac{1}{2}\delta, \frac{1}{2}\delta]}(x_j)$, where $\mathbb{I}$ denotes the indicator function.

For $h > 0, \delta > 0, F \in L^p(\mathbb{R}^n)$, we will study the quantity $\frac{1}{h} \int_0^h (\theta_\delta * F)(x + te_i)dt = \theta_{\delta,h}^i * F(x)$, where $\theta_{\delta,h}^i(x) = \frac{1}{h} \int_0^h \theta_\delta(x + te_i)dt$.

For $x = (x_1, \cdots, x_n) \in \mathbb{R}^n$, we have

$|\theta^i_{\delta,h}(x)| \leq \int_0^h \frac{1}{h} \cdot [\frac{C}{\delta^n} \prod_{j \neq i} \mathbb{I}_{[-\frac{\delta}{2},+\frac{\delta}{2}]}(x_j) \cdot \mathbb{I}_{[-\frac{\delta}{2},\frac{\delta}{2}]}(x_i + te_i)]dt \leq \frac{C'}{|I|}\mathbb{I}_I(x)$, where $I = \prod_{j=1}^n I_j$ and

$$I_j = (-\frac{\delta}{2}, +\frac{\delta}{2}) \qquad\qquad j \neq i$$

$$I_j = (-(\frac{\delta}{2}+h), +(\frac{\delta}{2}+h)) \qquad\qquad j = i$$

Consequently,

$$|\frac{1}{h}\int_0^h (\theta_\delta * F)(x+te_i)dt| \leq \int_{\mathbb{R}^n} |\theta^i_{\delta,h}(y)||F(x-y)|dy \leq \frac{C'}{|I|}\int_{y \in I} |F(x-y)|dy \leq C'MF(x).$$

Thus, we have shown that

$$|\frac{1}{h}\int_0^h (\theta_\delta * F)(x+te_i)dt| \leq CMF(x)$$

for all $\delta, h > 0, i = 1, \cdots, n, F \in L^p(\mathbb{R}^n)$.

More generally, the same argument shows that

$$|\frac{1}{h}\int_0^h (\theta_\delta * F)(x+te_i)dt| \leq CMF(x+se_i)$$

for any $s \in (0,h)$ and therefore that

$$|\int_0^h (\theta_\delta * F)(x+te_i)dt| \leq C\int_0^h MF(x+se_i)ds$$

for any $\delta, h > 0, i = 1, \cdots, n$ and $F \in L^p(\mathbb{R}^n)$.

The Analogue of Problem 15 on $\mathbb{R}^n$

The operators T, T_j, T_{ij} commute with convolution with θ_δ when acting on C_0^∞, hence also when acting on $L^p_{comp}(\mathbb{R}^n)$.

Let $f \in L^p_{comp}(\mathbb{R}^n)$. For $x \in \mathbb{R}^n, h > 0$ and $j = 1, \cdots, n$ we argue as follows.

$$h^{-1}[Tf(x+he_j) - Tf(x)] =$$

$$\lim_{\delta \to 0} h^{-1}[(\theta_\delta * Tf)(x+he_j) - (\theta_\delta * Tf)(x)]$$

(since Tf is continuous)

$$= \lim_{\delta \to 0} h^{-1}(T(\theta_\delta * f)(x + he_j) - T(\theta_\delta * f)(x))$$

(since T commutes with convolution with θ_δ)

$$= \lim_{\delta \to 0} \{ h^{-1} \int_0^h \partial_j(T(\theta_\delta * f))(x + te_j)dt \}$$

(by the fundamental theorem of calculus; note that $T(\theta_\delta * f)$ is smooth since $\theta_\delta * f \in C_0^\infty(\mathbb{R}^n)$)

$$= \lim_{\delta \to 0} h^{-1} \int_0^h T_j(\theta_\delta * f)(x + te_j)dt$$

(by definition of T_j).
   Therefore,
(*)

$$h^{-1}[Tf(x + he_j) - Tf(x)] - T_jf(x) =$$

$$\lim_{\delta \to 0} h^{-1} \int_0^h T_j(\theta_\delta * f)(x + te_j)dt - \lim_{\delta \to 0} \theta_\delta * T_jf(x)$$

(since T_jf is continuous)

$$= \lim_{\delta \to 0} h^{-1} \int_0^h T_j(\theta_\delta * f)(x + te_j)dt - \lim_{\delta \to 0} T_j(\theta_\delta * f)(x)$$

(since T_j commutes with convolution with θ_δ)

$$= \lim_{\delta \to 0} h^{-1} \int_0^h [T_j(\theta_\delta * f)(x + te_j) - T_j(\theta_\delta * f)(x)]dt.$$

For each $\delta > 0$, the Sobolev inequality (see (2A)) gives

$$|T_j(\varphi_\delta * f)(x + te_j) - T_j(\varphi_\delta * f)(x)| \le C|t|^\alpha ||\theta_\delta * f||_{L^p(\mathbb{R}^n)} \le C|t|^\alpha ||f||_{L^p(\mathbb{R}^n)}.$$

Hence,

$$|h^{-1} \int_0^h [T_j(\theta_\delta * f)(x + te_j) - T_j(\theta_\delta * f)(x)]dt| \le Ch^\alpha ||f||_{L^p(\mathbb{R}^n)},$$

and therefore (*) tells us that

$$|h^{-1}[Tf(x+he_j) - Tf(x)] - T_jf(x)| \le Ch^{\alpha}||f||_{L^p(\mathbb{R}^n)}.$$

Consequently,

$$T_jf(x) = \lim_{h \to 0+} h^{-1}[Tf(x+he_j) - Tf(x)].$$

Similarly,

$$T_jf(x) = \lim_{h \to 0-} h^{-1}[Tf(x+he_j) - Tf(x)].$$

Thus,

$$T_jf(x) = \lim_{h \to 0} h^{-1}[Tf(x+he_j) - Tf(x)]$$

for all $x \in \mathbb{R}^n, j = 1, \cdots, n, f \in L^p(\mathbb{R}^n)$.

We have shown that $T_jf = \partial_j Tf$ for all $f \in L^p_{comp}$, not just for $f \in C_0^{\infty}$.

Next, for $f \in L^p_{comp}, x \in \mathbb{R}^n, h > 0$, and $i, j = 1, \cdots, n$, we study the quantity

$$h^{-1}[T_jf(x+he_i) - T_jf(x)] =$$

$$\lim_{\delta \to 0} h^{-1}[(\theta_\delta * T_jf)(x+he_i) - (\theta_\delta * T_jf)(x)]$$

(since T_jf is continuous)

$$= \lim_{\delta \to 0} h^{-1}[T_j(\theta_\delta * f)(x+he_i) - T_j(\theta_\delta * f)(x)]$$

(since T_j commutes with convolution with θ_δ)

$$= \lim_{\delta \to 0} h^{-1} \int_0^h \partial_i T_j(\theta_\delta * f)(x+te_i)dt$$

(by the fundamental theorem of calculus; note that $T_j(\theta_\delta * f)$ is smooth, since $\theta_\delta * f \in C_0^{\infty}(\mathbb{R}^n)$)

$$= \lim_{\delta \to 0} h^{-1} \int_0^h T_{ij}(\theta_\delta * f)(x+te_i)dt$$

(by definition of T_{ij})

$$= \lim_{\delta \to 0} h^{-1} \int_0^h \theta_\delta * T_{ij}f(x+te_i)dt$$

(since T_{ij} commutes with convolution with θ_δ).

Therefore, by the results of section 3, we have

$$|\frac{1}{h}[T_jf(x+he_i) - T_jf(x)]| \leq CM(T_{ij}f)(x)$$

and

$$(!)\qquad |T_jf(x+he_i) - T_jf(x)| \leq C \int_0^h M(T_{ij}f)(x+se_i)ds.$$

The strong maximal inequality and the Calderón-Zygmund inequality now show that the function

$$x \mapsto \sup_{h>0}|h^{-1}[T_jf(x+he_i) - T_jf(x)]|$$

is pointwise less than or equal to $CM(T_{ij}f)(x)$ and thus has L^p-norm at most

$$C||M(T_{ij}f)||_{L^p(\mathbb{R}^n)} \leq C'||T_{ij}f||_{L^p(\mathbb{R}^n)} \leq C''||f||_{L^p(\mathbb{R}^n)}.$$

Now define

$$\varepsilon_{ij}f(x) = \limsup_{h\to 0}|h^{-1}[T_jf(x+he_i) - T_j(x)] - T_{ij}f(x)|.$$

Then our previous inequality and the boundedness of T_{ij} on L^p yields the inequality

$$(**)\qquad ||\varepsilon_{ij}f||_{L^p(\mathbb{R}^n)} \leq C||f||_{L^p(\mathbb{R}^n)}$$

for all $f \in L^p_{comp}(\mathbb{R}^n)$.

On the other hand, the definition of ε_{ij} shows that

$$0 \leq \varepsilon_{ij}(f+g)(x) \leq \varepsilon_{ij}(f) + \varepsilon_{ij}(g).$$

Moreover, for functions $\theta \in C_0^\infty(\mathbb{R}^n)$ we have $\varepsilon_{ij}(\theta) = 0$, since $T_j\theta$ is smooth and $T_{ij}\theta = \partial_i T_j\theta$.

Given $\varepsilon > 0$ and $f \in L^p_{comp}(\mathbb{R}^n)$, we find a function $\theta \in C_0^\infty(\mathbb{R}^n)$ with $||f - \theta||_{L^p(\mathbb{R}^n)} < \varepsilon$, and write $f = \theta + (f - \theta)$.

Therefore,

$$0 \leq \varepsilon_{ij}(f) \leq \varepsilon_{ij}(\theta) + \varepsilon_{ij}(f - \theta) = \varepsilon_{ij}(f - \theta).$$

and consequently, by $(**)$, we have

$$||\varepsilon_{ij}(f)||_{L^p(\mathbb{R}^n)} \leq ||\varepsilon_{ij}(f - \theta)||_{L^p(\mathbb{R}^n)} \leq C||f - \theta||_{L^p(\mathbb{R}^n)} \leq C\varepsilon.$$

Since $\varepsilon > 0$ is arbitrarily small, we see that $\varepsilon_{ij}(f) = 0$ almost everywhere. That is, for almost every $x \in \mathbb{R}^n$, we have

$$T_{ij}f(x) = \lim_{h \to 0+} h^{-1}[T_jf(x + he_i) - T_jf(x)].$$

Similarly, for almost every $x \in \mathbb{R}^n$, we have

$$T_{ij}f(x) = \lim_{h \to 0-} h^{-1}[T_jf(x + he_i) - T_jf(x)].$$

$$\boxed{\text{Thus, for almost every } x \in \mathbb{R}^n, \text{we have } T_{ij}f(x) = \lim_{h \to 0} h^{-1}[T_jf(x + he_i) - T_jf(x)].}$$

Thus, at almost every $x \in \mathbb{R}^n$, we have $T_{ij}f(x) = \partial_i(T_jf)(x)$. We have shown this for all $f \in L^p_{comp}$, not just for $f \in C_0^\infty$.

For $\theta \in C_0^\infty(\mathbb{R}^n)$, we have

$$\sum_{i=1}^n T_{ii}\theta = \sum_{i=1}^n \partial_i^2 T\theta = \Delta T\theta = \theta,$$

since T inverts the Laplacian when acting on $\theta \in C_0^\infty$. Since T_{ij} is bounded on L^p, we conclude that

$$\boxed{\sum_{i=1}^n T_{ii}f = f \quad \text{(almost everywhere) for all} \quad f \in L^p(\mathbb{R}^n).}$$

Next, we study the absolute continuity of

$$x_i \mapsto T_jf(x_1, \cdots, x_n) \text{ for fixed } x_1, \cdots, x_{i-1}, x_{i+1}, \cdots, x_n.$$

Fix i and $j \in \{1, \cdots, n\}$, and $f \in L^p_{comp}(\mathbb{R}^n)$.
Given $x' := (x_1, \cdots, x_{i-1}, x_{i+1}, \cdots, x_n) \in \mathbb{R}^{n-1}$, we define

$$F^{x'}(x_i) := T_jf(x_1, \cdots, x_{i-1}, x_i, x_{i+1}, \cdots, x_n)$$

for $x_i \in \mathbb{R}$.
Let $I_v = [a_v, b_v](v = 1, \cdots, N)$ be pairwise disjoint intervals.
Applying our previous inequality (!), with $x = (x_1, \cdots, x_{i-1}, a_v, x_{i+1}, \cdots, x_n)$ and $h = b_v - a_v$, we learn that

$$|F^{x'}(b_v) - F^{x'}(a_v)| \leq C \int_{y \in I_v} M(T_{ij}f)(x_1, \cdots, x_{i-1}, y, x_{i+1}, \cdots, x_n)dy.$$

Summing over v, we find that

$$\sum_{v=1}^{N} |F^{x'}(b_v) - F^{x'}(a_v)| \leq C \int_{y \in \cup_v I_v} M(T_{ij}f)(x_1, \cdots, x_{i-1}, y, x_{i+1}, \cdots, x_n) dy$$

$$\leq C(\int_{-\infty}^{\infty} [M(T_{ij}f)(x_1, \cdots, x_{i-1}, y, x_{i+1}, \cdots, x_n)]^p dy)^{\frac{1}{p}} \cdot (\sum_v |I_v|)^{\frac{1}{p'}}$$

by Hölder's inequality, where p' is the exponent dual to p.
It follows that $F^{x'}(x_i)$ is an absolutely continuous function of x_i, provided

$$(\#) \qquad \int_{-\infty}^{\infty} [M(T_{ij}f)(x_1, \cdots, x_{i-1}, y, x_{i+1}, \cdots, x_n)]^p dy < \infty.$$

However, we have already seen that $M(T_{ij}f) \in L^p(\mathbb{R}^n)$, and therefore $(\#)$ holds for almost every $(x_1, \cdots, x_{i-1}, x_{i+1}, \cdots, x_n) \in \mathbb{R}^{n-1}$.

Therefore, the function $x_i \mapsto T_jf(x_1, \cdots, x_n)$ is absolutely continuous on $\mathbb{R}$, for almost every $(x_1, \cdots, x_{i-1}, x_{i+1}, \cdots, x_n)$; this holds for any $f \in L^p_{comp}(\mathbb{R}^n)$.

This concludes our study of the analogue of Problem 15 in $\mathbb{R}^n$.

Passing from $\mathbb{R}^n$ to the Unit Cube

Let $f \in L^p(Q), Q = (-\frac{1}{2}, \frac{1}{2})^n$.
We introduce a function $\tilde{f} \in L^p_{comp}(\mathbb{R}^n)$, defined by

$$\tilde{f}(x_1, \cdots, x_n) =$$

$$\sum_{\substack{(i_1, i_2, \cdots, i_n) \in \mathbb{Z}^n \\ |i_1|, \cdots, |i_n| \leq 100}} (-1)^{i_1 + \cdots + i_n} f((-1)^{i_1}(x_1 - i_1), \cdots, (-1)^{i_n}(x_n - i_n)) \cdot \prod_{v=1}^{n} \mathbb{I}_{(-\frac{1}{2}, \frac{1}{2})}(x_v - i_v).$$

In particular, $\tilde{f} = f$ on Q.
Let $R_{j,\sigma} : \mathbb{R}^n \to \mathbb{R}^n$ be the reflection of $\mathbb{R}^n$ with respect to the hyperplane $H_{j,\sigma} = \{(x_1, \cdots, x_n) : x_j = \sigma\} (\sigma = \pm\frac{1}{2})$.
Thus, $H_{j,\sigma}(x_1, \cdots, x_n) = (x_1, \cdots, x_{j-1}, 2\sigma - x_j, x_{j+1}, \cdots, x_n)$.
Then $\tilde{f} + \tilde{f} \circ R_{j,\sigma} = 0$ on the cube $(-10, +10)^n$, as one checks by elementary computation.
The operator T commutes with $R_{j,\sigma}$ when acting on C_0^∞, hence also when acting on L^p_{comp}. It therefore follows from the result given in section (2C) that

$$T\tilde{f} + T\tilde{f} \circ R_{j,\sigma}$$

is smooth on the cube $(-1, 1)^n$.

In particular, $T\tilde{f}$ is smooth on the face $\bar{Q} \cap H_{j,\sigma}$ of the unit cube since $R_{j,\sigma} =$ identity on that face.

Applying the results of section 4 to $\tilde{f}$, we now learn the following.

Let $U = T\tilde{f}|_{\bar{Q}}$. Then

- U is continuous on $\bar{Q}$
- For each $x \in Q$ and $j = 1, \cdots, n$, the limit $U_j(x) := \lim_{h \to 0} h^{-1} [U(x + he_j) - U(x)]$ exists.
- For almost every $x \in Q$, and for each $i,j = 1, \cdots, n$, the limit $U_{ij}(x) := \lim_{h \to 0} h^{-1} [U_j(x + he_i) - U_j(x)]$ exists.
- For each i, and for almost every $(x_1, \cdots, x_{i-1}, x_{i+1}, \cdots, x_n) \in (-\frac{1}{2}, \frac{1}{2})^{n-1}$, the function

$$x_i \mapsto U_j(x_1, \cdots, x_{i-1}, x_i, x_{i+1}, \cdots, x_n)$$

  is absolutely continuous on $[-\frac{1}{2}, +\frac{1}{2}]$.
- The restriction of U to $\{(x_1, \cdots, x_n) \in \bar{Q} : x_j = \sigma\}$ is smooth, for each $\sigma = \pm\frac{1}{2}$ and each $j = 1, \cdots, n$.
- $\sum_{i=1}^{n} U_{ii} = f$ almost everywhere in Q.

Let V be the solution of the Dirichlet problem

$$\begin{bmatrix} V \text{ continuous on } \bar{Q}, \\ V = U \text{ on } \partial Q, \\ V \text{ harmonic in } Q \end{bmatrix} \qquad \text{(see section (2D))}$$

Then

- V is continuous on $\bar{Q}$
- For each $x \in Q$ and $j = 1, \cdots, n$, the limit $V_j(x) = \lim_{h \to 0} h^{-1} [V(x + he_j) - V(x)]$ exists.
- For each $x \in Q$ and $i,j = 1, \cdots, n$, the limit $V_{ij}(x) = \lim_{h \to 0} h^{-1} [V_j(x + he_i) - V_j(x)]$ exists.
- The restriction of V to $\{(x_1, \cdots, x_n) \in \bar{Q} : x_j = \sigma\}$ is smooth for each $\sigma = \pm\frac{1}{2}, j = 1, \cdots, n$.
- $\sum_{i=1}^{n} V_{ii}(x) = 0$ for all $x \in Q$.

We will prove that

$$(\dagger) \quad \boxed{\begin{array}{l} x_i \mapsto V_j(x_1, \cdots, x_{i-1}, x_i, x_{i+1}, \cdots, x_n) \text{ is a smooth function on } [-\frac{1}{2}, +\frac{1}{2}], \text{ for} \\ \text{each } (x_1, \cdots, x_{i-1}, x_{i+1}, \cdots, x_n) \in (-\frac{1}{2}, \frac{1}{2})^{n-1}. \end{array}}$$

To see this, we may assume without loss of generality, that $i = n$. Fix $x^\circ = (x_1^\circ, \cdots, x_{n-1}^\circ) \in (-\frac{1}{2}, \frac{1}{2})^{n-1}$, and pick $\varepsilon > 0$ so small that $|x_j^\circ| + 10\varepsilon < \frac{1}{2}$ for each j.

Pick a smooth curve Γ as in Figure 15.1 containing vertical line segments starting at $(\pm\frac{1}{2},0)$.

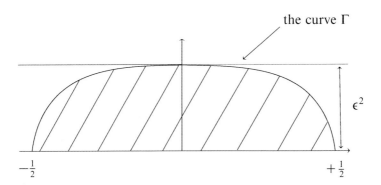

Fig. 15.1

Then define $\Omega \subset \mathbb{R}^n$ to consist of all $(x_1,\cdots,x_{n-1},x_n) \in \mathbb{R}^n$ s.t. the point $(\sum_1^{n-1}(x_i - x_i^0)^2, x_n) \in \mathbb{R}^2$ lies in the shaded region enclosed by Γ and the x-axis. (We include the interval $(-\frac{1}{2},\frac{1}{2}) \times \{0\}$ on the x-axis, but we do not include the curve Γ in the shaded region.)

Then

- Ω is a bounded domain with a smooth boundary in $\mathbb{R}^n$.
- $\partial\Omega$ contains a neighborhood of $(x_1^0,\cdots,x_{n-1}^0,\frac{1}{2})$ in $\mathbb{R}^{n-1} \times \{\frac{1}{2}\}$ and a neighborhood of $(x_1^0,\cdots,x_{n-1}^0,-\frac{1}{2})$ in $\mathbb{R}^{n-1} \times \{-\frac{1}{2}\}$.
- $\Omega \subset Q$.

Now V is a harmonic function on Ω, V is continuous on $\bar{\Omega}$, and $V|_{\partial\Omega}$ is smooth in a neighborhood of

$$(\cdot) \qquad (x_1^0,\cdots,x_{n-1}^0,\frac{1}{2}) \text{ and } (x_1^0,\cdots,x_{n-1},-\frac{1}{2}).$$

According to the result given in section (2D), V extends to a smooth function on the union of Ω with small balls centered at the two points $(\cdot)$. In particular, the function

$$(\cdot\cdot) \qquad x_n \mapsto V(x_1^0,\cdots,x_n^0,x_n) \text{ is smooth on } [-\frac{1}{2},-\frac{1}{2}+2\delta] \text{and on } [\frac{1}{2}-2\delta,\frac{1}{2}] \text{for small enough} \delta.$$

In $[-\frac{1}{2}+\delta,+\frac{1}{2}-\delta]$, that function is smooth, simply because V is harmonic.

Therefore, the function $(\cdot\cdot)$ is smooth on $[-\frac{1}{2},+\frac{1}{2}]$, completing the proof of $(\dagger)$.

We now define $u = U - V$. The known properties of U and V immediately imply that u has all the properties asserted in the statement of the theorem on the 1st page.

This work was partially supported by NSF grant #: DMS-1265524. I am grateful to Will Crow for TeXing this note.

References

In this section we provide references and/or sketch of proof for the background results in Section 2.

(A) The proof of Theorem 2 part (iii) in Stein's SINGULAR INTEGRALS AND DIFFERENTIABILITY PROPERTIES OF FUNCTIONS, Princeton U. Press 1970 contains the essential ideas of the proof.

(B) See A.P. Calderon and A. Zygmund, On the existence of certain singular integrals, Acta Math 88,1952, pp 339–393

(C) follows from the integral formulas in our references for (A) and (B) together with the Lebesgue dominated convergence theorem.

(D) First Bullet point. If the restriction of f to each face is a trigonometric polynomial, then a solution is easily produced by explicit formulas. One can then pass to the general case by approximating every continuous f uniformly by "simple" f's as described above. The harmonic extensions of the approximating f's converge uniformly, thanks to the maximum principle for harmonic functions.

Second bullet point. Existence of solutions of the Dirichlet problem is proven e.g. in *Partial Differential Equations:Second Edition* by L. C. Evans, Graduate Studies in Math vol 19, AMS.

To obtain the smoothness of u near x_0 as stated at the end of (D) one extends u to be zero outside Ω. Let U be the resulting extension, and let F be the Laplacian of U (as a distribution). Then F is supported on the boundary of Ω, and involves terms containing u and its normal derivative (let's call it v) on the boundary of Ω. Since U must agree with f as we approach the boundary from the inside, we obtain a singular integral (or pseudodifferential) equation for v in terms of f. That equation is elliptic, hence v is smooth wherever f is smooth. In particular, f and v are smooth in a neighborhood of x_0 in the context of (D). Since u inside Ω is given by solving Laplacian $U = F$ on R^n, it follows that u is smooth up to the boundary of Ω and may therefore be extended to a neighborhood of x_0 as asserted in (D).

(E) See B. Jessen, J. Marcinkiewicz and A. Zygmund, Note on the differentiability of multiple integrals, Fundamenta Math. 25 (1935), pp 217–234.

CHARLIE FEFFERMAN

PROBLEM 15.1: MAZUR, ORLICZ

Prize: Two small beers, S. Mazur

Is a space E, of type F, for which there exists a sphere K which is bounded, necessarily of type (B)? (A sphere is bounded if and only if $\chi_n \in K$, and if the numbers $t_n \to 0$, then $t_n \chi_n \to 0$.)

Addendum. Negative answer: It suffices to take for E the space of numerical sequences

$$x = \sum_{n=1}^{\infty} \xi_n$$

such that

$$\sum_{n=1}^{\infty} |\xi_n|^p < \infty; \quad 0 < p < 1,$$

with ordinary operations, and

$$\|x\| = \left(\sum_{n=1}^{\infty} |\xi_n|^p \right)^{1/p}.$$

Instead of the space ℓ^p one can also take L^p which consists of real-valued functions $x = x(t)$ in $[0,1]$, measurable, and such that

$$\int_0^1 |x(t)|^p dt < \infty,$$

with ordinary algebraic operations and

$$\|x\| = \left[\int_0^1 |x(t)|^p dt \right]^{1/p}.$$

MAZUR

May 1, 1937

Remark A space of type (F) is the terminology from Banach's monograph, Théorie des Opérations Linéaires, for a Fréchet space, a completely metrizable topological vector space.

PROBLEM 16: ULAM

Find a Lebesgue measure in the space of all measurable functions satisfying the following conditions:

If $\{H_n\}$ are measurable sets contained on the line $\{x = x_n\}$, then the set of all measurable functions $ff(x)$, satisfying the condition $f(x_n) \in H_n$ has a measure equal to $|H_1| \cdot |H_2| \ldots$ where $|H_n|$ denotes the measure of the set H_n.

Addendum. Such a measure exists.

S. Banach

May 15, 1941

Solution

If one takes M to be the set of all measurable functions from $\mathbb{R}$ into $\mathbb{R}$, then there is no such measure. This is easily seen by considering the set $A = \{f \in M : 0 \leq f(1) \leq 1\}$ and the set $B = \{f \in M : 0 \leq f(1) \leq 1 \text{ and } 0 \leq f(2) \leq 2\}$. If there were such a measure, then the measure of A would be 1, while the measure of the subset B of A would be 2.

On the other hand, if one takes M to be the space of all measurable functions from $\mathbb{R}$ into I, the unit interval, then the answer is yes, even if one interprets the words "Lebesgue measure" to mean a regular Borel measure in M relative to the topology induced by the product topology of $I^{\mathbb{R}}$. (Obviously, it must have been this case which Banach considered.) To prove this we first construct the standard product measure μ in $I^{\mathbb{R}}$. Thus μ is defined over the σ-algebra Σ of subsets of $I^{\mathbb{R}}$ generated by all subsets of the form $\{f : f(t) \in A\}$, where $t \in \mathbb{R}$ and A is a measurable subset of I. Now we use the following properties of Σ and μ.

(a) If $t_1, \ldots, t_n$ are different real numbers and $A_1, \ldots, A_n$ are measurable subsets of I, then

$$\mu\left(\{f : f(t_1) \in A_1, \ldots, f(t_n) \in A_n\}\right) = \lambda(A_1) \cdot \ldots \cdot \lambda(A_n)$$

where λ is the Lebesgue measure.

This follows of course from the definition of μ.

(b) The completion of Σ relative to μ contains a Borel subsets of $I^{\mathbb{R}}$, in other words for every Borel set $B \subseteq I^{\mathbb{R}}$ there exist sets $B_0 \subseteq B \subseteq B_1$, with $B_0, B_1 \in \Sigma$ and $\mu(B_0) = \mu(B_1)$.

This follows from a theorem on Haar measures, see [1, §64, Theorem H, or 2]. For convenience of the reader we shall prove it directly, and our proof generalizes to products of second countable spaces.

Proof. Since the completion $\bar{\Sigma}$ of Σ is a σ-field, to prove (b) it is enough to show that all open sets $V \subseteq I^{\mathbb{R}}$ are in $\bar{\Sigma}$. Let $I_1, I_2, \ldots$ be a countable basis of open sets for I. Let $\mathbb{B}$ be a basis of open sets for $I^{\mathbb{R}}$ which consists of all cylinders over finite nonempty products of the I_n's. Then V is a union of some sets $C_s \in \mathbb{B}$ ($s \in S$). Let $S_0 \subseteq S$ be a countable set such that $\mu(\bigcup_{s \in S_0} C_s)$ is maximal. We put $B_0 = \bigcup_{s \in S_0} C_s$. Let $T \subseteq \mathbb{R}$ be a countable set such that B_0 is a cylinder over an open subset of I^T. For any $A \in \mathbb{B}$ we put $A^* = $ (the cylinder in $I^{\mathbb{R}}$ over the projection of A into I^T). Thus $A \subseteq A^*$, A^* is open, and the range of the function * is countable. Let us prove that if $A \in \mathbb{B}$ and $A \subseteq V$ then $\mu(A^* - B_0) = 0$. First it is clear that $A^* - B_0! \in \Sigma$. Then $A = A^* \cap U$, where $U \in \mathbb{B}$ and U is a cylinder over an open set in I^S, where $S \cap T = \emptyset$. By the definition of B_0 we have $\mu(A \cup B_0) = \mu(B_0)$. Hence $0 = \mu(A - B_0) = \mu((A^* - B_0) \cap U) = \mu(A^* - B_0) \cdot \mu(U)$. Since $\mu(U) > 0$ it follows that $\mu(A^* - B_0) = 0$. Now we put

$$B_1 = \bigcup\{A^* : A \in \mathbb{B}, A \subseteq V\}.$$

Since the range of * is countable, we have $\mu(B_1) = \mu(B_0)$. Also it is clear that $B_0, B_1 \in \Sigma$ and $B_0 \subseteq V \subseteq B_1$.

Remarks. Since the above proof uses the separability of I in an essential way it may be interesting to recall the following facts related to (b) which hold for arbitrary compact spaces and Baire measures:

(1) If μ is a product of Baire measures μ_t in compact spaces C_t, then μ is a Baire measure in $\prod C_t$.

Proof. By the Tychonoff theorem and the Stone-Weierstrass theorem every continuous real-valued function $\phi(x)$ over $\prod C_t$ can be uniformly approximated by functions of the form $p(f_1(x(t_1)), \ldots, f_n(x(t_n)))$, where p is a polynomial and f_i is a continuous function over C_i. Hence ϕ is measurable relative to μ and hence μ is a Baire measure.

(2) Every Baire measure (in Halmos' sense [1]) is a locally compact space and has a unique extension to a regular Borel measure (again in Halmos' sense). This is proved in [1, § 54, Theorem D]. However, this is not true for the more widely used definitions. This has been demonstrated by Fremlin in a preprint.

If μ is a regular Borel measure in a compact space C and μ_0 is the Baire restriction of μ, then, by (1), μ_0^2 is a Baire measure in C^2. However, Fremlin [4] has given an example which shows that it is not necessarily true that μ^2 be consistent with the unique Borel extension of μ_0^2. This solves a problem of Bledsoe and Morse [3]. (If μ is a Borel measure in C which is not regular, then μ^2 need not be a Borel measure in C^2 (e.g., let $C = \{\alpha : \alpha \leq \omega_1\}$ with the interval topology. Thus C is compact. For any Borel set $B \subseteq C$, we put $\mu(B) = 1$, if B has a closed uncountable subset, and $\mu(B) = 0$ otherwise. Thus μ is a nonregular Borel measure, and, as is easily seen, the diagonal of C^2 is not μ^2-measurable.)

(c) The inner measure of M in $I^{\mathbb{R}}$ is 0 and the outer measure of M is 1.

Proof. For any $f \in I^{\mathbb{R}}$ let

$$A_f = \{g : |\{t : f(t) \neq g(t)\}| \leq \aleph_0\},$$

then A_f intersects every nonempty set in Σ. Of course if f is measurable then $A_f \subseteq M$ and if f is not measurable then $A_f \cap M = \emptyset$. And (c) follows.

In connection with this proof of (c) we have the following more special question. Let μ_0 be a probability Baire measure in a compact space C. Let $\bar{\mu}_0^S$ be the regular Borel extension of the product measure μ_o^S in C^S. Let $M \subset C^S$ be any set invariant under countable changes, i.e.,

$$|\{t : f(t) \neq g(t)\}| \leq \aleph_0 \implies (f \in M \iff g \in M),$$

and $\emptyset \neq M \neq C^S$ (in particular $|S| > \aleph_0$ follows). Is it true that the inner μ_0^S measure of M is 0 and the outer μ_0^S measure of M is 1?}

By (a), (b), and (c) the outer measure μ^* restricted to the class of sets of the form $M \cap X$, where $X \in \Sigma$ (= the completion of Σ), has all the properties required in Problem 16, restricted to functions in $I^{\mathbb{R}}$.

We are indebted to A. Hajnal, A. Mate, A. Ramsay, and R.J. Gardner for their help in writing this commentary.

1. P.R. Halmos, *Measure theory,* Van Nostrand Publ. Co., New York, 1950.
2. E. Hewitt and K.A. Ross, *Abstract Harmonic Analysis, Vol. I,* Springer-Verlag, New York, 1963.
3. W.W. Bledsoe and A.P. Morse, Product measures, *Trans. Amer. Math. Soc.,* 79 (1955), 173–215.
4. D.H. Fremlin, Products of Radon Measures: A counterexample, *Canad. Math. Bull.* 19, (1976), no. 3, 285–289.

R. Daniel Mauldin and J. Mycielski

PROBLEM 17: ULAM

Prove a converse of Poisson's theorem: that is, given a sequence of urns containing white balls (1) and black ones (0), with unknown composition $\{p_n\}$ and given also the result x_n of drawing from each urn in turn, prove that with probability 1,

$$\lim_{n \to \infty} \frac{1}{n} \sum_{i=1}^{n} x_i = p$$

implies that

$$\lim_{n \to \infty} \frac{1}{n} \sum_{i=1}^{n} p_i = p.$$

(See the Commentary to Problem 94.)

PROBLEM 17.1 ULAM

Let f be a continuous function defined for all $0 \le x \le 1$. Does there exist a perfect set of points C and an analytic function ϕ so that for all points of the set C we have $f \equiv \phi$?

Remark

Z. Zahorski (Sur l'ensemble des points singuliers d'une fonction d'une variable réelle admettant les derivées de tous les ordres, *Fund. Math.* 34 (1947), 183–245) showed that the answer is no and also raised a number of related problems.

Second Edition Commentary

A natural question is: What happens if one relaxes the requirement on the interpolant g, asking it to be not analytic but just smooth? In particular, in the paper mentioned above Z.Zahorski asked: Is it true that there is a function $f \in C(I), I = [0,1]$, such that for every g in $C^{\infty}(I)$ (or in $C^p(I)$) the set

$$E(f,g) = \{t : f(t) = g(t)\}$$

is at most countable?

Here are some results in the subject, obtained in the 1980s - 1990s.

1. S. Agronski, A. Bruckner, M. Laczkovich, and D. Preiss proved that for every $f \in C(I)$ there exists $g \in C^{\infty}(I)$ such that $E(f,g)$ is infinite [ABLP-85]. These authors proved also that for every $f \in C(I)$ there exists $g \in C^1(I)$ such that $E(f,g)$ is a perfect set. The following result of Z.Buczolich [B-88] provides a jump to "almost" C^2-functions: For every $f \in C(I)$ there is a convex function g such that $E(f,g)$ is perfect.

2. However , it is not possible, in general, to get the interpolation by C^2- function [O-94]: There is a Lipshitz function f on I, such that the set $E(f,g)$ is at most countable whenever $g \in C^2(I)$. This gives the positive answer to Zahorski's question above. In fact, one cannot "jump" over any integer in the following sense: Given $n \in N$ there exists f of smoothness $(n - \varepsilon)$ so that $E(f,g)$ is at most countable whenever g is of smoothness $(n + \varepsilon)$.

3. On the other hand, one can "jump" from 1 to 2: For every $f \in C^1(I)$ one can find $g \in C^2(I)$ such that $E(f,g)$ is a perfect set ([O-94]). As a contrast , let us mention the following result from that paper: There exists $f \in C^2(I)$ such that no function $g \in C^3(I)$ interpolates f on a perfect set. It can be stated in a more general form: Given $n \geq 2$ there exists $f \in C^n(I)$ such that every function $g \in C^{n+1}(I)$ may have only finitely many points of tangency of maximal order with f,

References

[ABLP-85] S.Agronsky, A.M.Bruckner, M.Laczkovich and D.Preiss, Convexity conditions and intersections with smooth functions, Trans. Amer. Math. Soc. 298 (1985),659–677.

[B-88] Z.Buczolich , Sets of convexity of continuous functions, Acta Math.Hungar 52 (1988), 291–303.

[O-94] A.Olevskii, Ulam-Zahorski problem on free interpolation by smooth functions, Trans. Amer. Math. Soc. 342 (1994) , 713–727.

A. Olevskii

PROBLEM 18: ULAM

Let a steady current flow through a curve in space which is closed and knotted. Does there exist a line of force which is also knotted (knotted = nonequivalent through any homeomorphism of the whole space R^3 with the circumference of a circle)?

Second Edition Commentary

We refer to the curve through which the current flows as the *wire*, and to lines of force as *(magnetic) field lines*. Wires are understood to be simple closed curves. This problem asks whether there always, for every knotted wire, exists a knotted field line. That question remains unanswered. One much weaker interpretation of the question, though, is whether there *can* exist a knotted field line, i.e., whether there is some knotted wire which has some field line which is a simple closed knotted curve.

That question has been answered in the affirmative using computer-assisted proof techniques.

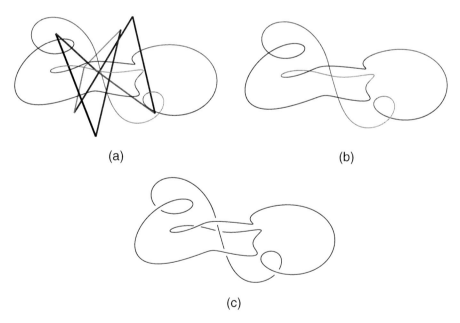

Fig. 18.1 A knotted field line, shown (a) with the wire and (b) without the wire. The wire is drawn in bold. A planar knot diagram is shown in (c).

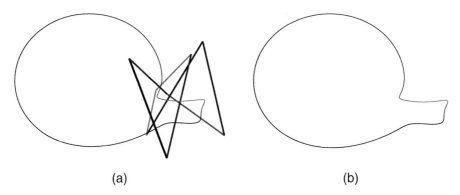

Fig. 18.2 An unknotted field line, shown (a) with the wire and (b) without the wire. The wire is drawn in bold.

Theorem 4 (Minton 2015 [9]). *Figures 18.1 and 18.2 depict two closed field lines for the same wire. The wire is piecewise-linear and forms a trefoil knot. The field line in Figure 18.1 is a trefoil knot, and the field line in Figure 18.2 is an unknot.*

The piecewise-linear trefoil knot was chosen for simplicity and computational efficiency; certainly, other choices could have been made. In fact, the computer proof yields the following strengthening essentially for free.

Corollary 1. *There is a wire with any desired knot type admitting a simple, closed field line equivalent to the trefoil knot (respectively, to the unknot).*

Magnetic field lines are generically closed for planar wires, but for nonplanar wires they are not. The topology of field lines can be quite nontrivial [11], even for simple examples [8]. Because they are not all closed, field lines can exhibit interesting topological behavior like interlocking [15]. Studying the linking of field lines has applications in physics ranging from solar dynamics [18] to tokamaks and fusion energy [12]. There are explicit examples known of electromagnetic fields in which every two field lines are linked [13, 6]. There are even explicit examples known of electromagnetic fields in which every torus knot appears as a field line [7]. These are solutions of Maxwell's equations, but they do not correspond to magnetic fields arising from current through a single wire. A modern perspective on magnetic field lines is that they are governed by a Hamiltonian system with one and a half degrees of freedom [4, 16]. This allows one to invoke the theory of Hamiltonian systems, implying properties like chaos and existence of certain periodic solutions (i.e., closed field lines) [12]. This argument does not *a priori* give us control over the topological structure of the closed field lines, though.

Computer assistance has been profitably used in rigorous mathematical proof for some time now, with perhaps the most notable examples being the Four-Color Theorem [1] and Hales' proof of the Kepler Conjecture [5]. It has also proven useful for showing existence of certain periodic solutions to differential equations [17, 2, 3]. This is essentially our problem.

Because magnetic field lines obey an action principle [4], they can be viewed in an existing computer-assisted proof framework [10]. Theorem 4 was proven with the following basic procedure: find an approximate solution using (nonrigorous) numerical calculations — this part uses standard scientific computing techniques — and then prove existence of a true solution near to the approximate solution using the following fixed-point theorem.

Existence Theorem. *Let U and V be Banach spaces, let $f: U \to V$ be a differentiable function, and let $T: V \to U$ be a linear operator. Suppose $x_0 \in U$ and $\varepsilon > 0$ are such that*

$$|f(x_0)| \cdot \frac{||T||}{1 - ||\mathrm{id}_V - Df(x) \circ T||} < \varepsilon$$

for all $x \in B(x_0, \varepsilon)$. Then there exists $x_* \in B(x_0, \varepsilon)$ such that $f(x_*) = 0$.

This theorem converts the problem of finding a continuous object that *exactly* satisfies a set of conditions into a problem of computing *bounds* on continuous objects. Computers are relatively ill-suited to the former, but (e.g., using interval arithmetic [14]) they are well-suited to the latter. We apply the Existence Theorem to our problem by looking for a zero of the function $f: \mathbf{x}(t) \mapsto \mathbf{x}'(t) - \mathbf{B}(\mathbf{x}(t))$, where $\mathbf{B}$ is the magnetic field in question. The main complications arise from the fact that the natural domain and codomain of this function f are infinite-dimensional; one must reduce to a finite-dimensional problem by controlling for the effects of higher Fourier modes [10, §III.4]. The details of the proof can be found in the report [9].

In the context of Theorem 4, what the computer actually proves is this: there is a closed field line which is within 10^{-10} relative units of the provided approximate curve. But in particular this means that, up to the precision of the diagram, the true solution is identical to the figure shown.

Note that the conditions of the Existence Theorem are not necessarily satisfied near every zero x_* of every function f: the derivative $Df(x_*)$ must be right-invertible. So this technique is widely but not universally applicable. Generally speaking, for a problem such as this, invertibility can be arranged by suitable choice of f if the solution is isolated. So the fact that field lines are not generically closed — and thus a closed field line is not likely to be in a continuous family of closed field lines — is actually critical for the proof to work.

Finally, note also that this proof technique is designed to handle individual cases; it does not give us a way to prove that knotted field lines always exist. If true, proving that will require new techniques.

1. Kenneth Appel and Wolfgang Haken. *Every planar map is four colorable*, volume 98 of *Contemporary Mathematics*. American Mathematical Society, Providence, RI, 1989. With the collaboration of J. Koch.
2. Gianni Arioli and Hans Koch. Integration of dissipative partial differential equations: A case study. *SIAM Journal on Applied Dynamical Systems*, 9(3):1119–1133, 2010.
3. CAPD - computer assisted proofs in dynamics, a package for rigorous numerics. http://capd.ii.uj.edu.pl/. Accessed: 2015-04-14.

4. John R Cary and Robert G Littlejohn. Noncanonical Hamiltonian mechanics and its application to magnetic field line flow. *Annals of Physics*, 151(1):1–34, 1983.
5. Thomas C. Hales. A proof of the Kepler conjecture. *Ann. of Math. (2)*, 162(3):1065–1185, 2005.
6. William T. M. Irvine and Dirk Bouwmeester. Linked and knotted beams of light. *Nature Physics*, 4:716–720, Aug 2008.
7. Hridesh Kedia, Iwo Bialynicki-Birula, Daniel Peralta-Salas, and William T. M. Irvine. Tying knots in light fields. *Phys. Rev. Lett.*, 111:150404, Oct 2013.
8. Keith Leon McDonald. Topology of steady current magnetic fields. *American Journal of Physics*, 22(9):586–596, 1954.
9. Gregory T. Minton. Knotted field lines and computer-assisted proof. Preprint.
10. Gregory T. Minton. *Computer-assisted proofs in geometry and physics*. PhD thesis, Massachusetts Institute of Technology, Sep 2013.
11. Tomoshige Miyaguchi, Makoto Hosoda, Katsuyuki Imagawa, and Katsuhiro Nakamura. Topology of magnetic field lines: Chaos and bifurcations emerging from two-action systems. *Phys. Rev. E*, 83:016205, Jan 2011.
12. P. J. Morrison. Magnetic field lines, Hamiltonian dynamics, and nontwist systems. *Physics of Plasmas*, 7(6):2279–2289, 2000.
13. Antonio F. Rañada. A topological theory of the electromagnetic field. *Letters in Mathematical Physics*, 18(2):97–106, 1989.
14. Nathalie Revol and Fabrice Rouillier. Motivations for an arbitrary precision interval arithmetic and the MPFI library. *Reliable Computing*, 11(4):275–290, 2005.
15. Joseph Slepian. Lines of force in electric and magnetic fields. *American Journal of Physics*, 19(2):87–90, 1951.
16. G. F. Torres del Castillo. Debye potentials and Hamiltonians for magnetic field line flow. *Journal of Mathematical Physics*, 36(7):3413–3418, 1995.
17. Yoshitaka Watanabe. A computer-assisted proof for the kolmogorov flows of incompressible viscous fluid. *Journal of Computational and Applied Mathematics*, 223(2):953–966, 2009.
18. A R Yeates and G Hornig. Dynamical constraints from field line topology in magnetic flux tubes. *Journal of Physics A: Mathematical and Theoretical*, 44(26):265501, 2011.

Gregory T. Minton

PROBLEM 19: ULAM

Is a solid of uniform density which will float in water in every position a sphere?

Commentary

Only a few special cases have been solved. In two dimensions there are counterexamples for density 1/2. For the limiting case of density 0, in two or three dimensions the answer is a qualified yes. If the body is central symmetric of density 1/2 the answer is yes, in any dimension.

The two-dimensional version of the problem concerns a cylinder of uniform density which floats in every position, having the axis parallel to the water surface,

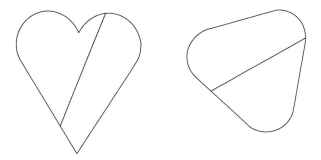

Figure 19.1 Two possible solutions. The line segment rotates within the curve, and in each position cuts off half the area and half the perimeter.

and compatible with Archimedes' law. H. Auerbach [1] showed that in the case of density 1/2, the cylinder need not be circular, or even convex, and gave a class of examples. We reproduce his illustration of two of them (Fig. 19.1).

For dimension 2 or 3 in the limiting case of density 0 the body must rest on a plane in every position. L. Montejano [2] showed that the shell of the body must be a sphere, and noted the example of a ball from which a smaller concentric ball had been removed. The proof given would seem to generalize to arbitrary dimension.

For arbitrary dimension d and density 1/2, if S is star-shaped, symmetric, bounded and measurable, then it differs from a ball by a set of measure 0. This follows from Theorem 1.4 in R. Schneider [3]:

Let Ω_d be the unit sphere $|u| = 1$, and $\langle , \rangle$ the inner product. "If ϕ is an even measure on Ω_d satisfying $\int_{\Omega_d} |\langle u, v \rangle| d\phi(u) = 0$ for each $v \in \Omega_d$, then $\phi = 0$." See also Problem 6331, *Am. Math. Monthly 88 (1981), 150.*

The simplest unsolved case seems to be that of dimension 2, central symmetry, and density other than 0 or 1/2.

1. H. Auerbach, Sur un problème de M. Ulam concernant l'équilibre des corps flottants, *Studia Math.* 7 (1938) 121–142.
2. L. Montejano, On a Problem of Ulam concerning a characterization of the sphere, *Studies in Applied Math.* 53, (1974), 243–248.
3. R. Schneider, Functional equations connected with rotations and their geometric applications, *L'Enseignement Math.* (2) 16 (1970), 297–305.

D. Hensley

Second Edition Commentary

The problem is still open in dimension three and only a few special cases have been solved.

In three dimensions, there are no solutions, other than the sphere, in the limit $\rho \to 0$ or 1, [5]. Also, there are no nontrivial solutions among star-shaped objects with central symmetry for density $\rho = \frac{1}{2}$, [4] and [6]. F. Wenger has recently proposed a perturbation expansion scheme starting from the sphere for objects with central symmetry and $\rho \neq \frac{1}{2}$ [11], as well as for bodies with arbitrarily shape and $\rho = \frac{1}{2}$ [12]. His results point out towards the existence of many nontrivial solutions in these wider classes of shapes, even though the proof is incomplete in that the convergence of the perturbation series has not been examined. Furthermore, no attempt to construct actual solutions of the problem, in dimension three, has been reported. In this dimension, Várkonyi [8], following the spirit of Auerbach [1], took a different approach to construct neutrally floating objects of density $\rho = \frac{1}{2}$ with cylindrical symmetry.

The two-dimensional version of the problem concerns a cylinder of uniform density ρ which floats in water in equilibrium in every position, having the axis parallel to the water surface. This problem is connected to several dynamical systems, as for example, the tire track problem [7], the problem of the existence of closed carrousels [3] and the problem of determining the trajectory of a charge moving in a perpendicular parabolic magnetic field.

H. Auerbach [1] showed that in the case of density $\rho = \frac{1}{2}$, a cylinder of uniform density ρ which floats in water in equilibrium in every position, having the axis parallel to the water surface, need not to be circular, or even convex and gave a class of examples. All these examples coincide with the Zindler curves [13] which, with a suitable geometric construction, are in one to one correspondence with the family of the figures of constant width. Auerbach's illustration of two of them are given in Figure 19.1.

If a figure D of density ρ floats in equilibrium in every position then the water surface divides the boundary of D in constant ratio, say, $\sigma : 1 - \sigma$. We call σ the perimetral density of D. In [9], F. Wenger was able to obtain noncircular solutions for $\rho \neq \frac{1}{2} \neq \sigma$, by a perturbative expansion around the circular solution. These figures have a p-fold rotational symmetry and have $(p - 2)$ different perimetral densities. On the other hand, Bracho, Montejano, and Oliveros [2] proved that if the perimetral density σ is $\frac{1}{3}$ or $\frac{1}{4}$, then the solution is circular. Later, in [10], F. Wegner was able to give nontrivial explicit solutions to this two-dimensional version.

1. H. Auerbach, Sur un probleme concernant l'equilibre des corps flottant, *Studia Math.* **7** (1938), 121–142.
2. J. Bracho, L. Montejano and D. Oliveros, A clasification theorem for Zindler carrousels, *Journal of Dynamical and Control Systems.* **7** (2001), 367–384.
3. J. Bracho, L. Montejano and D. Oliveros, Carrousels, Zindler curves and the floating body problem, *Period. Math. Hungar.* **49** (2004), 1–17.
4. K. J. Falconer, Applications on a result on spherical integration to the theory of convex sets, *Amer. Math. Monthly* **90** (1983), 690–693.
5. L. Montejano, On a problem of Ulam concerning the characterization of the sphere, *Studies Appl. Math.* **53** (1974), 243–248.
6. R. Schneider, Functional equations connected with rotations and their geometrical applications, *L'Enseignement Math.* **16** (1970), 297–305.

7. S. Tabachnkov, Tire track geometry; variations on a theme, *Israel Journal of Math.* **151** (2006), 1–28.
8. P. L. Varkonyi, Neutrally floating objects of density $\frac{1}{2}$, in three dimensions, *Studies Appl. Math.* **130** (2013), 295.
9. F. Wegner, Floating bodies of equilibrium *Studies Appl. Math.* **111** (2003), 167–183.
10. F. Wegner, Floating bodies of equilibrium in 2D, the tire track problem and electrons in a parabolic magnetic field, arXiv:physics/0701241v3 (2007).
11. F. Wegner, Floating bodies of equilibrium in three dimensions. The central symmetric case. arXiv:0803.1043 (2008).
12. F. Wegner, Floating bodies of equilibrium at density $\frac{1}{2}$ in arbitrary dimensions. arXiv:0902.3538 (2009).
13. K. Zindler, Über konvexe Gebilde II, *Monatsh. Math. Phys.* **31** (1921), 25–57.

Luis Montejano

PROBLEM 20: ULAM

Consider one-to-one and continuous transformations of the plane of the form $x' = x$, $y' = f(x,y)$ and $y' = y$, $x' = g(x,y)$ and also transformations which result from composing the above a finite number of times. Can every homeomorphic transformation be approximated by such?

(Analogous problem for the n-dimensional space)

Remark

H.G. Eggleston (A property of plane homeomorphisms, *Fund. Math.* 42 (1955), 61–74) has proved the answer is no. He also shows that if the plane $\mathbb{R} \times \mathbb{R}$ is replaced by the compact square $[0,1] \times [0,1]$ then the answer is yes. For related material see Problem 47 and the accompanying commentary.

Jan Mycielski

PROBLEM 20.1: MAZUR, ORLICZ

For every positive integer n determine the smallest positive integer k_n with the following property: If $f(x_1, \ldots, x_n)$ is an irreducible polynomial, there exist points

$$\left(x_{11}, \ldots, x_{1n}\right), \ldots, \left(x_{k_n 1}, \ldots, x_{k_n n}\right),$$

such that

$$f\left(\lambda_1 x_{11} + \ldots + \lambda_{k_n} x_{k_n 1}, \ldots, \lambda_1 x_{1n} + \ldots + \lambda_{k_n} x_{k_n n}\right),$$

considered as a polynomial of the variables $\lambda_1, \ldots, \lambda_{k_n}$, is irreducible. Is the sequence k_n bounded? ($x_{11}, \ldots, x_{k_n n}$ and $\lambda_1, \ldots, \lambda_{k_n}$ are real or complex variables.)

Commentary

According to Professor Orlicz, Problems 20.1, 27, and 56 emerged in connection with some problems which he and Mazur were considering [1, 2]. The exact meaning of the problems they were considering seems to have become obscured. Problems 20.1 and 56 still seem to be unsolved.

1. S. Mazur and W. Orlicz, Sur la divisibilité des polynomes abstraits, *C. R. Acad. Sci. Paris* 202 (1936), 621–623.
2. S. Mazur and W. Orlicz, Sur les fonctionnelles rationnelles, *C. R. Acad. Sci. Paris* 202 (1936), 904–905.

PROBLEM 21: ULAM

Can one make from the disc $x^2 + y^2 \leq 1$ the surface of a torus using transformations with arbitrary small counter-images? (That is to say, for every $\varepsilon > 0$ there should exist a transformation called $f(p)$ of the disc into the torus, such that if $|p_1 - p_2| \geq \varepsilon$ then $f(p_1) \neq f(p_2)$.)

Remark

This problem was solved negatively by T. Ganea in his paper "On ε-maps onto manifolds," *Fund. Math.* 47 (1959), 35–44. Also, compare with the commentary to Problem 97.

Jan Mycielski

PROBLEM 22: ULAM, SCHREIER

Is every set Z of real numbers a Borel set with respect to set G which are additive groups of real numbers? (Can any set Z be obtained through the operations Σ performed countably many times and through operations of forming differences of sets, starting with sets G such that if x, y belong to the set G, then $x - y$ also belongs to G?)

Remark

The answer to the question is no as it stands. This is because every set Z of real numbers which is in the Borel field generated from the additive groups has the property that if $z \in Z$, then $-z \in Z$. However, with this modification, the problem seems to be unsolved. Also, see Erdős' comments about this problem in his lecture.

R. Daniel Mauldin

PROBLEM 23: SCHAUDER

DEFINITION A.   A function defined in a certain n-dimensional region is called monotonic in this region if, in every subregion, it assumes its maximum and minimum on the boundary. A function is called a saddle function if, after subtracting an arbitrary linear function, it is always monotonic.

DEFINITION B. Let C be a plane region; $C =$ a Jordan curve which is its boundary; $K =$ a space curve over C with one-to-one projection. (That is to say, two different points of K have different projections on C.) I shall say that the curve K satisfies the triangle condition with a constant Δ, if the steepness of the plane defined by any three different points of K is always $\leq \Delta$. By the steepness of a plane $z = ax + by + c$, we mean the number $\sqrt{a^2 + b^2}$.

Rado (and later J. von Neumann) proved this theorem: The surface (function) defined in a convex region C, which is continuous $z = f(x,y)$, and is a saddle function, and whose boundary curve satisfies the triangle condition with the constant Δ satisfies a Lipschitz condition with the same constant Δ. That is to say, for any two points (x_1, y_1) and $(x_2, y_2) \in C$ we have:

$$|f(x_1, y_1) - f(x_2, y_2)| = \Delta\sqrt{(x_1 - x_2)^2 + (y_1 - y_2)^2}.$$

Problem A.   What can one say if the boundary curve is assumed to be merely continuous? For example, is a Lipschitz condition satisfied in every closed domain contained entirely in the interior of C? **Problem B.**   Can one prove anything analogous to Rado's theorem for functions of a greater number of variables $(n \geq 2)$?

PROBLEM 24: MAZUR

Prize: Two Small Beers, S. Mazur

In a space E of type (B), there is given an additive functional $F(x)$ with the following property: If $x(t)$ is a continuous function in $0 \leq t \leq 1$ with values in E, then $F(x(t))$ is a measurable function. Is $F(x)$ continuous?

Remark

The answer is yes (See I. Labuda and R.D. Mauldin, Problem 24 of the Scottish Book, *Coll. Math.* 48 (1984), 89–91).

PROBLEM 25: SCHAUDER

Recently the theory of integral equations was generalized for singular integral equations; that is to say, in which the integral expression $\int K(s,t)g(t)dt$ is considered as an improper integral in the sense of Cauchy. Under certain additional assumptions, the three well-known theorems of Fredholm (for equations with fixed limits) are also valid. In the sense of the theory of operations, equations of this type are probably not totally continuous in the corresponding spaces of type (B).

Problem. Find a new class of linear operations $F(x)$, which contains as special cases the integral equations of the above type (improper) and for which Fredholm theorems do not hold anymore. The equations are of type: $y = x + F(x)$.

PROBLEM 26: MAZUR, ORLICZ

Prize: One small beer, S. Mazur

Let E be a space of type (F_0) and $\{F_n(x)\}$ a sequence of linear functionals in E converging to zero uniformly in every bounded set $R \subset E$. Is the sequence then convergent to zero uniformly in a certain neighborhood of zero? [E is a type (F_0) means that E is a space of type (F) with the following condition: If $x_n \in E$, $x_n \to 0$ and the number series $\sum_{n=1}^{\infty} |t_n|$ is convergent, then the series $\sum_{n=1}^{\infty} t_n x_n$ is convergent. $R \subset E$ is a bounded region if $x_n \in R$, and if the numbers $t_n \to 0$, then $t_n x_n \to 0$.]

Addendum The answer is negative.

M. Eidelheit

June 4, 1938

Commentary

The answer is really negative, even in locally convex spaces. It follows from Słowikowski's example of a Montel space which is not a Schwartz space [1]. The example is also presented in [2, p. 149]. The space is the following.

Let k, m, n_1, n_2 be positive integers. Let $n = (n_1, n_2)$ and let

$$a_{k,m,n_1,n_2} = n^k \max(1, 2^{m-n_1}).$$

Let X be the space of all double sequences

$$x = \{x_{n_1,n_2}\}$$

such that

$$\|x\|_{k,m} = \sup_{n_1,n_2} a_{k,m,n_1,n_2} |x_{n_1,n_2}|$$

with the topology determined by the pseudonorms $\|x\|_{k,m}$. We put

$$F_n(x) = a_{1,1,n_1,n_2} x_{n_1,n_2}.$$

The functionals F_n do not tend uniformly to 0 on any neighborhood of zero, because

$$\limsup_{n \to \infty} \frac{a_{1,1,n_1,n_2}}{a_{k,m,n_1,n_2}} > 0.$$

On the other hand, for an arbitrary bounded set A we have

$$\lim a_{k,m,n_1,n_2} \sup_{x \in A} |x_{n_1,n_2}| = 0$$

for every k and m. In particular for $m = k = 1$, this implies that the functionals F_n tend uniformly to 0 on each bounded set A.

1. W. Słowikowski, On (S)- and (DS)-spaces, *Bull. Acad. Pol. Sci. Cl. III* 5 (1957), 599–600.
2. S. Rolewicz, *Metric Linear Spaces,* Monografie Matematyczne Vol. 56, Polish Scientific Publishers, Warszawa, Poland 1972.

Stefan Rolewicz

PROBLEM 27: MAZUR, ORLICZ

Prize: Five small beers, S. Mazur

Let E be a complex space of type (B); $F(x)$, $G(x)$ complex polynomials defined in E. Let us assume that there exist elements $x_n \in E$ such that $|x_n| \leq 1$ and $F(x_n) \to 0$, $G(x_n) \to 0$. Does there exist then an element x_0 such that $F(x_0) = 0$, $G(x_0) = 0$? *Addendum* The answer is positive. If there is no $x_0 \in E$, such that $F(x_0) = 0$, $G(x_0) = 0$, then there exist complex polynomials $\phi(x)$, $\psi(x)$ in E with the property that

$$F(x)\phi(x) + G(x)\psi(x) \equiv 1.$$

Mazur, Orlicz

April 4, 1939

Commentary

According to Professor Orlicz, Problems 20.1, 27, and 56 emerged in connection with some problems which he and Mazur were considering [1, 2]. The exact meaning of the problems they were considering seems to have become obscured. Problems 20.1 and 56 still seem to be unsolved.

1. S. Mazur and W. Orlicz, Sur la divisibilité des polynomes abstraits, *C. R. Acad. Sci. Paris* 202 (1936), 621–623.
2. S. Mazur and W. Orlicz, Sur les fonctionnelles rationnelles, *C. R. Acad. Sci. Paris* 202 (1936), 904–905.

PROBLEM 28: MAZUR
Prize: Bottle of wine, S. Mazur
  Let

$$\sum_{n=1}^{\infty} a_n,$$

be a series of real terms and let us denote by R the set of all numbers a for which there exists a series differing only by the order of terms from

$$\sum_{n=1}^{\infty} a_n,$$

summable by the method of the first mean to a. Is it true that if the set R contains more than one number but not all the numbers, then it must consist of all numbers of a certain progression $\alpha x + \beta$ $(x = 0, \pm 1, \pm 2, \ldots)$?

The same question for other methods of summation. [It is known that

(1) There exists a series

$$\sum_{n=1}^{\infty} a_n$$

   such that R consists of all terms of a sequence given in advance $\alpha x + \beta$ $(x = 0, \pm 1, \pm 2, \ldots)$;
(2) If the sequence $\{a_n\}$ is bounded, then R consists of either one number or contains all the numbers—the first case occurs only when the series $\sum a_n$ is absolutely convergent.]

Commentary

Let A be a linear method of summation (for example, a matrix method) which for some real series $\sum_0^{\infty} a_n$ produces a sum $s = A - \sum a_n$, while other series are perhaps not A-summable. For a given method A and a series $\sum a_n$ we consider all rearrangements $\sum a_{n_k}$ of the series, single out those among them that are A-summable, and consider the A-sums s. The set of all s may be called the "Riemann set" of the method A and the series $\sum a_n$. For example, Riemann's theorem is that for ordinary convergence, the Riemann set of a series is either a point or the whole real line $\mathbb{R}$. Steinitz' theorem asserts that the Riemann set of ordinary convergence for a series with complex terms is either one point, or a line, or the whole complex plane. Bagemihl and Erdős [*Acta Math.* 92 (1954), 35–53] answered the first part of Problem 28: for $(C, 1)$ summability, the Riemann sets are precisely as described there.

Lorentz and Zeller [*Acta Math.* 100 (1963), 149–169] answered the second part of Problem 28 negatively. They proved that Riemann sets of matrix methods *A* may be almost absolutely arbitrary. More exactly: a subset *S* of $\mathbb{R}$ is a Riemann set of some regular matrix method and some series $\sum a_n$ is and only if *S* is an analytic set. (Thus, all Borel sets *S* are Riemann sets.)

In particular, *S* may be any countable set. [But it is not known that this can be combined with the additional property for series of bounded terms mentioned in the addendum to Problem 28.] Obviously, there are many open problems here of different types. The exact determination of all Riemann sets of some "natural" summation method (such as the Abel or Euler method) is probably quite difficult.

G.G. Lorentz
The University of Texas at Austin

PROBLEM 29: ULAM

Is the group H_n of all homeomorphisms of the surface of an *n*-dimensional sphere simple? (In the following sense: the component of identity does not contain a nontrivial normal subgroup.) It is known (Schreier-Ulam) that the theorem holds for $n = 2$ and the component of identity of H_n does not contain any closed, normal proper subgroups for any *n*.

Commentary

The problem for the orientation-preserving homeomorphisms of S^1 was solved by Schreier and Ulam [4], and in 1947, Ulam and von Neumann showed the comparable result for S^2 [5].

The more general problem for S^m, $m > 1$ was partially solved in 1958 [1], with conditions on a space *X* and group *G* of autohomeomorphisms of *X* guaranteeing that every element of *G* is the product of six elements of *G*, each of the six being a conjugate of an arbitrary non-identity element of *G* or its inverse. Examples of spaces and groups satisfying the conditions include all *stable* autohomeomorphisms of S^m ($m > 1$) and all autohomeomorphisms of the Cantor set, the rationals, the irrationals, the Hilbert cube, the universal curve, etc. (An autohomeomorphism of a manifold is called *stable* if it is the product of finitely many autohomeomorphisms each supported on a cell.) It follows from the well-known 1968 results of Kriby and Siebenmann that the annulus conjecture is true and thus all orientation-preserving autohomeomorphisms of S^m are stable ($m > 5$) (later shown for $m = 5$). Thus with the similar earlier known results for $m = 1, 2, 3$, Problem 29 is settled in the affirmative for $m \neq 4$ (and the case $m = 4$ would follow from the annulus conjecture for S^4).

Later the author, in unpublished work, and then Nunnally [3] showed that for a slightly different class of spaces and groups including all the examples cited above, every element of *G* is the product of at most three conjugates of any non-identity element of *G*. Note that the inverse need not be used. It is not hard to see that in general two conjugates are not sufficient. But Nunnally showed that for any "dilation" g (as, for example, a motion from one pole toward the other on S^m), two conjugates of g do suffice.

Other related papers dealing with inner automorphisms of G are by Fine and Schweigert [2] and several papers by Whittaker in the early 1960s [6].

1. R.D. Anderson, The algebraic simplicity of certain groups of homeomorphisms, *Am. J. Math.* 80 (1958), 955–963.
2. N.J. Fine and G.E. Schweigert, On the group of homeomorphisms of an arc, *Ann. of Math.* 62 (1955), 237–253.
3. Ellard Nunnally, Dilations on Invertible Spaces, *Trans AMS* 123 (1966), 437–448.
4. J. Schreier and S. Ulam, Eine Bemerkung über die Gruppe der topolischen Abbildungen der Kreislinie auf sich selbst, *Studia Math.* 5 (1934), 155–159.
5. S.M. Ulam and J. von Neumann, On the group of homeomorphisms of the surface of a sphere, (abstract), *Bull. AMS* vol. 53 (1947), 506.
6. J.V. Whittaker, On isomorphic groups and homeomorphic spaces, *Ann. of Math.* 78 (1963), 74–91, MR27#737.

R.D. Anderson

PROBLEM 30: ULAM

Two elements a and b of a group H are equivalent if there exists $h \in H$ such that there is a relation $a = hbh^{-1}$. Two pairs of elements: a', a'' and b', b'' are called simultaneously equivalent if there exists $h \in H$ such that we have $a' = hb'h^{-1}$ and $a'' = hb''h^{-1}$.

Question: For which groups does it suffice for simultaneous equivalence of two pairs of elements a', a'' and b', b'' that every combination of the elements a' and a'' be equivalent to the corresponding combination of the elements b' and b''? (The necessity of this condition is obvious.)

PROBLEM 31: ULAM

June 18, 1936

In a metric group which is complete and compact, is the set of elements equivalent to a given element always of first category? Does this theorem hold under the additional assumptions that the group is connected or simple?

Addendum. Banach, Mazur counterexample:

$$S_a = e^{ix} \mapsto e^{i(x+a)}, \quad T_b = e^{ix} \mapsto e^{-i(x+b)}.$$

Commentary

The solution of Banach and Mazur is a misunderstanding, since by the set of conjugates of an element a in a group G, Ulam means $\{xax^{-1} : x \in G\}$. Of course in every matrix group $\mathrm{trace}(xax^{-1}) = \mathrm{trace}(a)$. Hence in most matrix groups the set of conjugates of any element is of dimension smaller than the dimension of the group. This observation need not *a priori* extend to all groups mentioned in the problem and so it remains unsolved.

(For the transfer of some results from matrix groups to all compact groups see J. Mycielski, Some properties of connected compact groups, *Coll. Math.* 5 (1958), 162–166.)

J. Mycielski

PROBLEM 32: ULAM

Let G be a compact metric group (the group operation we shall denote by $\times$). Does there exist for every $\varepsilon > 0$ a finite number of elements of the group: $a_1, a_2, \ldots, a_N$ for which we can define a group operation (denoted by the symbol o) so that with respect to this operation the given finite system forms a group and:

(1) $(a_i \times a_j, a_i o a_j) < \varepsilon$; $i,j = 1,2,\ldots,N$ [(a,b) denotes the distance between the elements a,b].
(2) The inverses of the elements a_i ($i = 1,2,\ldots,N$) with respect to the two operations are distant from each other by less than ε?

Remark

A.M. Turing [*Annals of Mathematics*, 39 (1938), 105–111] showed that the only finitely approximate Lie groups are the compact abelian groups. Thus $SU(3)$ would be a counterexample.

PROBLEM 33: ULAM

Two sequences of sets of real numbers A_n and B_n are called equivalent if there exists an arbitrary function f mapping the set of all numbers into itself in a one-to-one way and such that $f(A_n) = B_n$. **Questions:**

(α) Is every sequence A_n of projective sets equivalent to a certain sequence of Borel sets?
(β) Is every sequence of measurable sets—in the sense of Lebesgue—equivalent to a sequence of Borel sets? Can one prove that there exists a sequence not equivalent to any sequence of sets which are Lebesgue measurable?

Addendum. There exist sequences of projective sets and sequences of measurable sets nonequivalent to sequences of Borel sets. (Communicated by Mr. Szpilrajn, who obtained additional results concerning this notion of equivalence of sequences of sets.) (*Fund. Math.* 26)

Ulam
August 1, 1935

Commentary

The answer to both α and β is no. In fact, Szpilrajn (Sur l'équivalence des suites d'ensembles et l'équivalence des fonctions, *Fund. Math.* 26 (1936), 302–326) showed that there is a sequence of $(PCA) \cap (CPA) = \Delta_2^1$ sets which is not equivalent to any sequence of Borel sets.

S. Ulam

PROBLEM 34: ULAM

A class K of sets of real numbers has the following properties:

(1) The class K contains all sets measurable in the sense of Lebesgue.
(2) If $A \in K$ and $B \in K$, then $A - B \in K$.
(3) If $A_n \in K$, then $\Sigma A_n \in K$.
(4) If the whole space is decomposed into a noncountable number of sets A_γ all disjoint, each noncountable and each belonging to K, then there exists in the class K a set which contains exactly one element from each of the sets A_γ.

Question: Is the class K the class of all subsets of our space?

Commentary

The answer is negative (see [1]). Under natural additional assumptions however, the answer is positive. Those additional assumptions are that K be invariant under translation and that for some integer $n > 1$ for every partition of $\mathbb{R}$ into sets of cardinality n there exists a selector which belongs to K. This was proved in [2]. It was also proved in [2] that under the supposition of *CH* (more precisely the supposition that every union of less than $2^{\aleph_0}$ sets of measure zero is of measure zero), there exists a σ-field F of subsets of $\mathbb{R}$ such that (1) it includes the field of Lebesgue measurable sets, (2) it does not contain all subsets of $\mathbb{R}$, (3) for every partition of $\mathbb{R}$ into sets of cardinality $2^{\aleph_0}$ there exists a selector which belongs to F, (4) F is closed under images by all rational functions with real coefficients. The proof of [2] depends on the algebraic structure of $\mathbb{R}$ and it is not known if one could achieve invariance of F, e.g., under all homeomorphisms of $\mathbb{R}$ onto itself.

1. E. Grzegorek and B. Weglorz, Extensions of filters and fields of sets (I), *Journal Austral. Math. Soc.* 25, Series A, (1978), 275–290.
2. B. Weglorz, Large invariant ideals on algebras, *Alg. Univ.* 13 (1981), 41–55.

Jan Mycielski

PROBLEM 35: ULAM

Is projective Hilbert space (that is to say, the set of all diameters of the unit sphere in Hilbert space metrized by the Hausdorff formula) homeomorphic to the Hilbert space itself?

Commentary

The answer is no because Hilbert space is simply connected and projective Hilbert space is not simply connected (because it has a double covering).

W. Holsztynski

PROBLEM 36: ULAM

Can one transform continuously the full sphere of a Hilbert space into its boundary in such a way that the transformation should be identity on the boundary? *Addendum.* There exists a transformation with the required property given by

Tychonoff.

Commentary

This was answered affirmatively by a construction of Kakutani [2]. It was later shown by Klee [3] that there is a homeomorphism h of the unit ball $\{x : \|x\| \leq 1\}$ onto the punctured unit ball $\{x : 0 < \|x\| \leq 1\}$ such that h is the identity on the boundary $\{x : \|x\| = 1\}$. By results of Bessaga [1], h can even be made a diffeomorphism. With $f(x) = h(x)/\|h(x)\|$, f is a very nice transformation of the desired sort.

1. C. Bessaga, Every infinite-dimensional Hilbert space is diffeomorphic with its unit sphere, *Bull. Acad. Polon. Sci.,* Ser. Sci. Math. Astr. et Phys. 14 (1966), 27–31.
2. S. Kakutani, Topological properties of the unit sphere of a Hilbert space, *Proc. Imp. Acad. Tokyo* 19 (1943), 269–271.
3. V. Klee, Convex bodies and periodic homeomorphisms in Hilbert space, *Trans. Amer. Math. Soc.* 75 (1953), 10–43.

V. Klee

PROBLEM 37: ULAM

A class of sets K is called a ring if: whenever $A \in K$, $B \in K$, then both $(A+B)$ and $(A-B) \in K$. Two rings of sets K and L are isomorphic if one can make correspond to every set of the ring K, in a one-to-one fashion, a set of the ring L so that the sum of sets goes over into the sum, the difference into the difference, and the counterimage contains all the sets of the ring K. **Questions:**

(α) How many nonisomorphic rings of sets of real numbers exist?
(β) How many nonisomorphic rings of sets of integers exist?
(γ) Is the ring of projective sets isomorphic to the ring of Borel sets?

Analogous questions for rings in the sense of countable addition, i.e., countable summation of sets which belong to K also belongs to K.

Commentary

With no loss of generality assume that the problem is formulated for Boolean
algebras of sets, *BA*s. Concerning (α), it is known that there are at least 2^c such
*BA*s [1]. It is still open whether there can be 2^{2^c} such (the maximum possible).
Similarly, concerning (β), there are at least c such (by the folklore result that there
are c non-isomorphic denumerable *BA*s); the question is open whether there are 2^c
such (the maximum possible).

The answer to (γ) is clearly no: since both *BA*s are atomic, any isomorphism
between them must preserve infinite unions which exist in one or the other. But the
BA of projective sets is not closed under countable unions [2, p. 12].

The same questions were asked for σ-fields. For part (α), the same remark
and open question holds. For part (β), there are exactly $\aleph_0$ such, by an obvious
argument. Part (γ) is still open.

1. J.D. Monk and R.M. Solovay, On the number of complete Boolean algebras, *Alg. Univ.,* 2 (1972),
   365–368.
2. W. Sierpiński, *Les ensembles projectifs et analytiques,* Mémorial des Sciences Mathématiques,
   Fascicule CXII, Gauthier-Villars, Paris, 1950.

J.D. Monk

PROBLEM 38: ULAM

Let there be given N elements (persons). To each element we attach k others
among the given N at random (these are friends of a given person). What is the
probability P_{kN} that from every element one can get to every other element through
a chain of mutual friends? (The relation of friendship is not necessarily symmetric!)
Find $\lim_{N \to \infty} P_{kN}$ (0 or 1?).

Solution

First, if $k \geq 2$, the resulting graph is connected with probability tending to 1. Here a
is joined to b if a knows b or b knows a. This may be seen as follows.

Suppose that the graph, G, which has N vertices, is not connected. Then it must
be possible to split G into two parts, G_1 and G_2, so that $|G_1| = r$, $|G_2| = N - r$ where
$3 \leq r \leq N - 3$ and there is no edge connecting an element of G_1 to an element of
G_2. The probability that we do not join any of the r points of G_1 to any of the $N - r$
points of G_2 does not exceed

$$\left[\binom{r-1}{k} \bigg/ \binom{N-1}{k} \right]^r \left[\binom{N-r-1}{k} \bigg/ \binom{N-1}{k} \right]^{N-r}.$$

Since $k \geq 2$, this last estimate is less than

$$\left(\frac{r}{N-1}\right)^{2r}\left(\frac{N-r}{N-1}\right)^{2(N-r)}.$$

Thus, the probability that there is a split is less than

$$\sum_{3 \leq r \leq N-3}\binom{N}{r}\left(\frac{r}{N-1}\right)^{2r}\left(\frac{N-r}{N-1}\right)^{2(N-r)}. \tag{1}$$

To see that this sum goes to zero as $N \to \infty$ notice that if $3 \leq r < (N/8)$ and $N > 8e^2/(8-e^2)$, then $(er/N)^r > (e(r+1)/N)^{r+1}$. Thus,

$$\sum_{3 \leq r < N/8}\binom{N}{r}\left(\frac{r}{N-1}\right)^{2r}\left(\frac{N-r}{N-1}\right)^{2(N-r)}$$

$$\leq c\sum_{3 \leq r < N/8}\binom{N}{r}\left(\frac{r}{N}\right)^{2r} \leq c\sum e^r\left(\frac{r}{N}\right)^r$$

$$\leq k/N^2,$$

where k and c are constants. Also, for all N,

$$\sum_{\frac{N}{8} \leq r \leq \frac{N}{3}}\binom{N}{r}\left(\frac{r}{N-1}\right)^{2r}\left(\frac{N-r}{N-1}\right)^{2(N-r)} \leq \frac{N}{3}\left(\frac{e}{3}\right)^{N/3}$$

and

$$\sum_{\frac{N}{3} \leq r \leq \frac{N}{2}}\binom{N}{r}\left(\frac{r}{N-1}\right)^{2r}\left(\frac{N-r}{N-1}\right)^{2(N-r)} \leq \frac{N}{2}\left(\frac{e}{4}\right)^{N/2}.$$

These inequalities imply that the sum in (1) converges to zero.

Second, if $k = 1$, the resulting graph is connected with probability tending to zero. If $k = 1$, we may consider this as a problem on random mapping functions [1, p. 66]. Let f map $\{1,\ldots,n\}$ into itself by setting $f(i) = j$, provided i "knows" j. Katz and Rényi proved the following theorem: If $C(n)$ denotes the number of connected mapping functions, then

$$\lim_{n \to \infty}\frac{C(n)}{n^{n-1/2}} = (\pi/2)^{1/2}.$$

Thus,

$$\lim_{N \to \infty} P_{1N} = 0.$$

1. J.W. Moon, *Counting Labelled Trees,* Canadian Mathematical Monographs, No. 1, William Clowes and Sons, Limited, London, 1970.

P. Erdős

Second Edition Commentary

Stan Ulam's seminal question concerning the connectivity of random graphs with given (out-)degree foreshadowed the emergence of the subject of the evolution of random graphs, which was pioneered by Erdős and Renyi more than two decades later. This has subsequently become a major stream in the flow of probabilistic combinatorics. It is interesting that Erdős' comments on the problem only treated the (easier) problem for undirected graphs, whereas Ulam's question dealt with the (more difficult) case of directed graphs.

Ron Graham

More Second Edition Commentary and Solution

One interpretation of Ulam's question is that it asks not about the limiting behavior of the probability that certain random directed graphs are strongly connected, but about whether associated undirected graphs are connected. We now formalize the problem.

Consider a directed graph $D_{m,n}$ and an undirected graph $U_{m,n}$. $D_{m,n}$ and $U_{m,n}$ use the same set of n vertices, which we label $v_1, \ldots, v_n$. We construct $D_{m,n}$ as follows: independently for each vertex v_i, we randomly select m distinct vertices other than v_i and draw directed edge from v_i to each of these m vertices. Using $D_{m,n}$ we then construct $U_{m,n}$ as follows: we connect v_i and v_j by an undirected edge if and only if $D_{m,n}$ contains both $\overrightarrow{v_i v_j}$ and $\overrightarrow{v_j v_i}$. Let $P_{m,n}$ denote the probability that $U_{m,n}$ is connected. For large n, we wish to know how $P_{m,n}$ depends on m. We show that a transition occurs near $m = \sqrt{n \log n}$: for any $\varepsilon > 0$ and n sufficiently large, if $m < (1 - \varepsilon)\sqrt{n \log n}$, then $P_{m,n}$ is close to 0, whereas if $m > (1 + \varepsilon)\sqrt{n \log n}$, then $P_{m,n}$ is close to 1.

We remark that a related question is when $D_{m,n}$ is strongly connected. This question has a different transition that probably occurs around $m = \log n$, since $D_{m,n}$

is likely to have vertices with no in-edges when m is much smaller than $\log n$, but not when m is much bigger than $\log n$.

Define $S_i = \{v_j : D_{m,n} \text{ contains } \overrightarrow{v_i v_j}\}$.

<u>Lemma 1</u> *Fix any vertices $w_1, \ldots, w_k$ and sets of vertices $A_1, \ldots, A_k$ such that $w_i \notin A_j$ for any i, j. Then the probability that there is no directed edge in $D_{m,n}$ from A_i to w_i for any i is at most*

$$\left(1 - \frac{m}{n-1}\right)^{(|A_1| + \cdots + |A_k|)}.$$

Proof. Since the edges emanating from any one node are chosen independently of the edges emanating from any other node, the probability in question is:

$$\prod_{i=1}^{n} \Pr(\text{for all } j \text{ such that } v_i \in A_j, \text{ there is no directed edge from } v_i \text{ to } w_j). \qquad (1.1)$$

Let a_i denote the number of sets A_j that contain v_i. The probability that there is no edge from a particular v_i to the w_j's for which $v_i \in A_j$ is

$$\frac{\binom{n-m-1}{a_i}}{\binom{n-1}{a_i}} = \frac{(n-m-1)(n-m-2)\cdots(n-m-a_i)}{(n-1)(n-2)\cdots(n-a_i)} \leq \left(\frac{n-m-1}{n-1}\right)^{a_i}.$$

Thus, the probability in (1.1) is at most

$$\prod_{i=1}^{n} \left(\frac{n-m-1}{n-1}\right)^{a_i} = \left(\frac{n-m-1}{n-1}\right)^{(a_1 + \cdots + a_n)} = \left(\frac{n-m-1}{n-1}\right)^{(|A_1| + \cdots + |A_k|)}.$$

<u>Theorem 1</u> *Fix $\varepsilon, \delta > 0$. For n sufficiently large, if $m = \lfloor (1 - \varepsilon)\sqrt{n \log n} \rfloor$, then $P_{m,n} < \delta$.*

Proof. We say that a vertex is isolated if it has no incident edge in $U_{m,n}$. Obviously, $U_{m,n}$ cannot be connected if it has any isolated vertex. We will show that, as n gets large, if m has the specified value, then $U_{m,n}$ is likely to have isolated vertices. To do so, we compute the expected number of isolated vertices and the standard deviation; we then apply Chebyshev's Inequality.

Let p denote the probability that a particular vertex (say v_1) is isolated in $U_{m,n}$, and let $X_{m,n}$ denote the number of isolated vertices in $U_{m,n}$. The vertex v_1 is isolated if and only if, for each $v_j \in S_1$, the vertex v_1 is not in S_j. These events are independent and all have probability $1 - \frac{m}{n-1}$, so

$$p = \left(1 - \frac{m}{n-1}\right)^{m}.$$

We deduce that

$$EX_{m,n} = np = n \left(1 - \frac{m}{n-1}\right)^m.$$

Next we consider the variance of $X_{m,n}$. Define q to be the probability that a particular pair of vertices (say v_1 and v_2) are both isolated. In order to compute q, we consider cases depending on whether or not $v_2 \in S_1$ and $v_1 \in S_2$.

Case (1): $v_1 \notin S_2$ and $v_2 \notin S_1$. This occurs with probability $\left(1 - \frac{m}{n-1}\right)^2$. Next observe that v_1 and v_2 are both isolated if and only if the following two conditions hold: for every $v_j \in S_1$, we have $v_1 \notin S_j$, and for every $v_j \in S_2$, we have $v_2 \notin S_j$. By Lemma 1, this occurs with probability at most

$$\left(1 - \frac{m}{n-1}\right)^{|S_1|+|S_2|} = \left(1 - \frac{m}{n-1}\right)^{2m}.$$

Thus case (1) contributes at most

$$\left(1 - \frac{m}{n-1}\right)^{2m+2}$$

to q.

Case (2): $v_1 \notin S_2$ but $v_2 \in S_1$. This occurs with probability

$$\left(1 - \frac{m}{n-1}\right)\left(\frac{m}{n-1}\right).$$

Now v_1 and v_2 are both isolated if and only if the following two conditions hold: for every $v_j \in S_1 - \{v_2\}$, we have $v_1 \notin S_j$, and for every $v_j \in S_2$, we have $v_2 \notin S_j$. By Lemma 1, this occurs with probability at most

$$\left(1 - \frac{m}{n-1}\right)^{|S_1|-1+|S_2|} = \left(1 - \frac{m}{n-1}\right)^{2m-1}.$$

Thus case (2) contributes at most

$$\left(1 - \frac{m}{n-1}\right)^{2m}\left(\frac{m}{n-1}\right)$$

to q.

Case (3): $v_1 \in S_2$ but $v_2 \notin S_1$. By symmetry, case (3) makes the same contribution to q as case (2).

We need not worry about the case $v_1 \in S_2$ and $v_2 \in S_1$, because then v_1 and v_2 could not be isolated. Summing cases (1), (2), and (3),

$$q \le \left(1 - \frac{m}{n-1}\right)^{2m} \left(1 + \left(\frac{m}{n-1}\right)^2\right).$$

We next deduce the variance of $X_{m,n}$.

$$EX_{m,n}^2 = np + n(n-1)q \le n\left(1 - \frac{m}{n-1}\right)^m + n^2\left(1 - \frac{m}{n-1}\right)^{2m}\left(1 + \left(\frac{m}{n-1}\right)^2\right).$$

Thus

$$EX_{m,n}^2 - (EX_{m,n})^2 \le n\left(1 - \frac{m}{n-1}\right)^m + n^2\left(1 - \frac{m}{n-1}\right)^{2m}\left(\frac{m}{n-1}\right)^2.$$

Let σ denote the standard deviation of $X_{m,n}$. The value 0 is $EX_{m,n}/\sigma$ standard deviations away from the mean, so by Chebyshev's inequality, the probability that $X_{m,n} = 0$ is at most $\sigma^2/(EX_{m,n})^2 = (EX_{m,n}^2 - (EX_{m,n})^2)/(EX_{m,n})^2$. Thus the probability that there are no isolated vertices is at most

$$\frac{EX_{m,n}^2 - (EX_{m,n})^2}{(EX_{m,n})^2} \le \frac{n\left(1 - \frac{m}{n-1}\right)^m + n^2\left(1 - \frac{m}{n-1}\right)^{2m}\left(\frac{m}{n-1}\right)^2}{n^2\left(1 - \frac{m}{n-1}\right)^{2m}}$$

$$\le \frac{1}{n}\left(1 - \frac{m}{n-1}\right)^{-m} + \left(\frac{m}{n-1}\right)^2. \tag{1.2}$$

Recall that $m = \lfloor (1-\varepsilon)\sqrt{n\log n}\rfloor$. Thus for large n, the first term in (1.2) is approximately $e^{m^2/(n-1)}/n \approx e^{(1-\varepsilon)^2 \log n}/n \approx n^{(1-\varepsilon)^2 - 1}$. Thus for sufficiently large n, both terms in (1.2) are less than $\delta/2$, and so the probability that there are no isolated vertices (and hence $P_{m,n}$) is less than δ.

<u>Lemma 2</u> *For integers* $m > i \ge 0$, $a > j \ge 0$, *and* $b > k \ge 0$, *suppose* $m = a + b$ *and* $i = j + k$. *Then*

$$\frac{(a-1)(a-2)\cdots(a-j)\,(b-1)(b-2)\cdots(b-k)}{(m-1)(m-2)\cdots(m-i)} \le \frac{a^j b^k}{m^i}. \tag{1.3}$$

Proof. We use induction on the pair j, k. For the base cases, we note that since $m > a, b$, we have

$$\frac{(a-1)(a-2)\cdots(a-j)}{(m-1)(m-2)\cdots(m-j)} \le \frac{a^j}{m^j}$$

and

$$\frac{(b-1)(b-2)\cdots(b-k)}{(m-1)(m-2)\cdots(m-k)} \le \frac{b^k}{m^k}.$$

Now suppose that j and k are both nonzero. Suppose $\frac{j}{a}$ and $\frac{k}{b}$ are both less than $\frac{i}{m}$. Then

$$i = j+k = \left(\frac{j}{a}\right)a + \left(\frac{k}{b}\right)b < \left(\frac{i}{m}\right)a + \left(\frac{i}{m}\right)b = \frac{im}{m} = i,$$

which is a contradiction. Thus either $\frac{j}{a}$ or $\frac{k}{b}$ must be at least as big as $\frac{i}{m}$. We can assume without loss of generality that $\frac{j}{a} \ge \frac{i}{m}$. Then $\frac{a-j}{a} \le \frac{m-i}{m}$. Therefore, if we replace $(a-j)$ and $(m-i)$ by a and m, respectively, in the left-hand side of (1.3), then the value does not decrease. The result now follows by induction.

___Theorem 2___ *Fix* $1/4 > \varepsilon > 0$ *and* $\delta > 0$. *For* n *sufficiently large, if* $m = \lceil (1+\varepsilon)\sqrt{n\log n}\,\rceil$, *then* $P_{m,n} > 1-\delta$.

Proof. Fix a set S of i vertices, where $i \le n/2$, and let S^c denote the complement of S. We will say that S is "isolated" if there are no edges in $U_{m,n}$ from S to S^c. We consider the probability that S is isolated. Let $w_1,\ldots,w_i$ denote the vertices of S, and for $0 \le k \le i$, let A_k be the set of vertices $x \in S^c$ such that there is an edge $\overrightarrow{w_k x}$. For some k, if exactly j of the m out-edges from w_k end in S, then A_k has size $m-j$. This occurs with probability

$$\frac{\binom{i-1}{j}\binom{n-i}{m-j}}{\binom{n-1}{m}}$$

$$= \frac{m!}{j!(m-j)!}\frac{(i-1)(i-2)\cdots(i-j)(n-i)(n-i-1)\cdots(n-i-m+j+1)}{(n-1)(n-2)\cdots(n-m)}$$

$$= \binom{m}{j}\frac{(i-1)(i-2)\cdots(i-j)(n-i)(n-i-1)\cdots(n-i-m+j+1)}{(n-1)(n-2)\cdots(n-m)}.$$

$$= \left(\frac{n-i}{n-i-m+j}\right)\binom{m}{j}\frac{(i-1)(i-2)\cdots(i-j)(n-i)(n-i-2)\cdots(n-i-m+j)}{(n-1)(n-2)\cdots(n-m)},$$

which, for large n, is less than

$$2\binom{m}{j}\frac{(i-1)(i-2)\cdots(i-j)(n-i-1)(n-i-2)\cdots(n-i-m+j)}{(n-1)(n-2)\cdots(n-m)}.$$

By Lemma 2, this is at most

$$2\binom{m}{j}\frac{i^j(n-i)^{m-j}}{n^m}.$$

S is isolated if and only if there is no directed edge from A_k to w_k for any k. By Lemma 1, the probability that this occurs is at most

$$\left(\sum_{j=0}^{m} 2 \binom{m}{j} \frac{i^j (n-i)^{m-j}}{(n)^m} \left(1 - \frac{m}{n-1}\right)^{m-j} \right)^i$$

$$= 2^i \left(\frac{(n-i)(n-m-1)}{n(n-1)} \right)^{mi} \left(\sum_{j=0}^{m} \binom{m}{j} \left(\frac{i(n-1)}{(n-i)(n-m-1)} \right)^j \right)^i$$

$$= 2^i \left(\frac{(n-i)(n-m-1)}{n(n-1)} \right)^{mi} \left(1 + \frac{i(n-1)}{(n-i)(n-m-1)} \right)^{mi}$$

$$= 2^i \left(\frac{(n-i)(n-m-1) + i(n-1)}{n(n-1)} \right)^{mi}$$

$$= 2^i \left(\frac{(n-i)(n-1) + i(n-1) - (n-i)m}{n(n-1)} \right)^{mi}$$

$$= 2^i \left(1 - \frac{(n-i)m}{n(n-1)} \right)^{mi} < \left(2e^{-(n-i)m^2/n^2} \right)^i. \tag{1.4}$$

Now using the fact that $m = \lceil (1+\varepsilon)\sqrt{n \log n} \rceil$, the right-hand side of (1.4) is less than

$$\left(2e^{-(1+\varepsilon)^2 (\log n)(n-i)/n} \right)^i = \left(2n^{-(1+\varepsilon)^2 (n-i)/n} \right)^i.$$

We now multiply by $\binom{n}{i}$ to bound the expected number of isolated sets of size i:

$$\binom{n}{i} \left(2n^{-(1+\varepsilon)^2 (n-i)/n} \right)^i$$

$$\leq \left(\frac{ne}{i} \right)^i \left(2n^{-(1+\varepsilon)^2 (n-i)/n} \right)^i \leq \left(\left(\frac{2e}{i} \right) n^{1-(1+2\varepsilon)(n-i)/n} \right)^i = \left(\left(\frac{2e}{i} \right) n^{-2\varepsilon + (1+2\varepsilon)i/n} \right)^i. \tag{1.5}$$

We now consider two cases. First suppose $i \leq \varepsilon n$. Then the quantity in the right-hand side of (1.5) is less than

$$\left(\left(\frac{2e}{i} \right) n^{-2\varepsilon + (1+2\varepsilon)\varepsilon} \right)^i < \left(2en^{-\varepsilon/2} \right)^i$$

(since $\varepsilon < 1/4$). For any $\varepsilon > 0$, we can choose n sufficiently large that $2en^{-\varepsilon/2}$ is arbitrarily small, and so we can make

$$\sum_{i=1}^{\varepsilon n} \left(2en^{-\varepsilon/2}\right)^i$$

arbitrarily small.

Finally consider the case $\varepsilon n < i \le n/2$. Let $x = i/n$. Then the quantity in the right-hand side of (1.5) is less than

$$\left(\left(\frac{2e}{xn}\right)n^x\right)^i \doteq \left(\left(\frac{2e}{\varepsilon}\right)n^{x-1}\right)^i \le \left(\frac{2e}{\varepsilon\sqrt{n}}\right)^i.$$

Again, for any $\varepsilon > 0$, we can choose n sufficiently large that $\frac{2e}{\varepsilon\sqrt{n}}$ is arbitrarily small, and so we can make

$$\sum_{i=\varepsilon n}^{n/2} \left(\frac{2e}{\varepsilon\sqrt{n}}\right)^i$$

arbitrarily small.

Thus for $m = \lceil(1+\varepsilon)\sqrt{n\log n}\rceil$, as n goes to infinity, the expected number of isolated sets in $U_{m,n}$ of size $\le n/2$ goes to 0. But if $U_{m,n}$ is disconnected, then there must be an isolated set of size $\le n/2$, so the probability that $U_{m,n}$ is connected goes to 1.

Corollary 1 *Fix $\varepsilon, \delta > 0$. For n sufficiently large, if $m \ge (1+\varepsilon)\sqrt{n\log n}$, then $P_{m,n} > 1 - \delta$, and if $m \le (1-\varepsilon)\sqrt{n\log n}$, then $P_{m,n} < \delta$.*

Proof. A straightforward coupling argument shows that (for fixed n), as m increases, the probability that $U_{m,n}$ is connected cannot decrease. Therefore, this corollary follows from Theorems 1 and 2.

Douglas S. Jungreis

PROBLEM 39: AUERBACH

The absolute value of a number x satisfies the following conditions:

(1) $\phi(x) \ge 0$, $\phi(x) \not\equiv 0, 1$
(2) $\phi(x+y) \le \phi(x) + \phi(y)$
(3) $\phi(xy) = \phi(x)\phi(y)$.

The only continuous functions satisfying these conditions are: $\phi(x) = |x|^\alpha$, where α is constant and $0 < \alpha \le 1$. Do there exist discontinuous functions with the above properties?

Addendum This follows from Lebesgue's theorem [See for example, E. Kamke, Zur Definition der affinen Abbildung, *Jahresb. d.D.M.V.*, 36 (1927): There exists a complex function of a complex variable $w = f(z)$ discontinuous and such that: $f(z_1 + z_2) = f(z_1) + f(z_2), f(z_1 z_2) = f(z_1)f(z_2)$; $\phi(x) = |f(x)|$ satisfies (1), (2), (3), and is discontinuous.]

S. Mazur

April 10, 1937

Commentary

That the only continuous functions satisfying the conditions are: $\phi(x) = |x|^\alpha$, $0 < \alpha \leq 1$, follows of course from considering the functional equation $\phi(xy) = \phi(x)\phi(y)$ and Cauchy's equation into which it may be transformed: $f(z_1 + z_2) = f(z_1) + f(z_2)$. For a discussion of Cauchy's equation, see [1].

The page references to the Kamke paper is pp. 145–156. Slightly more precisely than is stated in the addendum, Kamke considered the following problem. Suppose $f(z)$ has the following properties:

(1) $f(z)$ is defined for each z.
(2) $f(z)$ takes on each value exactly once.
(3) $f(z_1 + z_2) = f(z_1) + f(z_2)$.
(4) $f(z_1 z_2) = f(z_1)f(z_2)$.

In the real case, Darboux showed that $f(z) = z$. In the complex case there are also solutions $f(z) = z$ and $f(z) = \bar{z}$. There are discontinuous solutions in the complex case, as was proved by Steinitz (1910), Ostrowski (1913) and Noether (1916). Kamke gives a construction for the discontinuous solutions using the well-ordering principle.

It is now well known that there are 2^c ring automorphisms of the complex numbers [2, p. 157]. Since each such automorphism f must be the identity on the positive rational numbers, it follows that if f is continuous, then f must be the identity on the real numbers.

Thus, if $|f|$ is continuous, then f must be either the identity or the conjugation map. So, if f is any of the 2^c discontinuous ring automorphisms of the complex numbers, then $\phi(x) = |f(x)|$ is discontinuous and satisfies (1), (2), and (3).

1. J. Aczel, *Lectures on functional equations and their applications,* Academic Press, New York-London, 1966.
2. N. Jacobson, *Lectures in Abstract Algebra, III,* Van Nostrand, New York, 1964.

W.A. Beyer and R. Daniel Mauldin

PROBLEM 40: BANACH, ULAM

Can one define a completely additive measure function for all the projective sets on the interval $(0,1)$ which, for Borel sets, coincides with Lebesgue measure? In particular, can one define such a measure on the ring of sets of the sets $P(A)$ (projective)? All this with the additional requirement that congruent sets should have equal measure.

Commentary

It is now known that this problem is connected with the axioms of set theory. For example, if $ZF+$ "there is an inaccessible cardinal" is consistent, then it is consistent that all sets are Lebesgue measurable [5]. If the axiom of projective determinacy is consistent, then all projective sets are Lebesgue measurable [3]. If there is a projective well-ordering of the real numbers into type ω_1 (which is the case under Gödel's axiom of constructibility), then there is no countably additive measure defined on all projective sets which coincides with Lebesgue measure [4].

Kakutani and Oxtoby (2) and Hulanicki [1] obtained some absolute results concerning extensions of Lebesgue measure.

1. A. Hulanicki, Compact abelian groups and extensions of Haar measures, *Rozprawy Mat.* 38 (1964), 58 pp. (MR31 #270.)
2. S. Kakutani and J.C. Oxtoby, Construction of a non-separable invariant extension of the Lebesgue measure space, *Ann. of Math.*, (2) 52 (1950), 580–590. (MR12-246.)
3. D.A. Martin, Descriptive Set Theory: Projective Sets, in *Handbook of Mathematical Logic*, Edited by John Barwise, Studies in Logic, volume 90, North-Holland Publ. Co., New York, 1977.
4. R.D. Mauldin, Projective well orderings and extensions of Lebesgue measure, *Coll. Math.*, 46 (1982), 185-188.
5. R.M. Solovay, A model of set theory in which every set of reals is Lebesgue measurable, *Ann. of Math.* 92 (1970), 1–56.
6. S.M. Ulam, *Problems in Modern Mathematics,* John Wiley, New York, 1960.

R. Daniel Mauldin

PROBLEM 41: MAZUR

Does there exist a 3-dimensional space of type (B) with the property that every 2-dimensional space of type (B) is isometric to a subspace of it? This is equivalent to the question: Does there exist in the 3-dimensional Euclidean space a convex surface W which has a center 0 with the property that every convex curve with a center is affine to a plane section of W through 0? More generally, given an integer $k \geq 2$, does there exist an integer i and an i-dimensional space of type (B) such that every k-dimensional space of type (B) is isometric to a subspace of it; given k, determine the smallest i.

Commentary

By very simple reasoning, Grunbaum [2] showed that no 3-dimensional Banach space is isometrically universal for all 2-dimensional Banach spaces. Bessaga [1], with more complicated reasoning, obtained the result with "3" replaced by "finite." Further refinements were contributed by Melzak [5], Klee [3], and Lindenstrauss [4].

1. C. Bessaga, A note on universal Banach spaces of finite dimension, *Bull. Acad. Polon. Sci. Ser. Sci. Math. Astr. Phys.* 6 (1958), 249–250.
2. B. Grünbaum, On a problem of S. Mazur, *Bull. Res. Council Israel* (Sect F), 7 (1958), 133–135.
3. V. Klee, Polyhedral sections of convex bodies, *Acta Math.* 103 (1960), 243–267.
4. J. Lindenstrauss, Notes on Klee's paper: "Polyhedral sections of convex bodies," *Israel J. Math.* 4 (1966), 235–242.
5. Z. Melzak, Limit sections and universal points of convex surfaces, *Proc. Amer. Math. Soc.,* 9 (1958), 729–734.

V. Klee

PROBLEM 42: ULAM

To every closed, convex set X, contained in a sphere K in Euclidean space, there is assigned another convex, closed set $f(X)$, contained in K, in a continuous manner (in the sense of the Hausdorff metric); does there exist a fixed point, that is to say, a closed convex X_0 such that $f(X_0) = X_0$?

Theorem (Mazur).   Let E be a class of convex closed sets contained in a sphere K with the properties:

(1) If $A \in E$, $B \in E$, then also $\lambda A + (1 - \lambda)B \in E$, for $0 \le \lambda \le 1$ [$\lambda A + (1 - \lambda)B$ denotes the set of points $\lambda x + (1 - \lambda)y$ for $x \in A$ and $y \in B$];
(2) If $A_n \in E$ and the sequence $\{A_n\}$ converges to A, then $A \in E$.

Suppose that $f(x)$ is a continuous function in E whose values belong to E; then, there exists a fixed point; that is, an $X_0 \in E$ such that $f(X_0) = X_0$. Examples of such a class E are, for instance, the class of all closed, convex sets contained in K with diameter not greater than a given number $\phi > 0$.

Second Edition Commentary

There is a canonical embedding of the set of all nonempty convex compact subsets of a linear space into another linear space. The latter is normed if so is the former. This embedding preserves the convex combinations. In problem 42, the image of this embedding would be a compact convex subset in a normed space and the result (existence of a fixed point) follows from Schauder's theorem.

Michael Zarichnyi

PROBLEM 43: MAZUR

Prize: One bottle of wine, S. Mazur

Definition of a certain game. Given is a set E of real numbers. A game between two players A and B is defined as follows: A selects an arbitrary interval d_1; B then selects an arbitrary segment (interval) d_2 contained in d_1; then A in his turn selects an arbitrary segment d_3 contained in d_2 and so on. A wins if the intersection $d_1, d_2, \ldots, d_n, \ldots$ contains a point of the set E; otherwise, he loses. If E is a complement of a set of first category, there exists a method through which A can win; if E is a set of first category, there exists a method through which B will win. **Problem.** It is true that there exists a method of winning for the player

A only for those sets E whose complement is, in a certain interval, of first category; similarly, does a method of win exist for B if E is a set of first category?

Addendum. Mazur's conjecture is true.

S. Banach

August 4, 1935

Modifications of Mazur's Game

(1) There is given a set of real numbers E. Players A and B give in turn the digits 0 or 1. A wins if the number formed by these digits in a given order (in the binary system) belongs to E. For which E does there exist a method of win for player A (player B)?

Ulam

(2) There is given a set of real numbers E. The two players A and B in turn give real numbers which are positive and such that a player always gives a number smaller than the last one given. Player A wins if the sum of the given series of numbers is an element of the set E. The same question as for (1). [Ed. See Problem 67 for another modification.]

Banach

Commentary

The first published paper on general finite games with perfect information is Zermelo's [14]. Here, in Problem 43, we have the first interesting definition of an infinite one. A proof of the solution of Banach which is announced here was published by Oxtoby [5]. This theorem constitutes a very useful characterization of meager or comeager sets. E.g., this characterization immediately yields that the set of real numbers, such that in their binary representation 1 has a frequency, is meager (yet, by the strong law of large numbers, the set of those numbers in which 1 has frequency 1/2 has a complement of measure zero). For other applications of this criterion, see [4].

The theorem of Banach also has the following corollary (to be compared with the theorems which follow): If C is a class of subsets of the real line $\mathbb{R}$ which is closed under preimages by continuous functions $f : \mathbb{R} \to \mathbb{R}$, then every set $E \in C$ has the property of Baire iff for every set $E \in C$ the game of Mazur is determined (i.e., one of the players has a winning strategy).

A similar game was introduced by Morton Davis [1]. Here a set $E \subseteq \{0,1\}^\omega$ is given and Player I chooses a finite sequence of 0's and 1's, Player II chooses 0 or 1 and again I chooses a finite sequence of 0's and 1's and II chooses 0 or 1, etc., ω times. If the juxtaposition of the consecutive choices belongs to E then I wins, otherwise II wins. Davis has proven that I has a winning strategy iff E has a perfect subset, and II has a winning strategy iff E is countable or finite.

Ulam's modification (1) of Mazur's game is particularly important. Let C be a class of subsets of the unit interval $[0,1]$ which is closed under some natural operations, e.g., under finite unions and preimages by Borel measurable functions. Then, if for every set $E \in C$ the Ulam game is determined, every set $E \in C$ has the property of Baire, is Lebesgue measurable, and is either countable or has a perfect subset and the same is true for the complement of E (see [1, 11, 12, 13]).

The conjecture that all projective sets are determined is called the axiom of projective determinacy and is often used in the theory of projective sets since it has very natural or fitting consequences which cannot be proved otherwise (and are inconsistent with $V = L$), see [10].

Ulam's game is more general than Mazur's or Davis' in the sense that for every game of the latter kinds an equivalent game of the first kind can be defined (see [11]). In spite of this generality all known proofs of the existence of nondetermined games of Ulam require the axiom of choice for uncountable families of sets of reals. Therefore, and because of its excellent consequences, it is a plausible conjecture that if we remove from set theory the full axiom of choice and put in the axiom of determinacy (which tells that for *every* $E \subseteq [0,1]$ the Ulam game is determined) the resulting theory is consistent. But this conjecture is beyond the reach of present methods because the axiom of determinacy yields the consistency of set theory with very strong axioms of infinity [Solovay, Martin, Harrington], whence (by Gödel) one cannot prove the opposite.

The best one can do, therefore, is to prove for larger and larger classes C that for all $E \in C$ the Ulam game is determined. The strongest results in this direction were obtained by D.A. Martin. In [7] he proves this for the class of all Borel sets. This theorem is outstanding because, although it pertains to sets of low set-theoretic rank, it still uses the full power of Zermelo-Skolem set theory (often incorrectly called Zermelo-Fraenkel set theory). Friedman had proven earlier [3] that this could not be demonstrated in Zermelo's set theory. In [6] Martin proves that if there exists a cardinal κ having the Erdős-Rado partition property $\kappa \to (\omega_1)_2^{<\omega}$ then for all analytic sets E the Ulam game is determined. Finally, in [8] he proves, under the assumption of the existence of some extremely large cardinals (called iterable), that for all $E \in \Sigma_1^2$ (continuous images of complements of analytic sets) the Ulam game is determined. Those results of Martin constitute brilliant examples of the impact of strong axioms of infinity upon the theory of projective sets. The problem of proving the full axiom of projective determinacy from some strong axiom of infinity is still open, and is perhaps the most outstanding problem of set theory.

Why don't we abandon the axiom of choice and accept in its place the axiom of determinacy, in spite of the fact that this removes such artificial phenomena as paradoxical decompositions of the sphere and yields so many "positive" consequences? The answer is that it seems more natural to restrict those consequences (and the axiom) to a suitable class C as above, e.g., the class of projective sets, without violating our basic intuitions about the class of all sets (of which the axiom of choice is a part) and without wrecking the unicity of our fundamental theory.

Banach's modification (2) is of a more special character than that of Ulam. Some work has been done on such modifications (see [2, 9]). It appears that the specific question of Ulam and Banach (the question of who is the winner) does not lead to nontrivial answers. One may observe that a countable set is avoidable by any of the players and that the existence of a winning strategy means that the corresponding set has a perfect subset of a particular shape. For studies of some other infinite games with perfect information, see [11] and the commentary to Problem 67. Other such games were recently studied by F. Galvin, R. McKenzie, R. Laver, J. Mycielski, K. Prikry, and many others. A paper of F. Galvin, J. Mycielski, R. Solovay and some others is in preparation.

1. Morton Davis, Infinite games of perfect information, *Advances in game theory,* Ann. Math. Studies No. 52, Princeton University Press, Princeton, N.J. 1964, 85–101.
2. A. Ehrenfeucht, G. Moran and S. Shelah, Size direction games over the real line, *Israel J. Math.* 14 (1973), Part I, 163–168, Part II, 418–441, and Part III, 442–449.
3. H. Friedman, Higher set theory and mathematical practice, *Ann. of Math. Logic* 2 (1971), 326–357.
4. J. F. Lynch, Almost sure theories, *Ann. of Math. Logic* 18 (1980), 91–135.
5. J.C. Oxtoby, The Banach-Mazur game and Banach category theorem, *Contributions to the theory of games,* vol. III Ann. of Math Studies no. 39, Princeton 1957, 159–163.
6. D. A. Martin, Measurable cardinals and analytic games, *Fund. Math.* 66 (1970), 287–291.
7. D. A. Martin, Borel determinacy, *Ann. of Math.* 102 (1975), 363–371.
8. D. A. Martin, Infinite games, *Proc. of the Int. Congress of Mathematicians in Helsinki, 1978,* 269–273, Acad. Sci. Fennica, Helsinki, 1980.
9. G. Moran, Existence of nondetermined sets for some two person games over the reals, *Israel J. Math.* 9 (1971), 316–329.
10. Y. Moschovakis, Descriptive Set Theory: A Foundational Approach, North-Holland, Amsterdam, 1980.
11. J. Mycielski, On the axiom of determinateness, Part I, *Fund. Math.* 53 (1964), 205–224, and Part II, *Fund. Math.* 59 (1966), 203–212.
12. J. Mycielski and H. Steinhaus, A mathematical axiom contradicting the axiom of choice, *Bull. Acad. Polon. Sci. Sér. Math., Astronom. Phys.* 10 (1962), 1–3.
13. J. Mycielski and S. Świerczkowski, On the Lebesgue meaurability and the axiom of determinateness, *Fund. Math.* 54 (1964), 67–71.
14. E. Zermelo, Über eine Anwendungen der Mengen lehre auf die Theorie des Schachspiels, *Proc. Fifth Congress of Math.,* Cambridge 1912, vol. II, 501.

Jan Mycielski

Second Edition Commentary

> The investigation of the determinacy of infinite games is the most distinctive and
> intriguing development of modern set theory ....

> Akihiro Kanamori, Introduction to the *Handbook of Set Theory* [7, page 44]

Infinite games of various types – not only the three types of games in Problem
43 of the *Scottish Book* – now appear in many branches of set theory. In some cases,
the game is introduced for the reason indicated in the questions of Mazur, Ulam,
and Banach: the existence of a winning strategy for Player I (or II) is a useful way
of characterizing some interesting property. In other cases, the interest lies in the
determinacy of the game.

This commentary will be confined to the three types of games in Problem 43,
which will be discussed here in reverse order. Since Jan Mycielski's 1981 commentary there have been new results on each of the three. All of the results involve the
determinacy of the game, not the original question.

Before discussing these games, it is necessary to digress briefly for the benefit
of those readers who are unfamiliar with set theory (at the level of Jech [5] or
Kanamori [6]). To such a reader, there is much about this commentary which might
seem strange, for three reasons, which in increasing order of strangeness are as
follows. First, we work in ZF (not ZFC), and consider a number of propositions
which contradict the axiom of choice. Second, we will be concerned solely with
various propositions which are not decidable in ZF (and in some cases not in ZFC),
and the relationship between two such propositions. One type of relationship is
implication. The other type involves consistency. It may be that (ZF + P) and (ZF +
Q) are equiconsistent. And it may be that (ZF + P) implies the consistency of (ZF +
Q); if so, then the consistency of (ZF + P) implies the consistency of (ZF + Q) but,
by Gödel's Theorem, the consistency of (ZF + Q) does not imply the consistency of
(ZF + P). In this case, we say that P has greater *consistency strength* than Q. And
third, large cardinal axioms are involved.

A large cardinal axiom asserts that there exists a cardinal number satisfying a
particular property. That property is often technical and its motivation will not be
evident to the non-specialist. It will also not be evident that cardinals satisfying
that property are large. Large cardinal axioms are linearly ordered by consistency
strength, and it is standard practice in set theory to gauge the consistency strength of
a proposition by comparison to large cardinal axioms. Some – though certainly not
all – set theorists hold the following three beliefs: first, that it is meaningful to say
that a mathematical statement is true, even when there is no possibility of proving
it; second, large cardinal axioms are probably true; third, the collection of axioms
that constitute "the usual axioms of set theory" should be expanded to include large
cardinal axioms. For a more serious discussion of these matters, see Kanamori [6]
and Maddy [11].

For any set X with at least two members and any set $E \subset X^\omega$, the game $G_X(E)$ is the infinite game in which Players I and II alternately play elements $x_0, x_1, x_2, \ldots$ of X, and Player I wins iff the sequence $(x_0, x_1, x_2, \ldots)$ is in E. When $X = 2 \ (= \{0, 1\})$, this is the game of Ulam's question. For any such X, let AD_X be the proposition that for every $E \subset X^\omega$, $G_X(E)$ is determined. AD_2 and AD_ω are equivalent; this is the *axiom of determinacy*, denoted simply as AD. AD and $AD_\mathbb{R}$ are the only axioms of this form consistent with ZF. For a history of determinacy axioms, see Larson [10].

Clearly $AD_\mathbb{R}$ implies AD. It is also clear, since a strategy for a game on ω can be identified with a real number, that AD implies that AD holds in $L(\mathbb{R})$, the smallest model of ZF containing all reals and all ordinals. But $AD_\mathbb{R}$ does not hold in $L(\mathbb{R})$, so AD does not imply $AD_\mathbb{R}$. In fact, $AD_\mathbb{R}$ has greater consistency strength than AD. This follows from the fact that AD holds in $L(\mathbb{R})$, together with the theorem of Solovay [17] that $AD_\mathbb{R}$ implies that "the sharp" of $L(\mathbb{R})$ exists.

For any $E \subset \mathbb{R}$, the *Banach game*, $B(E)$, is the game of Banach's question. The *axiom of determinacy for Banach games*, ADB, is the proposition that every Banach game is determined.

One of the consequences of the 1981 publication of the *Scottish Book* is that Tony Martin learned about Banach games. He showed that one could simulate a game on 2 by a Banach game [4, 1.2]. That is, he defined a function $E \mapsto E^*$ from 2^ω into $\mathbb{R}$ and then proved that if Player I has a winning strategy for the game $B(E^*)$ then he also has a winning strategy for $G_2(E)$, and similarly for Player II. This clearly shows that ADB implies AD, and it also provides evidence that Banach's original question is unanswerable.

Since Banach games are games on reals, trivially $AD_\mathbb{R}$ implies ADB. Thus $AD_\mathbb{R} \Rightarrow ADB \Rightarrow AD$, and since $AD_\mathbb{R}$ is stronger than AD it is not possible that both arrows can be reversed. This leads Martin to ask how strong ADB is – as strong as AD, as strong as $AD_\mathbb{R}$, or somewhere in between? There are two questions here, one of implication, the other of consistency strength.

Note that a strategy for a game on reals is not a real, but rather is – modulo some coding – a set of reals. Freiling [4, 1.3 and 1.4] provided a partial answer to Martin's question by proving that if there is any winning strategy for the game $B(E)$ then there is a winning strategy (set of reals) which is fairly simple to define – so simple that it is in the model $L(\mathbb{R})$. Therefore, the relationship between $AD_\mathbb{R}$ and ADB is the same as the relationship between $AD_\mathbb{R}$ and AD: $AD_\mathbb{R}$ implies that ADB holds in $L(\mathbb{R})$ and the sharp of $L(\mathbb{R})$ exists, hence ZF + ADB is consistent. This still left open the question of how ADB is related to AD.

Using the work of Freiling and some unpublished work of Martin, Becker [2] then proved that AD implies ADB. Hence AD and ADB are equivalent, relative to ZF.

This equivalence has local versions, which we state using the notation of descriptive set theory (see [8] or [13]). Let Γ be a collection of pointsets in $\mathbb{R}$ and in 2^ω. Call Γ *nice* if every Δ_3^1 set is in Γ, Γ is closed under finite unions and intersections and Γ is closed under preimages by Δ_2^1 functions. For example, the projective pointclasses $\boldsymbol{\Pi}_n^1$, $\boldsymbol{\Sigma}_n^1$ and $\boldsymbol{\Delta}_n^1$ for $n \geq 3$ are nice, as is the class of projective sets and the class of all sets. For any nice $\boldsymbol{\Gamma}$ the following are equivalent: for any $E \subset \mathbb{R}$, if $E \in \boldsymbol{\Gamma}$ then $B(E)$ is determined; for any $E \subset 2^\omega$, if $E \in \boldsymbol{\Gamma}$ then $G_2(E)$ is determined. The latter fact is called $\boldsymbol{\Gamma}$-*determinacy*. The equivalence of AD and ADB is the special case in which $\boldsymbol{\Gamma}$ is all sets. For the other $\boldsymbol{\Gamma}$'s mentioned above, the equivalent propositions are compatible with the axiom of choice, and many set theorists would say that they are (probably) true.

Why should we believe that for projective $E \subset 2^\omega$, the determinacy of Ulam's game $G_2(E)$ is true (or even consistent)? One of the reasons for this belief is that large cardinal axioms imply that $G_2(E)$ is determined. Proving the determinacy of definable games from large cardinal axioms is arguably the most important development in set theory since Cohen's invention of forcing in 1963. In 1981, when Mycielski's commentary was published, this subject was still in its infancy. The strongest result mentioned there has now been improved in two ways: the large cardinal axiom has been weakened and the collection of definable games has been enlarged beyond $\boldsymbol{\Sigma}_2^1$. (There is a typo in that commentary: $\boldsymbol{\Sigma}_2^2$ should be $\boldsymbol{\Sigma}_2^1$.)

The relevant large cardinal axiom was discovered by Hugh Woodin and now bears his name. A cardinal κ is a *Woodin cardinal* if for each function $f : \kappa \to \kappa$ there exists an elementary embedding $j : V \to M$ with critical point $\lambda < \kappa$ closed under f such that $V_{(j(f))(\lambda)} \subset M$. Woodin was led to formulate this definition by previous work of Foreman, Magidor, Shelah and himself, work which was not directly related to proving determinacy.

Shortly after Woodin formulated the above definition, Martin-Steel [12] proved that if there exist n Woodin cardinals with a measurable cardinal above them then $\boldsymbol{\Pi}_{n+1}^1$-determinacy is true. Later Woodin proved that the existence of ω Woodin cardinals with a measurable above them implies that for every E in $L(\mathbb{R})$, $G_2(E)$ is determined [14, 8.24]; hence $L(\mathbb{R})$ is a model of AD. And he proved from a stronger large cardinal axiom that there is a model of $AD_{\mathbb{R}}$ [18, 8.3].

All proofs of definable determinacy from large cardinal axioms have the following form. The game $G_2(E)$ is simulated by a game of the form $G_\kappa(E^*)$, where κ is a large cardinal and E^* is a closed subset of κ^ω. Closed games are determined. What makes the proof work is that E^* is a very special type of closed set, namely the set of branches of a *homogeneous tree*. For information on homogeneous trees, see [6] and [10].

There is a known upper bound on the amount of definable determinacy that can be proved from large cardinal axioms. Work of Abraham, Shelah, and Woodin (see [1]) shows that one can always produce a forcing extension of the universe in which there is a nondetermined $\boldsymbol{\Delta}_1^2$ game. From this it follows that no large cardinal axiom can imply $\boldsymbol{\Delta}_1^2$-determinacy.

It is not possible for determinacy axioms, or any other statement about small sets, to imply the actual existence of large cardinals, since one can pass to the initial segment of the universe below the least inaccessible cardinal and thereby get a model with no large cardinals but no change in the small sets. Consistency, as opposed to implication, is another matter. Woodin proved that the theories (ZFC + there exist infinitely many Woodin cardinals) and (ZF + AD) are equiconsistent. (A proof of the forward direction is given in [18, 6.1]; a proof of the reverse direction is given in [9, 6.2].) He also proved some (rather technical) level-by-level consistency results; but there are still some open problems in this area.

Finally, we consider Mazur's game. Proving that it is *true* (as opposed to merely consistent) that for all projective $E \subset \mathbb{R}$, Mazur's game is determined, that is, every projective set has the property of Baire, requires axioms just as strong as those needed to prove that for every projective $E \subset 2^\omega$, $G_2(E)$ is determined. But in contrast to the games of Ulam and Banach, the determinacy of Mazur's game has very low consistency strength. Solovay [16] proved that assuming the consistency of an inaccessible cardinal – which is the weakest large cardinal axiom – the theory (ZF + every set has the property of Baire) is consistent. At the time Mycielski's commentary was written it was open whether the above theory was equiconsistent with ZFC. Shelah [15] later proved that it is.

1. U. Abraham and S. Shelah, Coding with ladders a well ordering of the reals, *Journal of Symbolic Logic* **67** (2002), 579–597.
2. H. Becker, Determinacy of Banach games, *Journal of Symbolic Logic* **50** (1985), 110–122.
3. M. Foreman and A. Kanamori (Eds.), *Handbook of Set Theory*, Springer, 2010.
4. C. Freiling, Banach games, *Journal of Symbolic Logic* **49** (1984), 343–375.
5. T. Jech, *Set Theory*, third edition, Springer, 2003.
6. A. Kanamori, *The Higher Infinite*, second edition, Springer, 2003.
7. A. Kanamori, Introduction, in [3], 1–92.
8. A. S. Kechris, *Classical Descriptive Set Theory*, Springer, 1995.
9. P. Koellner and W. H. Woodin, Large cardinals from determinacy, in [3], 1951–2119.
10. P. B. Larson, A brief history of determinacy, in: A. S. Kechris, B. Löwe and J. R. Steel (Eds.), *Large Cardinals, Determinacy and Other Topics: the Cabal Seminar*, volume IV, to appear.
11. P. Maddy, Believing the axioms, I and II, *Journal of Symbolic Logic* **53** (1988), 481-511 and 736–764.
12. D. A. Martin and J. R. Steel, A proof of projective determinacy, *Journal of the American Mathematical Society* **2** (1989), 71–125.
13. Y. N. Moschovakis, *Descriptive Set Theory*, second edition, American Mathematical Society, 2009.
14. I. Neeman, Determinacy in L($\mathbb{R}$), in [3], 1877–1950.
15. S. Shelah, Can you take Solovay's inaccessible away?, *Israel Journal of Mathematics* **48** (1984), 1–47.
16. R. M. Solovay, A model of set theory in which every set of reals is Lebesgue measurable, *Annals of Mathematics* **92** (1970), 1–56.
17. R. M. Solovay, The independence of DC from AD, in: A. S. Kechris and Y. N. Moschovakis (Eds.), *Cabal Seminar 76-77*, Springer, 1978.
18. J. R. Steel, The derived model theorem, http://www.math.berkeley.edu/$\thicksim$steel/papers/dm.ps

Howard Becker

From l to r: Stan Ulam, John Oxtoby, and Tony Martin

PROBLEM 44: H. STEINHAUS

A continuous function $z = f(x, y)$ represents a surface such that through every point of it there exist two straight lines contained completely in the surface. Prove that the surface is then a hyperbolic paraboloid. Do the same without assuming continuity of f.

Addendum. This problem was solved affirmatively by Banach—also without assuming continuity. The proof is based upon the remark: Any two straight lines on this surface either intersect or else their projection on the plane xy are parallel.

July 30, 1935

PROBLEM 45: BANACH

Let G be a metric group which is complete and non-Abelian; $U_1(x), U_2(x), \ldots,$ $U_n(x)$ multiplicative operations defined in G and with values belonging to G. Prove that if the operation $U(x) = U_1(x)U_2(x) \cdots U_n(x)$ is of a Baire class, then it must be continuous. This statement is true for $n = 2$.

Commentary

It is well known that if U is a homomorphism of G into G which has the Baire property, then U is continuous, so the statement is true for $n = 1$. I have been unable to locate a proof for $n = 2$.

R. Daniel Mauldin

PROBLEM 46: BANACH

Is the sphere in a space of type (B) unicoherent? (That is to say, in every decomposition of it into continua A, B, is the intersection AB connected?)

Addendum. An affirmative answer to Prof. Banach's problem follows from the following theorem of Borsuk: In every space which is connected, locally connected, complete and unicoherent, there exists a simple closed curve which is a retract. In general linear spaces, in which the multiplication is continuous, an affirmative answer to Prof. Banach's problem follows from my theorem in *Fund. Math.,* Vol. 26, p. 61.

S. Eilenberg

PROBLEM 47: BANACH

Can every permutation of a matrix $\{A_{ik}\}\, i, k = 1, 2, \ldots, \infty$ be obtained by composing a finite number of permutations in such a way that the rows go over into rows and columns into columns? (*Vide* Problem 20: Ulam)

Commentary

This problem was solved by M. Nosarzewska in 1951—the answer is yes. For further results and references, see E. Grzegorek, On axial maps of direct products II, *Coll. Math.* 34 (1976), 145–164.

J. Mycielski

PROBLEM 48: MAZUR, BANACH

Let E be a set of real numbers which is countable, closed, and bounded. W is the set of all continuous real-valued functions defined on E. Is the space W [if we define the norm of a function $f \in W$ as follows: $\|f\| = \max_{x \in E} f(x)$] isomorphic to the space c of all convergent sequences?

Addendum. The answer is affirmative. (The solution is unpublished.)

Mazur
February 15, 1939

Commentary

A negative response can be gleaned from a beautiful paper of Jozef Schreier (*Studia Math.* 2 (1930), 58–67). Though Schreier carries out his construction in $[0, 1]$ he actually shows that if the ωth derived set of the compact metric space K is nonvoid, then there exists a uniformly bounded sequence (f_n) of continuous real-valued functions on K which is weakly convergent to zero yet admits no subsequence whose arithmetic means are norm convergent to zero. In particular, if one considers the ordinal ray $[0, \omega^\omega + 1)$ in its order topology, the resulting compact countable metric space imbeds homeomorphically in $[0, 1]$ yet (because the ωth derived set of $[0, \omega^\omega + 1)$ is nonvoid) $C([0, \omega^\omega + 1))$ is not isomorphic to c (a Banach space is easily seen to have the delightful property that each weakly null sequence admits a subsequence whose arithmetic means are norm null).

Much more can be said here and perhaps the best way of indicating this was pointed out by C. Bessaga and A. Pełczyński (*Studia Math.* 19 (1960), 53-62), who classified the isomorphic types of spaces $C([0, \alpha + 1))$ for ordinals $\alpha < \omega_1$. Their result: if $\alpha < \beta < \omega_1$, then $C([0, \alpha + 1))$ is isomorphic to $C([0, \beta + 1))$ if and only if $\beta < \alpha^\omega$. It follows that the isomorphic types of spaces W as described in Problem 48 are uncountable in number.

Joseph Diestel
Kent, Ohio

Second Edition Commentary

Schreier showed that inside of $[0, 1]$, you can find a countable set K such that in $C(K)$, there is a sequence which while uniformly bounded and pointwise convergent to 0 has the property that the arithmetic means of any subsequence are at least $1/2$ in the sup norm. In fact, a close analysis of Schreier's construction uncovers the fact that if K is any compact metric space whose ω^{th} derived set is non-void, then there is a sequence in $C(K)$ which is uniformly bounded, converges pointwise to 0 but none of whose subsequences have norm convergent arithmetic means. Since any weakly null sequence in $C(K)$ has a subsequence whose arithmetic means are norm null, the answer to problem 48 is 'no'. A beautiful exposition on extension of Schreier's result can be found in the paper of N. Farnum (Canad. J.Math. 26(1974), 91–97).

Joseph Diestel
Kent, Ohio

PROBLEM 49: MAZUR, BANACH

Does there exist a space E of type (B) with the property (W) which is universal for all spaces of type (B) with the property (W)? One should investigate this question for the following properties (W):

(1) the space is separable and weakly compact (that is, from every bounded sequence one can select a subsequence weakly convergent to an element).
(2) The space contains a base (countable).
(3) The adjoint space is separable.

The space E is universal isometrically (or isomorphically) for spaces of a given class K if every space of this class is isometric (or isomorphic) to a linear subspace of the space E.

Commentary

This problem was solved negatively for property (1) in 1967 by W. Szlenk of Warsaw (*Studia Math.* 30 (1968), 53–61), who showed that if X is any separable Banach space that contains isomorphs of all separable reflexive Banach spaces, then X^* is nonseparable. Szlenk's solution makes heavy use of a mode of derivation the idea for which comes from the work of Z. Zalcwasser (*Studia Math.* 2 (1930), 63–67) and of D.C. Gillespie and W.A. Hurewicz (*Trans. AMS*, 32 (1930), 527–543).

More recently, using other work of W.A. Hurewicz (*Fundamenta Math.* 15 (1930), 4–17), Jean Bourgain (*Proc. Amer. Math. Soc.* 79 (1980), 241–246) proved the theorem that *if a separable Banach space B is universal for all separable reflexive Banach spaces, then B is universal for all separable Banach spaces!* We also refer the reader to Haskell Rosenthal's presentations of the result (entitled "On applications of the boundedness principle to Banach space theory according to J. Bourgain," in *Expose 5, Publ. Math. Univ. Pierre et Marie Curie, 29*, Univ. Paris VI, 1979.

(2) was solved by S. Banach and S. Mazur (cf. Banach's book *Theorie des operations lineaires*), who showed $C[0,1]$ is universal for all separable Banach spaces. $C[0,1]$ has a (in fact, the original) Schauder basis.

The problem was solved negatively for (3) by P. Wojtaszczyk (*Studia Math.* 37 (1970), 197–202). Modifying the method of Szlenk, he showed that if X is a separable Banach space for which every separable reflexive Banach space is isomorphic to a subspace of X, then X is not isomorphic to a dual space. Jean Bourgain noticed (*Bull Soc. Math. Belg. Ser. B* 31 (1979), 87–117) that if X is a Banach space that contains an isomorph of $C(K)$ for every countable compact subset of $[0, 1]$, then X contains an isomorph of every separable Banach space. Since $C(K)^*$ is isomorphic to ℓ_1 whenever K is a compact countable subset of $[0, 1]$, this gives a substantial improvement to Wojtaszczyk's results.

Joseph Diestel
Kent, Ohio

Second Edition Commentary

With regard to (2), A. Pełczyński (*Studia Math.* 32 (1969), 247-268) showed there exists a Banach space **U** with a (Schauder) basis such that every Banach space with such a basis is isomorphic to a complemented subspace of **U**; in the same vein M. I. Kadets (*Studia Math.* 40 (1971), 85-89) established the existence of a separable Banach space **V** with the bounded approximation property such that every separable Banach space with the bounded approximation property is isomorphic to a complemented subspace of **V**. Using the so-called Pełczyński decomposition technique, it can be shown that **U** and **V** are isomorphic. Very recently, calling on categorical idea developed by R. Fraisse (*Publ. Sci. Univ. Alger.* 1 (1954) 35-182), J. Garbulinska-Wegrazyn (*Banach J. Math. Anal.* 8 (2014), 211–220) has shown how to construct an isometric version of the Kadets- Pełczyński space.

Those interested in (3) and the mentioned work of Bourgain would be well advised to read P. Dodos (*J. Functional Anal.* 260 (2011), 1285–1303).

Joseph Diestel
Kent, Ohio

PROBLEM 50: BANACH

Prove that the integral of Denjoy is not [Editor note: The word not was left out in the Los Alamos edition] a Baire functional in the space $\mathscr{S}$ (that is to say, in the space of measurable functions).

Second Edition Commentary

The formulation of this problem is rather vague. However it was speculated that the results in the paper [1] could be construed as a solution to this problem; see, e.g., the review of H. Becker, MR0934228 (89g:03067) of [2] (which contains a summary of the results in [1]).

We present below a plausible precise interpretation of Problem 50 and explain how the results in the above paper provide a positive solution.

Denote by $\mathscr{S}$ the set of (Lebesgue) measurable functions $f : [0,1] \to \mathbb{R}$ (we use here the interval $[0,1]$ but of course everything below works for any interval $[a,b]$). We let for $f, g \in \mathscr{S}, f \sim g \iff f = g, a.e..$ Let $S = \mathscr{S}/\sim$ be the space of measurable functions (modulo equality a.e.). This is a topological vector space with the topology induced by the invariant (under translation), complete, separable metric

$$d([f]_\sim, [g]_\sim) = \int_0^1 \frac{|f(x) - g(x)|}{1 + |f(x) - g(x)|} dx.$$

We refer to Chapter 7 of the book [3], for a detailed introduction to the Denjoy integral. We will only need the following facts.

The Denjoy integral is defined on a subset $\mathscr{DI}$ of $\mathscr{S}$ (the set of *Denjoy integrable functions*) such that $f \in \mathscr{DI}, g \sim f \implies g \in \mathscr{DI}$. Let $DI = \{[f]_\sim : f \in \mathscr{DI}\}$. For $f \in \mathscr{DI}$ the Denjoy integral of f is a continuous function F on $[0,1]$, uniquely determined up to a constant, and we denote by $\mathscr{I}(f) = F(1) - F(0)$ the corresponding definite integral. Moreover $f \sim g \implies \mathscr{I}(f) = \mathscr{I}(g)$, so $\mathscr{I}$ descends to a unique function $I: DI \to \mathbb{R}$, given by $I([f]_\sim) = \mathscr{I}(f)$. The crucial property of the Denjoy integral is now the following: If $F: [0,1] \to \mathbb{R}$ is a differentiable function with $F' = f$, then $f \in \mathscr{DI}$ and $\mathscr{I}(f) = F(1) - F(0)$, i.e, the Denjoy integral recovers the primitive of any derivative.

We can now formulate a precise version of Banach's problem: Prove that the function $I : DI \to \mathbb{R}$ is not in the Baire class of functions (from the separable metrizable space DI into $\mathbb{R}$), i.e., the smallest class of functions containing the continuous functions and closed under limits of pointwise convergent sequences of functions. Equivalently this means that I is not a Borel function (i.e., the preimage of some open set is not Borel in DI).

Under this interpretation, it is a corollary of the results in [1] that this is true. This can be seen as follows.

Let $C = C([0,1])$ be the Banach space of continuous functions on $[0,1]$ and consider the infinite product space $C^{\mathbb{N}}$, a Polish space. Let CN be the subset of $C^{\mathbb{N}}$ consisting of all pointwise convergent sequences of continuous functions and for $\bar{f} = (f_n) \in CN$, let $\lim \bar{f}$ be its pointwise limit, which is clearly in $\mathscr{S}$. It is easy to check that the function $L(\bar{f}) = [\lim \bar{f}]_\sim$ from CN into $\mathscr{S}$ is Borel. Let now D be the subset of CN consisting of all sequences $\bar{f}$ such that $\lim \bar{f}$ is a derivative (of some differentiable function on $[0,1]$). Then $L_D = L|D$ is a Borel function from D into DI. Thus if I was a Borel function, so would be the composition $I \circ L_D$ from D to $\mathbb{R}$. In particular, the set of all $\bar{f} \in D$ such that $\mathscr{I}(\lim \bar{f}) > 0$ would be a Borel subset of D, contradicting Theorem 4 of [1].

On the other hand, it can be shown that the set DI is coanalytic in S and that the Denjoy integral $I : DI \to \mathbb{R}$ is Δ_1^1-measurable on DI, i.e., the preimage of any open set is both analytic and coanalytic in DI. This is due to Ajtai (unpublished). A proof can be also given using the techniques in [1].

References

[1] R. Dougherty and A.S. Kechris, The complexity of antidifferentiation, *Advances in Mathematics*, (88)(2) (1991), 145–169.
[2] A.S. Kechris, The complexity of antidifferentiation, Denjoy totalization, and hyperarithmetic reals, *Proceedings of the International Congress of Mathematicians, Vol. 1, 2* (Berkeley, Calif., 1986), 307–313, Amer. Math. Soc., Providence, RI, 1987.
[3] R.A. Gordon, *The Integrals of Lebesgue, Denjoy, Perron, and Henstock*, Graduate Studies in Mathematics, Volume 4, Amer. Math. Soc., 1994.

A. S. Kechris

PROBLEM 51: MAZUR

Is a set of functions, measurable in $[0,1]$ with the property that every two functions of the set are orthogonal, at most countable? (I do not assume that the functions are square-integrable!)

(b) An analogous question for sequences: Is the set of sequences with the property that any two sequences $\{\varepsilon_n\}$, $\{\eta_n\}$ of this set are orthogonal, that is

$$\sum_{n=1}^{\infty} \varepsilon_n \eta_n = 0,$$

at most countable?

Addendum. Solved by Mazurkiewicz.

PROBLEM 52: BANACH

Show that the class of functions which are continuous and defined in the interval $[0,1]$ and which have everywhere a derivative, does not form a Borel set in the space C of all continuous functions in $(0,1)$. One can show that it is not a set F_σ and also it is the complement of an analytic set.

Addendum. Solved by Mazurkiewicz.

Commentary

Mazurkiewicz proved that this set forms a coanalytic subset of C which is not a Borel set [3]. It is also true that this set is of the first category (meager) [6, p. 45].

Mazurkiewicz also showed that the set of all continuous functions f on the unit square for which there is some y so that $\partial f / \partial x$ exists at (x,y) for all x in $[0,1]$ forms a *PCA* set which is not a *CPCA* set [4].

It has been shown that the set of all continuous nowhere differentiable functions forms a coanalytic subset of C which is not a Borel set [2]. It can be shown that the set of functions of Besicovitch also forms a coanalytic subset of C [7] and is nonempty [5]. Of course, almost every path in Brownian motion is a continuous nowhere differentiable function [1].

1. A. Dvoretzky, P. Erdős, and S. Kakutani, Nonincrease everywhere of the Brownian motion process, *Proceedings of the Fourth Berkeley Symposium on Mathematical Statistics and Probability,* University of California Press (1961), 103–116.
2. R.D. Mauldin, The set of continuous nowhere differentiable functions, *Pac. J. Math.* 80 (1979), 199–205. Correction, *Pac.J. Math.* 121 (1986), 110-111.
3. S. Mazurkiewicz, Über die Menge die differenzierbaren Funktionen, *Fund. Math.* 27 (1936), 244–249.
4. S. Mazurkiewicz, Eine projektive Menge der Klasse PCA in Funktionalraum, *Fund. Math.* 28 (1937), 7–10.
5. A.P. Morse, A continuous function with no unilateral derivatives, *Trans. Amer. Math. Soc.,* 44 (1938), 496–507.
6. J.C. Oxtoby, *Measure and Category,* Springer-Verlag, New York, 1970.
7. S. Saks, On the functions of Besicovitch in the space of continuous functions, *Fund. Math.* 12 (1928), 244–253.

R. Daniel Mauldin

PROBLEM 53: BANACH

A surface element C (i.e., a one-to-one continuous image of a disc) has the following property: For every $\varepsilon > 0$ one can find $\eta > 0$ such that any two points of C with a distance less than η can be connected by an arc contained in C with a length less than ε. Show that C has a finite area and almost everywhere a tangent plane.

Addendum. There exists a surface element C of the form $z = f(x,y)$, $0 \leq x, y \leq 1$, satisfying the above conditions but without possessing a finite area.

Mazur

August 1, 1935

PROBLEM 54: SCHAUDER

A convex, closed, compact set H is transformed by a continuous mapping $U(x)$ on a part of itself. H is contained in a space of type (F). Does there exist a fixed point of the transformation?

(b) Solve the same problem for arbitrary linear topological spaces or such spaces in which there exist arbitrarily small convex neighborhoods.

[A solution exists for spaces of type (F_0); in the more general theorem, H need not be compact; only $U(H)$ is assumed compact.]

Remark

This problem has led to an incredible number of fixed point theorems. This topic is discussed in Andrzej Granas' lecture published in this edition of the Scottish Book (pp. 45–61)—Problem 54 is discussed in the last section of his talk.

The second and third parts of the problem have a positive solution; the first part of Problem 54 is still unsolved.

Second Edition Commentary

A. Tychonoff [Ein Fixpunktsatz, Math. Ann. 111 (1935), 767–776] provided a proof of the fixed point theorem for locally convex spaces. Versions of Schauder's fixed point theorem were proved by Glicksberg, Fan, Krasnoselskii, Schaefer, and many other authors. In 2001, Robert Cauty [Solution du problème de point fixe de Schauder, Fund. Math. 170 (2001) 231–246] considered Schauder's problem in the general case; however, later it was discovered that his proof (as well as its elaboration in [T.Dobrowolski, Revisiting Cauty's proof of the Schauder conjecture, Abstr. Appl. Anal. 2003, no. 7 , 407–433]) contained a gap.

In 2005, R. Cauty [Rétractes absolus de voisinage algébriques, Serdica Math. J. 31 (2005), no. 4, 309–354] introduced the notion of algebraic ANR. He applied the theory of algebraic ANRs to the fixed point theorem for compact upper semi-continuous multivalued maps with acyclic compact point images. In particular, this gave an affirmative solution of Schauder's problem.

M. Zarichnyi

PROBLEM 55: MAZUR
There is given, in an n-dimensional space E or, more generally, in a space of type (B), a polynomial $W(x)$ bounded in an ε-neighborhood of a certain nonbounded set $R \subset E$ (an ε-neighborhood of a set R is the set of all points which are distant by less than ε from R). Does there exist a polynomial $V(x)$ and a polynomial of first degree $\phi(x)$ such that

(1) $W(x) = V(\phi(x))$;
(2) The set $\phi(R)$, that is to say the image of the set R under the mapping $\phi(x)$, is bounded?

Addendum. In the case of Euclidean spaces, a solution for a finite system of polynomials:

There exists a linear substitution with determinant $\neq 0$ under which all the polynomials of the given system go over into polynomials depending on a smaller number of the given variables. (*Studia Math.* 5.) [See also Problem 75.]

Auerbach

August 3, 1935

PROBLEM 56: MAZUR, ORLICZ

In a space E of type (B) there is given a functional $F(x)$ of degree m and discontinuous. "F is of degree m" means that for $x_0, h_0 \in E$ there exist numbers $a_0, \ldots, a_m$ such that $F(x_0 + th_0) = a_0 + ta_1 + \ldots + t^m a_m$ for rational t. Do there then exist points $x_n \in E$ such that $x_n \to 0$ and $|F(x_n + x)| \to +\infty$ or even only

$$\varlimsup_{n \to \infty} |F(x_n + x)| = +\infty$$

for all $x \in E$? Not solved even for finite-dimensional spaces E.

Commentary

According to Professor Orlicz, Problems 20.1, 27, and 56 emerged in connection with some problems which he and Mazur were considering [1, 2]. The exact meaning of the problems they were considering seems to have become obscured. Problems 20.1 and 56 still seem to be unsolved.

1. S. Mazur and W. Orlicz, Sur la divisibilité des polynomes abstraits, *C. R. Acad. Sci. Paris,* 202 (1936), 621–623.
2. S. Mazur and W. Orlicz, Sur les fonctionnelles rationnelles, *C. R. Acad. Sci. Paris* 202 (1936), 904–905.

PROBLEM 57: RUZIEWICZ

Given are two functions $w(h)$ and $\phi(h)$, decreasing with $|h|$ to 0, and satisfying the conditions

$$\lim_{h \to 0} \frac{w(h)}{|h|} = \infty,$$

and

$$\lim_{h \to 0} \frac{w(h)}{\phi(h)} = \infty.$$

Does there exist a function satisfying the conditions:

$$|f(x+h) - f(x)| < w(h) \tag{1}$$

and

$$\limsup_{h \to 0} \left| \frac{f(x+h) - f(x)}{\phi(h)} \right| = \infty? \tag{2}$$

Second Edition Commentary

We shall prove that the answer to Stanisław Ruziewicz's question is affirmative. We denote by $\langle x \rangle$ the distance from x to the nearest integer.

Lemma 1. *For every $x \in \mathbb{R}$ we have*

$$\max(|\langle x + (1/3) \rangle - \langle x \rangle|, |\langle x + (2/3) \rangle - \langle x \rangle|) \geq 1/6 \tag{3}$$

and

$$\max(|\langle x - (1/3) \rangle - \langle x \rangle|, |\langle x - (2/3) \rangle - \langle x \rangle|) \geq 1/6. \tag{4}$$

Proof. We may assume $x \in [0,1]$. Suppose $|\langle x + (1/3) \rangle - \langle x \rangle| < 1/6$ and $|\langle x + (2/3) \rangle - \langle x \rangle| < 1/6$. If $x \in [0, 1/2]$, then these inequalities imply $x + (1/3), x + (2/3) \in [1/2, 1]$. From these we infer $|(2/3) - 2x| < 1/6$ and $|(1/3) - 2x| < 1/6$, which is impossible. If $x \in [1/2, 1]$, then necessarily $x + (1/3), x + (2/3) \in [1, 3/2]$. Thus we have $|2x - (5/3)| < 1/6$ and $|2x - (4/3)| < 1/6$, which is also impossible. This proves (3).

Applying (3) with $-x$ in place of x and using the fact that the function $\langle x \rangle$ is even, we obtain (4). $\qquad\square$

Lemma 2. *Suppose that $v: [0, \infty) \to \mathbb{R}$ is continuous, increasing, concave, and satisfies $v(0) = 0$ and $\lim_{x \to +0} v(x)/x = \infty$. Then, for every sequence (c_n) of positive numbers converging to zero there exist a function $g: \mathbb{R} \to \mathbb{R}$ and a subsequence (c_{n_k}) of (c_n) such that*

(i) $|g(x+h) - g(x)| \leq 2v(|h|)$ *for every $x, h \in \mathbb{R}$, and*
(ii)

$$\max(|g(x + (c_{n_k}/3)) - g(x)|, |g(x + (2c_{n_k}/3)) - g(x)|) \geq \frac{1}{12} v(c_{n_k})$$

and

$$\max(|g(x - (c_{n_k}/3)) - g(x)|, |g(x - (2c_{n_k}/3)) - g(x)|) \geq \frac{1}{12} v(c_{n_k})$$

for every $x \in \mathbb{R}$ and $k = 1, 2, \ldots$.

Proof. We define the sequence (n_i) of positive integers and the sequences (λ_i), (a_i) of positive numbers by induction. We put $n_1 = 1$, $\lambda_1 = v(c_1)$ and $a_1 = 1/c_1$. Suppose that n_i, $\lambda_i = v(c_{n_i})$ and $a_i = 1/c_{n_i}$ have been defined. Since $\lim_{x \to +0} v(x)/x = \infty$ and $v(c_n) \to 0$, we can choose an index $n_{i+1} > n_i$ such that $v(c_{n_{i+1}})/c_{n_{i+1}} > 16v(c_{n_i})/c_{n_i}$ and $v(c_{n_{i+1}}) < v(c_{n_i})/16$. We put $\lambda_{i+1} = v(c_{n_{i+1}})$ and $a_{i+1} = 1/c_{n_{i+1}}$. In this way we defined the numbers n_i, λ_i and a_i for every $i = 1, 2, \ldots$.

It is clear from the construction that $\lambda_{i+1} < \lambda_i/16$ and $a_{i+1}\lambda_{i+1} > 16a_i\lambda_i$ hold for every i. Therefore, we have

$$\sum_{i=k+1}^{\infty} \lambda_i < \lambda_{k+1} \cdot \left(1 + \frac{1}{16} + \frac{1}{16^2} + \dots\right) = \frac{16}{15}\lambda_{k+1}$$

and

$$\sum_{i<k} \lambda_i a_i < \lambda_k a_k \cdot \left(\frac{1}{16} + \frac{1}{16^2} + \dots\right) = \lambda_k a_k/15.$$

for every k.

We define $g(x) = \sum_{i=1}^{\infty} \lambda_i \langle a_i x \rangle$, and prove that g satisfies the requirements. We have

$$g(x+h) - g(x) = \sum_{i=1}^{\infty} \lambda_i(\langle a_i x + a_i h \rangle - \langle a_i x \rangle) =$$

$$= A_k + \lambda_k(\langle a_k x + a_k h \rangle - \langle a_k x \rangle) + B_k \tag{5}$$

for every $x, h \in \mathbb{R}$ and $k = 1, 2, \dots$, where

$$A_k = \sum_{i<k} \lambda_i(\langle a_i x + a_i h \rangle - \langle a_i x \rangle) \qquad \text{and} \qquad B_k = \sum_{i=k+1}^{\infty} \lambda_i(\langle a_i x + a_i h \rangle - \langle a_i x \rangle).$$

Since $|\langle x+h \rangle - \langle x \rangle| \leq |h|$ and $|\langle x+h \rangle - \langle x \rangle| \leq 1/2$ for every x and h, it follows that

$$|A_k| \leq \sum_{i<k} \lambda_i a_i |h| < \lambda_k a_k |h|/15,$$

$$|B_k| \leq \frac{1}{2} \sum_{i=k+1}^{\infty} \lambda_i < \frac{8}{15}\lambda_{k+1} < \frac{1}{30}\lambda_k$$

and

$$|g(x+h) - g(x)| \leq |A_k| + \lambda_k a_k |h| + |B_k|. \tag{6}$$

If $|h| > c_1$, then

$$|g(x+h) - g(x)| \leq \sum_{i=1}^{\infty} \lambda_i < (16/15)\lambda_1 = (16/15)v(c_1) \leq (16/15)v(|h|),$$

and thus (i) is true in this case.

If $0 < |h| \leq c_1$, then there exists an index k such that $c_{n_{k+1}} \leq |h| \leq c_{n_k}$. By (6) we have

$$|g(x+h) - g(x)| \leq \frac{16}{15}\lambda_k a_k |h| + \frac{8}{15}\lambda_{k+1}. \tag{7}$$

We have $\lambda_{k+1} = v(c_{n_{k+1}}) \leq v(|h|)$. Since v is concave and $v(0) = 0$, it follows that the function $v(x)/x$ is decreasing on $(0,\infty)$. Therefore, by $|h| \leq c_{n_k}$ we have

$$\frac{v(c_{n_k})}{c_{n_k}} \leq \frac{v(|h|)}{|h|};$$

that is, $\lambda_k a_k |h| \leq v(|h|)$. Thus, by (7), we obtain $|g(x+h) - g(x)| \leq (24/15)v(|h|)$, which proves (i).

Now we prove (ii). Let $x \in \mathbb{R}$ and $k \geq 1$ be fixed. By Lemma 1, there are $h_1, h_2 \in \{c_{n_k}/3, 2c_{n_k}/3\}$ such that

$$|\langle a_k x + a_k h_1 \rangle - \langle a_k x \rangle| \geq 1/6 \quad \text{and} \quad |\langle a_k x - a_k h_2 \rangle - \langle a_k x \rangle| \geq 1/6.$$

(Note that $a_k = 1/c_{n_k}$.) Thus, by (5), we obtain

$$|g(x + (-1)^{j-1}h_j) - g(x)| \geq -\frac{1}{15}\lambda_k a_k h_j + \frac{1}{6}\lambda_k - \frac{1}{30}\lambda_k \geq$$

$$\geq -\frac{2}{45}\lambda_k + \frac{1}{6}\lambda_k - \frac{1}{30}\lambda_k > \frac{1}{12}\lambda_k = \frac{1}{12}v(c_{n_k})$$

for $j = 1, 2$, which completes the proof. $\qquad\square$

Lemma 3. *Let* $u\colon [0,\infty) \to \mathbb{R}$ *be an increasing function such that* $\lim_{x\to+0} u(x) = u(0) = 0$ *and* $\lim_{x\to+0} u(x)/x = \infty$. *Then there exists a function* $v\colon [0,\infty) \to \mathbb{R}$ *such that* $\lim_{x\to+0} v(x)/x = \infty$, v *is increasing, continuous, concave,* $v \leq 2u$ *on* $[0,\infty)$, *and* $v = u$ *at the points of a sequence converging to zero.*

Proof. Let $x_0 > 0$ be an arbitrary point, and put $s_0 = u(x_0)/x_0$. Since u is increasing and $\lim_{x\to+0} u(x)/x = \infty$, it follows that for every $s \geq s_0$ there is a smallest positive number $d(s)$ such that $u(d(s))/d(s) = s$. Indeed, it is easy to check that $d(s) = \inf\{x > 0\colon u(x) \leq sx\}$ has this property. Note that $u(x)/x > s$ for every $0 < x < d(s)$.

We put $d_0 = d(s_0)$. Since $\lim_{x\to+0} u(x) = 0$, there is a point $0 < x_1 < \min(d_0, 1)$ such that $u(x_1) < u(d_0)$. We put $s_1 = u(x_1)/x_1$ and $d_1 = d(s_1)$. Let ℓ_1 be the linear function whose graph connects the points $(d_1, u(d_1))$ and $(d_0, u(d_0))$. Then $\ell_1(0) > 0$, and thus we can select a point $0 < x_2 < \min(d_1, 1/2)$ such that $u(x_2) < \ell_1(0)$. We put $s_2 = u(x_2)/x_2$ and $d_2 = d(s_2)$. Let ℓ_2 be the linear function whose graph connects the points $(d_2, u(d_2))$ and $(d_1, u(d_1))$. Then $\ell_2(0) > 0$, and thus we can select a point $0 < x_3 < \min(d_2, 1/3)$ such that $u(x_3) < \ell_2(0)$. We put $s_3 = u(x_3)/x_3$ and $d_3 = d(s_3)$. Continuing this process we define the points d_n for every $n \geq 0$ and the linear functions ℓ_n for every $n \geq 1$.

It is clear that (d_n) is a strictly decreasing sequence converging to zero, and that $u(x)/x > u(d_n)/d_n = s_n$ for every $x \in [0, d_n)$ and $n = 0, 1, \ldots$. Thus the sequence (s_n) is strictly increasing.

We define $v(0) = 0$, $v(x) = u(d_0)$ for every $x \geq d_0$, and $v(x) = \ell_n(x)$ for every $x \in [d_n, d_{n-1}]$ and $n = 1, 2, \ldots$. It is easy to check that v is continuous, increasing, and concave on $[0, \infty)$, satisfies $\lim_{x \to +0} v(x)/x = \infty$, and equals u at the points d_n.

We prove that $v(x) \leq 2u(x)$ for every $x \geq 0$. This is clear for $x \geq d_0$ (since u is increasing). Let $n \geq 1$ be fixed; we prove that $v(x) = \ell_n(x) \leq 2u(x)$ for every $x \in [d_n, d_{n-1}]$. Since u is increasing and $u(x) > s_{n-1}x$ for every $x < d_{n-1}$, it follows that $u(x) \geq \max(u(d_n), s_{n-1}x)$ for every $x \in [d_n, d_{n-1}]$. Thus it is enough to show that

$$\ell_n(x) \leq 2 \cdot \max(u(d_n), s_{n-1}x) \tag{8}$$

if $x \in [d_n, d_{n-1}]$. Let $a = u(d_n)/s_{n-1}$. Then $a > u(d_n)/s_n = d_n$. Since the slope of ℓ_n is smaller than s_{n-1}, it follows that

$$\ell_n(a) < u(d_n) + s_{n-1}(a - d_n) < u(d_n) + s_{n-1}a = 2u(d_n).$$

Therefore $\ell_n(x) < 2u(d_n)$ in the interval $[d_n, a]$. Since $\ell_n(a) < 2u(d_n) = 2s_{n-1}a$, $\ell_n(d_{n-1}) = u(d_{n-1}) = s_{n-1}d_{n-1} < 2s_{n-1}d_{n-1}$ and ℓ_n is linear, it follows that $\ell_n(x) < 2s_{n-1}x$ for every $a \leq x \leq d_{n-1}$. This proves (8) and the lemma.   □

Now we turn to the **solution of the problem.** Let w and ϕ be given as in the problem. We put $w_1(h) = \min(w(-h), w(h))$ for every $h \geq 0$. Then w_1 is increasing on $[0, \infty)$, and satisfies $w_1(0) = 0$ and $\lim_{x \to +0} w_1(x)/x = \infty$. By Lemma 3, there exists a function v such that v is increasing, continuous, concave, $v \leq 2w_1$ on $[0, \infty)$, $\lim_{x \to +0} v(x)/x = \infty$, and $v = w_1$ at the points of a sequence (d_n) converging to zero.

For every n we have $v(d_n) = w_1(d_n) = \min(w(-d_n), w(d_n))$, and thus either $v(d_n) = w(d_n)$ or $v(d_n) = w(-d_n)$. Therefore, taking a suitable subsequence of (d_n) we can find a sequence (c_n) of positive numbers tending to zero such that either $v(c_n) = w_1(c_n) = w(c_n)$ for every n, or $v(c_n) = w_1(c_n) = w(-c_n)$ for every n.

Applying Lemma 2 we obtain a function g satisfying (i) and (ii). We prove that $f = g/5$ satisfies the requirements of the problem. For every x and $h \neq 0$ we have

$$|f(x+h) - f(x)| \leq 2v(|h|)/5 \leq 4w_1(|h|)/5 \leq 4w(h)/5 < w(h),$$

and thus (1) is satisfied.

If $v(c_n) = w(c_n)$ for every n, then we use the first inequality in (ii) of Lemma 2, and choose a sequence (h_k) such that $h_k \in \{c_{n_k}/3, 2c_{n_k}/3\}$ and

$$|g(x+h_k) - g(x)| \geq v(c_{n_k})/12 = w(c_{n_k})/12 \geq w(h_k)/12$$

for every k. Then

$$\lim_{k \to \infty} \frac{|f(x+h_k) - f(x)|}{\phi(h_k)} \geq \frac{1}{60} \cdot \lim_{k \to \infty} \frac{w(h_k)}{\phi(h_k)} = \infty. \tag{9}$$

Therefore, in this case we have

$$\limsup_{h \to +0} \left| \frac{f(x+h) - f(x)}{\phi(h)} \right| = \infty \tag{10}$$

for every x.

In the case when $v(c_n) = w(-c_n)$ for every n, we use the second inequality in (ii) of Lemma 1, and choose a sequence (h_k) such that $h_k \in \{-c_{n_k}/3, -2c_{n_k}/3\}$ and

$$|g(x+h_k) - g(x)| \geq v(c_{n_k})/12 = w(-c_{n_k})/12 \geq w(h_k)/12$$

for every k. Then we have (9) again, proving that

$$\limsup_{h \to -0} \left| \frac{f(x+h) - f(x)}{\phi(h)} \right| = \infty \tag{11}$$

holds for every x. This shows that f satisfies the requirements. □

Remark 1. We proved slightly more than what was asked in the problem. We showed that there exists a function f satisfying (1), and either (10) for every x or (11) for every x. It is clear from the proof that *if the function ϕ is even, then f satisfies*

$$\limsup_{h \to +0} \left| \frac{f(x+h) - f(x)}{\phi(h)} \right| = \limsup_{h \to -0} \left| \frac{f(x+h) - f(x)}{\phi(h)} \right| = \infty \tag{12}$$

for every x.

In the general case, however, we cannot expect (12) to be true. Moreover, it is possible that *whenever a function f satisfies* (1), *then*

$$\lim_{h \to +0} \left| \frac{f(x+h) - f(x)}{\phi(h)} \right| = 0. \tag{13}$$

Consider the following simple example. Let $0 < \alpha < \beta < \gamma < \delta < 1$ be given numbers, and define w and ϕ as follows: $w(h) = h^\alpha$, $\phi(h) = h^\beta$ if $h \geq 0$, and $w(h) = (-h)^\gamma$, $\phi(h) = (-h)^\delta$ if $h < 0$. It is clear that w and ϕ satisfy the conditions of the problem.

Suppose (1). If $h > 0$, then

$$|f(x+h) - f(x)| = |f((x+h) - h) - f(x+h)| \leq w(-h) = h^\gamma,$$

and thus

$$\limsup_{h \to +0} \left| \frac{f(x+h) - f(x)}{\phi(h)} \right| \leq \limsup_{h \to +0} \frac{h^\gamma}{h^\beta} = 0.$$

M. Laczkovich

PROBLEM 58: RUZIEWICZ

A set E_1 of real numbers precedes the set E_2, which we denote by E_1pE_2, if:

(1) E_1 is of a lower homoie class than E_2 ($E_1 < E_2$),
(2) There does not exist a set E_3 so that $E_1 < E_3 < E_2$.

 (a) Do there exist sets A, B, C, and $\{A_n\}$, $n = 1, 2, \ldots, N$, ($N > 1$), such that
 $ApBpC$ and $ApA_1pA_2p \ldots pA_npC$?
 (Remark: For $n = 2$ such sets exist; cf. *Fund. Math.*, 15, p. 95.)
 (b) Do there exist sets A, B, C, and $\{A_n\}$, $n = 1, 2, \ldots$ ad inf. such that $ApBpC$
 and $ApA_1pA_2pA_3p \ldots$ ad inf., and $A_n < C$ for $n = 1, 2, 3, \ldots$?

Commentary

This problem concerns dimensional types as defined by Fréchet [1, p. 30]. If X and Y are topological spaces, then the type of X is $\leq$ the type of Y (symbolized by $dX \leq dY$) provided there is a homeomorphism of X into Y. The spaces X and Y are of the same type ($dX = dY$) provided $dX \leq dY$ and $dY \leq dX$. If X and Y are homeomorphic, then $dX = dY$. It is easily seen that the converse is not true. The type of X is less than the type of Y provided $dX \leq dY$ but there is no homeomorphism of Y into X. This is what is meant by the expressions "$E_1 < E_2$" or "E_1 is of lower homoie class than E_2" in this problem. There is a discussion of this concept and some early results in [2].

Apparently, the problems posed here are still open.

1. M. Fréchet, Les espaces abstraits, Gauthier-Villars, Paris, 1928.
2. W. Sierpiński, General Topology, University of Toronto Press, Toronto, 1956.

R. Daniel Mauldin

PROBLEM 59: RUZIEWICZ

Can one decompose a square into a finite number of squares all different?

Commentary

A square dissected into a finite number of squares no two the same is now referred to as a perfect squared square or simply a perfect square. The earliest published mention of the problem which has been found is in a paper of 1925 by Zbigniew Moroń [11]. Moroń was concerned, however, with the dissection of a rectangle into squares. He stated that Ruziewicz had asked if a rectangle could be made up of different squares and presented two such rectangles in answer. A letter from Prof. Władysław Orlicz in 1977 gives some details [9]. He and Moroń were schoolmates studying mathematics at the University of Lwów and in about 1923–1924 both were junior assistants in the Institute of Mathematics. Stanisław Ruziewicz, professor of mathematics, proposed the problem (presumably in his mathematical seminar) of dissecting a rectangle into unequal squares, which he said he had heard of from mathematicians of the University of Cracow. The students worked diligently on the problem without success until they were all surprised by Moroń's solutions.

Moroń observed that if there were two different dissections of the same rectangle such that each square of one dissection is different from each square of the other, they can be put together with two added squares to form a square dissected into squares which are all different provided that neither of the two dissected rectangles contained a square equal to one of the two added squares (two examples are given in Figures 59.1 and 59.2).

It may be that Ruziewicz also mentioned dissecting a square at that time; in any event, he did so later. The problem was believed by some to be impossible. Kraitchik in 1930 stated that the Russian mathematician N.N. Lusin had communicated to him the proposition (believed to be true though not demonstrated) that it was not possible to dissect a square into a finite number of different squares [10].

Figure 59.1

Figure 59.2

Moroń did not state how he obtained his two dissected rectangles; obviously, they must have been found empirically. Some others were found over a dozen years later by Sprague. He succeeded in making up two different dissected rectangles the same size and forming a perfect square according to Moroń's observation, published in 1939 [12]. The square has 55 component squares and is 4205 units in size; its structure is shown in Figure 59.1. It contains five disjoint subsets of squares arranged into five rectangles, two pairs of which are Moroń's two rectangles magnified different amounts.

A second perfect square was published, by listing the sides of the component squares, only a few months later [13]. This is shown in Figure 59.2; it has 28 elements (the component squares) and is 1015 units in size. This square was the result of the work of four students at Trinity College, Cambridge, R.L. Brooks, C.A.B. Smith, A.H. Stone, and W.T. Tutte, who had been working on the problem of squaring rectangles and squares during the years 1936–1938. Their paper [4] is a classic; see also [14] for a very interesting expository account of their work.

A distinction is made between a squared rectangle (square) which contains a subset of elements, more than one and less than all, which are themselves arranged into a rectangle, and one which does not; the former are designated compound and the latter simple. The Cambridge group founded and developed the theory of simple squared rectangles. They proved that every simple squared rectangle with n elements can be produced from the complete set of 3-connected planar graphs of order $n + 1$. This is done by considering the graph to be an electrical network with an emf placed in one branch and the other branches having unit resistance. The relative values of the currents in these branches are calculated from Kirchhoff's laws. These values represent the relative values of the sides of the component squares of a dissection and the connectivities of the branches of the network represent the manner in which the component squares are arranged. They also produced a number of perfect squares, including several departing from Moroń's suggested type, and developed some theory concerning them. The treatment utilizes electrical network theory and graph theory, and makes contributions in each of these fields.

Bouwkamp described the method of [4] in more physical terms in 1946–1947 [1] and listed the 3-connected planar graphs with up to 14 edges (by drawings) and the simple squared rectangles with up to 13 elements (by giving their elements in a certain order). This reached the practical limit of what could be done systematically and completely by hand. Later, in 1960, complete sets of 3-connected planar graphs with up to 19 edges were produced by computer [3], and from these complete sets of simple squared rectangles, also by computer. A catalogue of the rectangles with up to 15 elements was published [2]. The methods used are described in Duijvestijn's 1962 thesis [5].

One relevancy of simple squared rectangles to Problem 59 is that compound perfect squares can be produced from them. A summary of various methods of combining squared rectangles was given in [8], which also introduced another method of producing compound perfect squares of the type shown in Figure 59.3, by means of which many compound perfect squares of order higher than 24 were produced. In 1948 Willcocks had constructed the compound perfect square with only 24 elements [15, 16] which is shown in Figure 59.3. It remained the lowest order perfect square until 1978 and is the lowest order compound perfect square. This was shown by a computer search by Duijvestijn and Leeuw based on a method proposed by Federico [17]. Over two thousand perfect squares of order higher than 24 have been produced.

On the other hand, the production of simple perfect squares is still not possible by any general direct method. The Cambridge group in [4] and later papers developed a theoretical method by means of which some simple perfect squares of a special type but of a high order were produced. Simple perfect squares of another special type, of orders 25 and 26, were produced by Wilson in 1967 [17].

The other relevancy of the systematic production of simple perfect rectangles is that if carried out to a sufficiently high order one with equal sides might be found, a simple perfect square. This was done up to order 19 by Duijvestijn in his 1962

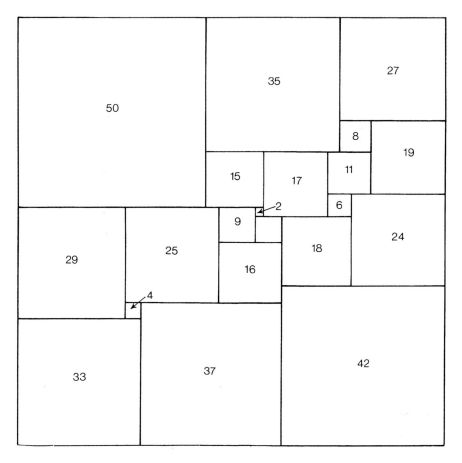

Figure 59.3

thesis [5]. He found that there were no simple perfect squares below order 20. The work of testing for order 19 squares, for example, required the construction of the complete set of nonisomorphic 3-connected planar graphs with 20 edges. However, further work on some of them was eliminated *a priori* by virtue of certain theorems in the basic paper [4]. Even so, the computer time was high. The construction of the graphs necessarily results in a considerable amount of duplication and to eliminate the duplicates and easily calculable numerical identifying characteristic of graphs, invariant under isomorphism, was developed. An incidental result of the construction of complete sets of 3-connected planar graphs was an advance in Euler's problem of the enumeration of convex polyhedra, since these graphs are isomorphic with the graphs of convex polyhedra.

Duijvestijn returned to the problem in recent years. With improvements in details of the methods and the availability of a faster computer he was able to complete the search through order 21. In 1978 he announced that there was one, and only one,

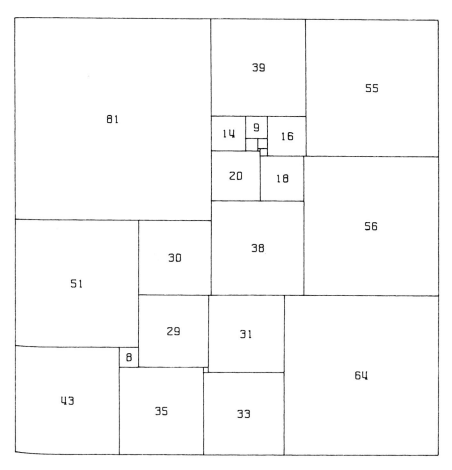

Figure 59.4

simple perfect square of order 21 and none of lower order [6]. The square is shown in Figure 59.4. *Scientific American* called this a pluperfect square, which term is deserved not only because it is the lowest order perfect square possible but also on account of the elegance of its construction.

An extended historical review of this subject is available in [9].

Acknowledgements

I am indebted to Michael Goldberg for making the drawings for Figures 59.1 and 59.4 and to P. Leeuw for drawing Figures 59.2 and 59.3 by computer.

1. C.J. Bouwkamp, On the dissection of rectangles into squares. I., *Nederl. Akad. Wetensch. Proc.*, AA49 (1946), 1176–1188; II, III, A50 (1947), 58-71, 72–78.
2. C.J. Bouwkamp, A.J.W. Duijvestijn, and P. Medema, *Tables Relating to Simple Squared Rectangles of Orders Nine through Fifteen,* Technische Hogeschool, Eindhoven, Netherlands, Aug. 1960, 360 pages.

3. C.J. Bouwkamp, A.J.W. Duijvestijn, and P. Medema, *Tables of c-nets of Orders 8–19 inclusive,* Phillips Research Laboratories, Eindhoven, Netherlands, 2 vols. 1960; unpublished, available in UMT file of Mathematics Computation; see Math. Comp., 24 (1970), 995-997 for description. (The basic paper [4] and others refer to 3-connected planar graphs as *c*-nets.)

4. R.L. Brooks, C.A.B. Smith, A.H. Stone, and W.T. Tutte, The dissection of rectangles into squares, *Duke Math. J.* 7 (1940), 312–340.

5. A.J.W. Duijvestijn, *Electronic Computation of Squared Rectangles,* Dissertation, Technische Hogeschool, Eindhoven, Netherlands, 1962, 96 pages; also in *Phillips Res. Rep.* 17 (1962), 523–612.

6. A.J.W. Duijvestijn, Simple perfect squared square of lowest order, *J. Comb. Theory,* B25 (1978), 240-243. See also *Sci. Amer.,* 238 (June 1978), 86–88

7. A.J.W. Duijvestijn, P.J. Federico, and P. Leeuw, Compound Perfect Squares, *Amer. Math. Monthly* 89 (1982), 15-32.

8. P.J. Federico, Note on some low-order perfect squared squares, *Canad. J. Math.* 15 (1963), 350–362.

9. P.J. Federico, Squaring Rectangles and Squares; A Historical Review with Annotated Bibliography, in *Graph Theory and Related Topics,* J.A. Bondy and U.S.R. Murty, Eds., Academic Press, 1979, 173–196.

10. M. Kraitchik, *La Mathématique des Jeux ou Récréations Mathématiques,* Stevens Frères, Brussels, 1930, 272.

11. Z. Moroń, O rozkladach prostokatow na kwadraty (On the dissection of a rectangle into squares), *Przeglad Mat. Fiz.,* 3 (1925), 152-153.

12. R. Sprague, Beispiel einer Zerlegung des Quadrats in lauter verschiedene Quadrate, *Math. Zeit.,* 45 (1939), 607–608.

13. A.H. Stone, Problem E401, *Amer. Math. Monthly,* 47 (Jan. 1940), 48.

14. W.T. Tutte, Squaring the square, *Sci. Am.,* 199 (Nov. 1958), 136–142, 166. Reprinted with addendum and enlarged bibliography in Martin Gardner, *The 2nd Scientific American Book of Mathematical Puzzles and Diversions,* pp. 186–209, 250, Simon & Schuster, New York, 1961; also paperback edition.

15. T.H. Willcocks, Problem 7795 and solution, *Fairy Chess Rev.* 7 (1948), 97 (Aug.), 106 (Oct.).

16. T.H. Willcocks, A note on some perfect squares, *Canad. J. Math.,* 3 (1951), 304–308.

17. John C. Wilson, *A method for finding simple perfect squared squarings,* Ph.D. Thesis, University of Waterloo, Waterloo, Ontario, Canada, 1967, 80 pages plus 72 pages computer output.

P. J. Federico

PROBLEM 60: RUZIEWICZ

Can one, for every $\varepsilon > 0$, represent the surface of a sphere as a sum of a finite number of regions which are smaller in diameter than ε, closed, connected, congruent, and have no interior point in common? We assume that the boundaries of these sets are: (a) polygons, (b) curves of finite length, (c) sets of measure zero.

PROBLEM 61: STEINHAUS

(a) Determine the surfaces $z = f(x, y)$ such that in each of their points there intersect two plane curves congruent to each other.

(b) Determine the surface $z = f(x, y)$ such that in each point there intersect two plane curves congruent to one of them (for every point the same curve).

(cf. Problem 44)

Addendum. All surfaces of revolution have this property; whether these are the only ones is not known.

Ruziewicz
July 31, 1935

PROBLEM 62: MAZUR, ULAM

In a group G there are given groups G_n, $n = 1, 2, \ldots$ ad inf. with the following properties: $G = G_1 + G_2 + \ldots + G_n + \ldots$, $G_n \subset G_{n+1}$, G_n is isomorphic to G_1. Is G isomorphic to G_1?

Addendum. As R. Baer remarked, the answer is trivial: G_n is the group of numbers with the denominator n, $G = \sum G_n =$ the group of rational numbers.

PROBLEM 63: MAZUR, ULAM

The set E of elements of a group G we call a base if E spans a group which is identical with G, but no proper subset of the set E has this property. If there is a base in a group G, does there exist a base for every subgroup H of it?

Commentary

The answer is no. Several examples of groups with a minimal set of generators having subgroups without minimal sets of generators were given by V. Dlab. (On a problem of Mazur and Ulam about irreducible generating systems in groups, *Coll. Math.* 7 (1959), 171–176).

Jan Mycielski

PROBLEM 64: MAZUR

In a space E of type (B) there are given two convex bodies A and B and their distance from each other is positive. (A convex body is a convex set which is closed, bounded, and possesses interior points.) Does there exist a hyperplane H which separates the two bodies, A, B? That is to say, a hyperplane which has the property that one of the bodies lies on one, the other on the other side of this hyperplane. [Hyperplane means a set of all points x satisfying the equation $F(x) - c = 0$, where $F(x)$ is a linear functional $\neq 0$, and c a constant.]

Addendum The theorem is true even when the two bodies are not disjoint, but do not have common interior points.

Eidelheit
January 11, 1936

Commentary

The result of Eidelheit is, of course, now well known. The theorem stated here was published in M. Eidelheit, "Zur Theorie der konvexen Mengen in linearen normierten Räumen," *Studia Math.* 6 (1936), 104–111. See also S. Kakutani, "Ein Beweis des Satzes von Eidelheit über konvexe Mengen," *Proc. Imp. Acad. Japan,* 13 (1937), 93–94. I do not know if there now is a stronger theorem for normed linear spaces. For the strongest form of the separation theorem in finite-dimensional Euclidean space, see F.A. Valentine, *Convex Sets,* McGraw-Hill, New York (1964), p. 66.

W.A. Beyer

PROBLEM 65: MAZUR
In a space E of type (B) there is given a convex set W, containing 0 and nowhere dense. Is the smallest convex set containing W, symmetric with respect to 0 (that is to say, the set generated by elements $x - y$, where $x \in W$, $y \in W$), also nowhere dense?

Addendum. False—the set W composed of functions which are nondecreasing, in the space (C) of all continuous functions is convex and nowhere dense. It contains 0. The convex set containing W and symmetric with respect to zero contains all functions of bounded variation and is not nowhere dense.

Mazur

PROBLEM 66: MAZUR
The real-valued function $z = f(x, y)$ of real variables x, y possesses the first partial derivatives $\partial f / \partial x$, $\partial f / \partial y$ and the second partial derivatives $\partial^2 f / \partial x^2$, $\partial^2 f / \partial y^2$. Do there exist then almost everywhere the mixed second partial derivatives $\partial^2 f / \partial x \partial y$, $\partial^2 f / \partial y \partial x$. According to a remark by Prof. Schauder, this theorem is true with the following additional assumptions: The derivatives $\partial f / \partial x$, $\partial f / \partial y$ are absolutely continuous in the sense of Tonelli, and the derivatives $\partial^2 f / \partial x^2$, $\partial^2 f / \partial y^2$ are square integrable. An analogous question for n variables.

Remark

This problem was just settled in the negative by V. Mykhaylyuk and A. Plichko in their paper *On a Mazur problem from the "'Scottish Book" concerning second partial derivatives.* The paper will appear in Colloquium Mathematicum. April 2015.

PROBLEM 67: BANACH
August 1, 1935

A modification of Mazur's game [See Problem 43].)

We call a half of the set E [in symbols, $(1/2)E$] an arbitrary subset $H \subset E$ such that the sets $E, H, E - H$ are of equal power.

(1) Two players A and B give in turn set $E_i, i = 1, 2, \ldots$ ad inf. so that $E_i = (1/2)E_{i-1}$, $i = 1, 2, \ldots$, where E_0 is a given abstract set. Player A wins if the product $E_1 E_2 \cdots E_i E_{i+1} \cdots$ is vacuous.
(2) The game, similar to the one above, with the assumption that $E_i = 1/2[E_0 - E_1 - \ldots - E_{i-1}]$, $i = 2, 3, \ldots$ ad inf., and $E_1 = (1/2)E_0$. Player A wins if $E_1 + E_2 + \ldots = E_0$.

Is there a method of win for player A? If E_0 is of power cofinal with $\aleph_0$, then player A has a method of win. Is it only in this case? In particular, solve the problem if E_0 is the set of real numbers.

Addendum. There exists a method of play which will guarantee that the product of the sets is vacuous. The solution was given by J. Schreier.

August 24, 1935

Commentary

The solution of J. Schreier was published in [2].

The strategies for both games are easy to describe: One well orders E with a relation $<$. Then for game (1), given any $X \subset E$ with $|X| = |E|$ one always chooses the subset of all those elements of X which have an immediate predecessor in X relative to $<$. Then, after ω steps the intersections is empty. For game (2) one chooses the subset of all elements of X which have no immediate predecessors in X. After ω steps E is covered.

Many related games were studied in [1], e.g., the following game: A set S is given. Player I cuts S into two pieces, Player II chooses one of them, then again I cuts the chosen piece and II chooses one of them, etc., for ω steps. Player II wins if the intersection of the chosen parts is empty and Player I wins otherwise. Then it is proved that I has a winning strategy iff $|S| \geq 2^{\aleph_0}$ and II has a winning strategy iff $|S| \leq \aleph_0$. For more material on games, see the commentary to Problem 43.

1. F. Galvin, J. Mycielski, R.M. Solovay, and others, a long paper in preparation.
2. J. Schreier, Eine Eigenschaft abstrakter Mengen, *Studia Math.* 7 (1938), 155–156.

J. Mycielski

PROBLEM 68: ULAM

There is given an n-dimensional manifold R with the property that every section of its boundary by a hyperplane of $n-1$ dimensions gives an $n-2$ dimensional closed surface (a set homeomorphic to a surface of the sphere of this dimension). Prove that R is a convex set. This question was settled affirmatively for $n = 3$ by Schreier. (That is to say, a manifold contained in E_3, such that every section by a plane gives a single simple closed curve, must be convex.)

Commentary

This problem remains unsolved. Of course, one could restate the problem to take into account the possible lower dimensional intersections of the manifold with an $(n-1)$-dimensional hyperplane.

Schreier [3] showed that each 2-dimensional surface in R^3, each of whose nondegenerate planar sections is a Jordan curve, is convex. Aumann [1] proved that a continuum K in R^3 is convex provided that for each plane P, $P \cap K$ and $P - K$ is connected. Aumann also showed that a continuum K in R^n is convex if and only if $R^n - K$ is connected and the intersection of K with each $(n-1)$-dimensional hyperplane is convex.

Aumann [2] later claimed the following theorem. A closed bounded subset K of R^n is convex if and only if $K \cap P$ is simply connected for all 2-dimensional hyperplanes P.

It is rather interesting that a problem formulated and partially solved by Schreier and Ulam should have been included in the first volume of *Deutsche Mathematik,* now available in an expurgated edition.

1. G.G. Aumann, Eine einfache Characterisierung der convexen Kontinuen im R_3, *Deutsche Mathematik* 1 (1936), 108.
2. G. Aumann, Über Schnitteigenschaften convexer Punktmengen im R_3, *Deutsche Mathematik* 1 (1936), 162–165; Swets and Zeitlinger N.V. Amsterdam, 1966.
3. J. Schreier, Über Schnitte convexer Flächen, *Bull. Int. Ac. Pol.* Series A (1933), 155–157.

R. Daniel Mauldin

Second Edition Commentary

We may interpret this problem as follows: "if R is a compact, $(n+1)$-dimensional manifold with boundary in $\mathbb{R}^{n+1}$ for which every n-dimensional hyperplane H that meets R in more than a point has $H \cap \partial R$ an $(n-1)$-sphere, is R convex?"

Schreier [3] showed first that a two-dimensional surface in $\mathbb{R}^3$, each of whose nondegenerate planar sections is a Jordan curve, is the boundary of a convex body. Montejano [4] generalized Schreier's theorem as follows: "Let N be a closed, connected n-manifold topologically embedded in $\mathbb{R}^{n+1}$. Suppose that for every n-dimensional hyperplane H that meets N in more than a point, $H - N$ has exactly two components. Then N is the boundary of a convex $(n+1)$-body."

As a corollary of this theorem, Problem 68 was settled affirmatively.

The following interpretation of Ulam's problem was also proved by Montejano in [4]: Let $1 \leq k \leq n$. If R is a compact $(n+1)$-manifold with boundary in $\mathbb{R}^{n+1}$ for which every k-dimensional plane H that meets the interior of R has $H \cap \partial R$ a $(k-1)$-sphere, then R is a convex (n+1)-body.

In [4], the following homological characterization of convexity which is closely related with Problem 68 is considered. Let $1 \leq k \leq n$ and let K be a compact subset of $\mathbb{R}^{n+1}$, suppose that for every k-dimensional plane H, $H \cap K$ is acyclic, then K is convex. This theorem was also obtained by Kosinski [7] and generalizes the following results of Aumann [1, 2], mentioned in the first edition of this book: a continuum K in $\mathbb{R}^3$ is convex provided that for each plane P, $P \cap K$ and $P - K$ is connected; a continuum K in $\mathbb{R}^n$ is convex if and only if $\mathbb{R}^n - K$ is connected and the intersection of K with each $(n-1)$-dimensional hyperplane is convex; and a closed subset K of $\mathbb{R}^n$ is convex if and only if $K \cap P$ is simply connected for all 2-dimensional hyperplanes P. Later Montejano and Shchepin [5] and [6] gave homological characterizations of convex sets in terms of acyclic support sets.

1. G. Aumann, Eine einfache Charakterisierung der konvexen Kontinuen in R^3, *Deutsche Mathematik* **1** (1936), 108.
2. G. Aumann, Über Schnitteigenschaften konvexer Punktmengen im R^3, *Deutsche Mathematik* **1** (1936), 162-165. Swets and Zeitlinger N.V. Amsterdam, 1966.
3. J. Schreier, Über Schnitte konvexen Flächen, *Bull. Int. Ac. Pol.* Series A (1933), 155-157.
4. L. Montejano, About a problem of Ulam concerning flat sections of manifolds, *Comment. Math. Helvetici* **65** (1990), 462-473.
5. L. Montejano and E. Shchepin, A characterization of convex sets in terms of acyclic support sets, *Bull. London Math. Soc.*, **28** (1996), 501-504.
6. L. Montejano and E. Shchepin, Topological tomography in convexity, *Bull. London Math. Soc.* 34 (2002), 353-358.
7. A. Kosinski, A theorem on families of acyclic sets and its applications, *Pacific J. Math.*, **12** (1962), 317-325

Luis Montejano

PROBLEM 69: MAZUR, ULAM

The problem of characterizing the spaces of type (B) among the metric spaces. There is given a complete metric space E with the following properties:

(1) If $p, q \in E$, there exists exactly one $x \in E$, such that x is a metric center of the couple (p, q);
(2) If $p, q \in E$, there exists exactly one $x \in E$, such that q is a metric center of the couple (p, x).

Is the space E isometric to a certain space of type (B)? [Every space of type (B) has the properties of (1) and (2).]

Definition of a metric center of a couple of points (p, q): We take the set of all points $x \in E$ such that $px + xq = pq$; we denote it by R. By R_1 we denote the set of all points $r \in R$ such that $rx \leq d(R)/2$ for every $x \in R$, where $d(R)$ is the diameter of the set R; we denote by R_{n+1} the set of all points $r \in R_n$ such that $rx \leq d(R_n)/2$ for all $x \in R_n$. One can show that the intersection $R_1 R_2 \cdots R_n \cdots$ contains at most one point; if such a point exists, we call it the metric center of the pair (p, q).

Addendum. The answer is negative.

S. Mazur
December 21, 1936

Commentary

Lobaczewski's geometry is a counterexample. A more elementary example is a hyperbolic paraboloid with distances measured on geodesics.

Stefan Rolewicz

PROBLEM 70: ULAM

Prove the following lemma: Let $f_1(x_1, x_2, \ldots, x_n; t_1, t_2, \ldots, t_r)$; $0 \leq x_i \leq 1$; $0 \leq t_j \leq 1$; $i = 1, \ldots, n$; $j = 1, \ldots, r$ be a polynomial with variables x_i and t_j real-valued and vanishing identically at the point $X = (0, 0, \ldots, 0; t_1, t_2, \ldots, t_r)$; ε a positive number. There exists then a polynomial f_2 in the same variables and constants K and ρ both positive and independent from ε (both K and $\rho = 1$?) such that the following conditions are satisfied.

(1) $|f_1(x_1, \ldots, x_n; t_1, \ldots, t_r) - f_2(x_1, \ldots, x_n; t_1, \ldots, t_r)| < \varepsilon$.

(2) The derivatives with respect to the variables x at the point $x_i = 1$; $i = 1, \ldots, n$ imitate the behavior of the polynomial; that is to say, if T' and T'' denote two sets of variables $t'_1, \ldots, t'_r$ and $t''_1, \ldots, t''_r$ so that $|f_2(x_1, \ldots, x_n; T') - f_2(x_1, \ldots, x_n; T'')| < \varepsilon$, then we have for every i:

$$\left| \left[\frac{\partial f_2(x_1, \ldots, x_n; t'_1, \ldots, t'_r)}{\partial x_i} \right]_{x_1 = x_2 = \ldots = x_n = 0} - \left[\frac{\partial f_2(x_1, \ldots, x_n; t''_1, \ldots, t''_r)}{\partial x_i} \right]_{x_1 = x_2 = \ldots = x_n = 0} \right| < K\varepsilon.$$

(3) The derivatives with respect to at least one of the variables x at the point $x_1 = x_2 = \ldots = 0$ are *essentially* different from zero. That is to say, there exist points T^* and T^{**} such that

$$\left| \left[\frac{\partial f_2(x_1, \ldots, x_n; T^*)}{\partial x_i} \right]_{x_1 = x_2 = \ldots = x_n = 0} - \left[\frac{\partial f_2(x_1, \ldots, x_n; T^{**})}{\partial x_i} \right]_{x_1 = x_2 = \ldots = x_n = 0} \right| > \rho.$$

From an affirmative solution, i.e., from this lemma, there would follow an affirmative answer to Hilbert's problem concerning introduction of analytic parameters in n-parameter groups. (The problem was solved for compact groups by von Neumann.)

PROBLEM 71: ULAM

Find all the permutations $f(n)$ of the sequence of natural integers which have the property that if $\{n_k\}$ is an arbitrary sequence of integers with a density α, then the sequence $f(n_k)$ has also a density α, in the set of all integers.

PROBLEM 72: MAZUR

Let E be a space of type (F) with the following property: If $Z \subseteq E$ is a compact set, then the smallest closed convex set containing Z is also compact. Is E then a space of type (F_0)? [See Problem 26 for a definition of (F_0).]

Commentary

Mazur's theorem states that the closed convex hull of a compact subset of a Banach space is compact [1]. Problem 72 then asks for a partial converse to this result. It was answered by Mazur and Orlicz [2] who showed that an F-space X is locally convex if and only if whenever $x_n \to 0$ in X, $t_n \geq)$ and $\sum t_n \leq 1$ then the series $\sum t_n x_n$ is bounded (i.e., has bounded partial sums). Thus if the convex hull of every compact set is bounded, then X is locally convex. In particular, spaces of type (F_0) are locally convex and the answer to Problem 72 is affirmative.

1. S. Mazur, Über die kleinste konvexe Menge, die eine gegebene kompakte Menge enthält, *Studia Math.* 2 (1930), 7–9.
2. S. Mazur and W. Orlicz, Sur les espaces métriques linéares (1), *Studia Math.* 10 (1948), 184–208.

N.J. Kalton

PROBLEM 73: MAZUR, ORLICZ

Let c_n be the smallest number with the property that if $F(x_1, \ldots, x_n)$ is an arbitrary symmetric n-linear operator [in a space of type (B) and with values in such a space], then

$$\sup_{\substack{\|x_i\| \leq 1, \\ i=1,2,\ldots,n}} \|F(x_1, \ldots, x_n)\| \leq c_n \sup_{\|x\| \leq 1} \|F(x, \ldots, x)\|.$$

It is known (Mr. Banach) that c_n exists. One can show that the number c_n satisfies the inequalities

$$\frac{n^n}{n!} \leq c_n \leq \frac{1}{n!} \sum_{k=1}^{n} \binom{n}{k} \cdot k^n.$$

Is $c_n = n^n/n!$?

Commentary

The answer to this problem is yes for any real normed linear spaces and it is now a standard fact in the field of infinite dimensional holomorphy. R.S. Martin proved that $c_n \le n^n/n!$ in his 1932 thesis [11] with the aid of an n-dimensional polarization formula. His argument was published a few years later in [15] by A.E. Taylor. Although this polarization formula was known to Mazur and Orlicz [12, p. 52], it appears that they used the case $x = 0$ rather than the case $x = -\sum_1^n h_k/2$, which gives the best estimate. Extremal examples in ℓ^1 and $L[0,1]$ showing that $c_n = n^n/n!$ are given in [5, 10], and [16]. For expositions, see [4, p. 48] and [13, p. 7]. Note that by the Hahn-Banach theorem, there is no loss of generality in this problem if all multilinear mappings are taken to be complex valued.

S. Banach [1] showed in 1938 that $c_n = 1$ when only real Hilbert spaces are considered. (He also assumed separability, though this assumption is not needed.) His result can be deduced quite easily from [3, Satz 9] or [9, Th. IV]. For modern expositions, see [2, p. 62], [6], or [8]. It is shown in [6] that Banach's result and an improvement by Szego of Bernstein's inequality for trigonometric polynomials are easily deduced from each other. For complex L^p-spaces, $1 \le p < \infty$, it is conjectured in [6] that

$$ c_n \le \left(\frac{n^n}{n!} \right)^{\frac{|p-2|}{p}}, $$

and this is proved when n is a power of 2. It also follows from [6] that

$$ c_n \le \frac{n^{n/2}(n+1)^{(n+1)/2}}{2^n n!} $$

holds for J^*-algebras [7]. (In particular, the space $C(S)$ of all continuous complex-valued functions on a compact Hausdorff space S and, more generally, any B^*-algebra is a J^*-algebra.) Since $c_n = 1$ for the space $C(S)$ with S a two point set [6, p. 154], it is natural to ask whether this holds for any compact Hausdorff space S. If so, then it is easy to deduce that the Bernstein inequality holds for polynomials on $C(S)$. (See [6, p. 149].)

A natural generalization of Problem 73 is the following: Let $k_1, \ldots, d_n$ be nonnegative integers whose sum is n and let $c(k_1, \ldots, k_n)$ be the smallest number with the property that if F is any symmetric n-linear mapping of one real normed linear space into another, then

$$ \sup_{\substack{\|x_i\| \le 1, \\ i=1,2,\ldots,n}} \|F(x_1^{k_1}, \ldots, x_n^{k_n})\| \le c(k_1, \ldots, k_n) \sup_{\|x\| \le 1} \|F(x, \ldots, x)\|. $$

where the exponents denote the number of coordinates in which the base variable appears. It is shown in [6] that if only complex normed linear spaces and complex scalars are considered, then

$$c(k_1, \ldots, k_n) = \frac{k_1! \cdots k_n!}{k_1^{k_1} \cdots k_n^{k_n}} \frac{n^n}{n!} \tag{1}$$

(where $0^0 = 1$) but there are many cases where (1) does not hold when real normed linear spaces are considered.

1. S. Banach, Über homogene Polynome in (L^2), *Studia Math.* 7 (1938), 36–44.
2. J. Bochnak and J. Siciak, Polynomials and multilinear mappings in topological vector spaces, *Studia Math.* 39 (1971), 59–76.
3. J.G. van der Corput and G. Schaake, Ungleichungen für Polynome und Trigonometrische Polynome, *Composito Math.* 2 (1935), 321–361; Berichtigung ibid., 3 (1963), 128.
4. H. Federer, *Geometric Measure Theory,* Springer-Verlag, Berlin-Heidelberg-New York 1969.
5. B. Grünbaum, Two examples in the theory of polynomial functionals, Riveon Lematematika 11 (1957), 56–60.
6. L.A. Harris, Bounds on the derivatives of holomorphic functions of vectors, *Analyse Fonctionnelle et Applications,* Leopoldo Nachbin editor, Hermann, Paris 1975.
7. _____, *Bounded symmetric homogeneous domains in infinite dimensional spaces,* Lecture Notes in Math. 364, Springer-Verlag, Berlin, 1974, 13–40.
8. L. Hörmander, On a theorem of Grace, *Math. Scand.* 2 (1954), 55–64.
9. O.D. Kellogg, On bounded polynomials in several variables, *Math. Zeit.* 27 (1928), 55-64.
10. J. Kopeć and J. Musielak, On the estimation of the norm of the n-linear symmetric operation, *Studia Math.* 15 (1955), 29–30.
11. R.S. Martin, Thesis, Cal. Inst. of Tech., 1932.
12. S. Mazur and W. Orlicz, Grundlegende Eigenschaften der polynomischen Operationen. I–II, *Studia Math.* 5 (1934), 50–68, 179-189.
13. L. Nachbin, *Topology on Spaces of Holomorphic Mappings,* Springer-Verlag, New York, 1969.
14. T.J. Rivlin, *The Chebyshev Polynomials,* Wiley, New York, 1974.
15. A.E. Taylor, Additions to the theory of polynomials in normed linear spaces, *Tôhoku Math. J.* 44 (1938), 302–318.
16. _____, (Review of a paper of Kopéc and Musielak), *Math. Reviews* 17 (1956), 512.
17. D.R. Wilhelmsen, A Markov inequality in several dimensions, *J. Approx. Theory* 11 (1974), 216–220.

Lawrence A. Harris

Mathematics Department

University of Kentucky

Lexington, Kentucky 40506

PROBLEM 74: MAZUR, ORLICZ

Given is a polynomial

$$W(t_1, \ldots, t_n) = \sum_{k_1 + \ldots + k_n = n} a_{k_1, \ldots, k_n} t_1^{k_1} \cdots t_n^{k_n}$$

in real variables $t_1, \ldots, t_n$, homogeneous and of order n; let us assume that $|W(t_1, \ldots, t_n)| \leq 1$ for all $t_1, \ldots, t_n$ such that $|t_1| + \ldots + |t_n| \leq 1$. Do we then have

$$|a_{k_1, \ldots, k_n}| \leq \frac{n^n}{k_1! \cdots k_n!}?$$

Commentary

The answer to this problem is yes and the solution is an easy consequence of the solution to Problem 73. Indeed, let F be the symmetric n-linear map on ℓ_n^1 such that $W(x) = F(x,\ldots,x)$ for all $x \in \ell_n^1$. Then applying the multinomial theorem [12, p. 52] with $x = t_1 e_1 + \ldots + t_n e_n$ and the uniqueness of the representation for $W(t_1,\ldots,t_n)$, we obtain

$$a_{k_1,\ldots,k_n} = \frac{n!}{k_1!\cdots k_n!} F(e_1^{k_1} \cdots e_n^{k_n}), \tag{2}$$

where $e_1,\ldots,e_n$ is the standard basis for ℓ_n^1. The desired estimate follows. Note that the problem of determining the best estimate $\alpha(k_1,\ldots,k_n)$ in Problem 74 is equivalent to the problem of determining $c(k_1,\ldots,k_n)$ in Problem 73; for, if F is a symmetric n-linear map satisfying $\|F(x,\ldots,x)\| \le 1$ for all $\|x\| \le 1$ and if $\|x_1\| \le 1,\ldots,\|x_n\| \le 1$, then the polynomial $W(t_1,\ldots,t_n) = F(x,\ldots,x)$, where $x = t_1 x_1 + \ldots + t_n x_n$, satisfies the hypotheses of Problem 74 and

$$a_{k_1,\ldots,k_n} = \frac{n!}{k_1!\cdots k_n!} F(x_1^{k_1},\cdots,x_n^{k_n}).$$

Thus

$$\alpha(k_1,\ldots,k_n) = \frac{n!}{k_1!\cdots k_n!} c(k_1,\cdots,k_n). \tag{3}$$

The general problem of obtaining estimates on the coefficients of polynomials in m variables which satisfy a given growth condition on R^m can be solved with the aid of the following generalized polarization formula: Let $W(t_1,\ldots,t_m)$ be any homogeneous polynomial of degree n and let $a_{k_1,\ldots,k_m}$ be the coefficient of $t_1^{k_1} \cdots t_m^{k_m}$ in its expansion. For each $i = 1,\ldots,m$, choose distinct real numbers $x_{i0},\ldots,x_{ik_i}$ and put

$$\Gamma_{ij} = \prod_{\ell \ne j}(x_{ij} - x_{i\ell}), \quad 0 \le j \le k_i,$$

with $\Gamma_{ij} = 1$, if $k_i = 0$. Then

$$a_{k_1,\ldots,k_m} = \sum \frac{W(x_{1j_1},\ldots,x_{mj_m})}{\Gamma_{1j_1}\cdots\Gamma_{mj_m}}, \tag{4}$$

where the sum is taken over all $0 \le j_1 \le k_1,\ldots,0 \le j_m \le k_m$. Note that one can convert any polynomial p of degree $\le n$ in $m-1$ variables to a homogeneous polynomial W of degree n in m variables by defining

$$W(t_1,\ldots,t_m) = t_m^n p\left(\frac{t_1}{t_m},\ldots,\frac{t_{m-1}}{t_m}\right).$$

One can obtain estimates on the left-hand side of (4) by estimating the right-hand side of (4) and minimizing. (A reasonable first choice is $x_{ij} = k_i/2 - j$.) To prove (4), observe that if $p(t)$ is a polynomial of degree $\leq k_i$, then the coefficient of t^{k_i} in the Lagrange interpolation formula for p is $\sum_{j=0}^{k_i} p(x_{ij})/\Gamma_{ij}$ and apply this to each variable of W.

For example, we show that the improved estimate

$$|a_{k_1,\ldots,k_n}| \leq \frac{n^n}{k_1! \cdots k_n!} r^\ell \tag{5}$$

holds in Problem 74, where

$$r = \frac{1 + e^{-2}}{2}, \quad \ell = \sum_{i=1}^n \left[\frac{k_i}{2}\right].$$

Indeed, choose $x_{i0} = 2$, $x_{i1} = 0$, $x_{i2} = -2$ for $i = 1, \ldots, \ell$, $x_{i0} = 1$, $x_{i1} = -1$ for $i = \ell + 1, \ldots, n - \ell$, and $x_{i0} = 0$ for $i = n - \ell + 1, \ldots, n$. Then by (4),

$$|a_{2\ldots21\ldots10\ldots0}| \leq \frac{1}{4^\ell} \sum_{j=0}^\ell \binom{\ell}{j} (n - 2j)^n \leq 2^{-\ell} n^n r^\ell, \tag{6}$$

where the last inequality follows from $(1 - 2j/n)^n \leq e^{-2j}$. Clearly

$$c(k_1, \ldots, k_n) \leq c(2, \ldots, 2, 1, \ldots, 1, 0, \ldots, 0),$$

and this together with (3) and (6) implies (5).

A related problem of interest is to find a Banach space analogue of Markov's theorems; that is, to find the smallest number $M_{n,k}$ with the property that if P is any polynomial of degree $\leq n$ mapping one real normed linear space into another, then

$$\sup_{\|x\| \leq 1} \|\hat{D}^k P(x)\| \leq M_{n,k} \sup_{\|x\| \leq 1} \|P(x)\|,$$

where $\hat{D}^k p(x) y = d^k/dt^k p(x + ty)\,|_{t=0}$. It is not difficult to show that

$$T_n^{(k)}(1) \leq M_{n,k} \leq 2^{2k-1} T_n^{(k)}(1), \tag{7}$$

where T_n is the Chebyshev polynomial of degree n. (See [14, p. 119].) Indeed, let P be any real-valued polynomial of degree $\leq n$ on a real normed linear space and suppose $|P(x)| \leq 1$ for all $\|x\| \leq 1$. Let $\|x\| \leq 1$, $\|y\| \leq 1$ and $-1 \leq s \leq 1$. Define $q(t) = P(\phi(t))$, where $\phi(t) = [x - sy + t(x + sy)]/2$, and note that $\|\phi(t)\| \leq (1 + t)/2 + (1 - t)/2 = 1$ when $-1 \leq t \leq 1$. Then q is a polynomial of degree $\leq n$ satisfying $|q(t)| \leq 1$ for $-1 \leq t \leq 1$, so $|q^{(k)}(1)| \leq |T^{(k)}(1)|$ by [14, 1.5.11] and clearly $q^{(k)}(1) = 2^{-k} \hat{D}^k P(x)(x + sy)$. Hence the map $s \to \hat{D}^k T_n^{(k)}(1)$ on $[0, 1]$ and $\hat{D}^k P(x) y$ is the coefficient of s^k in this polynomial. Therefore,

$$|\hat{D}^k P(x)y| \le 2^{k-1}[2^k T_n^{(k)}(1)]$$

by [14, p. 57]. Thus (7) follows by the Hahn-Banach theorem.

Note that the value of $M_{n,k}$ is unchanged when only real-valued polynomials on ℓ_2^1 are considered. It is shown in [6] and [9] that $M_{n,1} = n^2$ when only real Hilbert spaces are considered and it would be interesting to know whether $M_{n,k} = T_n^{(k)}(1)$ for all $1 \le k \le n$ in this case. See also [17].

1. S. Banach, Über homogene Polynome in (L^2), *Studia Math.* 7 (1938), 36–44.
2. J. Bochnak and J. Siciak, Polynomials and multilinear mappings in topological vector spaces, *Studia Math.* 39 (1971), 59–76.
3. J.G. van der Corput and G. Schaake, Ungleichungen für Polynome und Trigonometrische Polynome, *Composito Math.* 2 (1935), 321–361; Berichtigung ibid., 3 (1963), 128.
4. H. Federer, *Geometric Measure Theory,* Springer-Verlag, Berlin-Heidelberg-New York 1969.
5. B. Grünbaum, Two examples in the theory of polynomial functionals, Riveon Lematematika 11 (1957), 56–60.
6. L.A. Harris, Bounds on the derivatives of holomorphic functions of vectors, *Analyse Fonctionnelle et Applications,* Leopoldo Nachbin editor, Hermann, Paris 1975.
7. _____, *Bounded symmetric homogeneous domains in infinite dimensional spaces,* Lecture Notes in Math. 364, Springer-Verlag, Berlin, 1974, 13–40.
8. L. Hörmander, On a theorem of Grace, *Math. Scand.* 2 (1954), 55–64.
9. O.D. Kellogg, On bounded polynomials in several variables, *Math. Zeit.* 27 (1928), 55-64.
10. J. Kopéc and J. Musielak, On the estimation of the norm of the n-linear symmetric operation, *Studia Math.* 15 (1955), 29–30.
11. R.S. Martin, Thesis, Cal. Inst. of Tech., 1932.
12. S. Mazur and W. Orlicz, Grundlegende Eigenschaften der polynomischen Operatoren I–II, *Studia Math.* 5 (1935), 50–68, 179-189.
13. L. Nachbin, *Topology on Spaces of Holomorphic Mappings,* Springer-Verlag, New York, 1969.
14. T.J. Rivlin, *The Chebyshev Polynomials,* Wiley, New York, 1974.
15. A.E. Taylor, Additions to the theory of polynomials in normed linear spaces, *Tôhoku Math. J.* 44 (1938), 302–318.
16. _____, (Review of a paper of Kopéc and Musielak), *Math. Reviews* 17 (1956), 512.
17. D.R. Wilhelmsen, A Markov inequality in several dimensions, *J. Approx. Theory* 11 (1974), 216–220.

Research supported in part by the National Science Foundation.

Lawrence A. Harris

University of Kentucky

PROBLEM 75: MAZUR

In Euclidean n-dimensional space E, or, more generally, in a space of type (B) there is given a polynomial $W(x)$. α is a number $\ne 0$. If the polynomial $W(x)$ is bounded in an ε-neighborhood of a certain set $R \subset E$, is it then bounded in a δ-neighborhood of the set αR (which is the set composed of elements αx for $x \in R$)? (See Problem 55.)

Addendum. From the solution of Problem 55, it follows that the theorem is true in the case of a Euclidean space.

Mazur

PROBLEM 76: MAZUR

Given in 3-dimensional Euclidean space is a convex surface W and point 0 in its interior. Consider the set V of all points P defined by the property that the length of the interval $P0$ is equal to the area of the plane section of W through 0 and perpendicular to this interval. Is the set V convex?

PROBLEM 77: ULAM

Prize for (a), one bottle of wine, S. Eilenberg

(a) Let A and B e two topological spaces such that the spaces A^2 and B^2 are homeomorphic. Is then the space A homeomorphic to the space B?
(b) Let A and B be two metric spaces such that A^2 is isometric to B^2. Is A isometric with B?
(c) Let A and B be two abstract groups such that A^2 and B^2 form isomorphic groups. Is A isomorphic with B?

[We understand by A^2 (resp. B^2) the set of ordered pairs of elements of the set A (or B).] A topology [or, in Problem (c), the group operation] in such sets is defined, for example, in the "Euclidean" manner: by the square root of the sum of squares of the distances between projections.

Commentary Parts (a) and (b)

A number of papers have been devoted to part (a). In 1947, R. H. Fox [7] gave an example of two nonhomeomorphic compact 4-dimensional manifolds whose cartesian squares are homeomorphic. In 1960, J. Glimm [10] noticed that the cartesian square of the contractable open manifold, which is not homeomorphic to E^3, described by J.H.C. Whitehead in [17], is homeomorphic to E^6. This result was generalized to a class of contractable open 3-manifolds by D.R. McMillan, Jr. [13]. In 1964, K.N. Kwun proved [11] that if α is a simple arc in E^n and β is a simple arc in E^m, then $E^n/\alpha \times E^m/\beta$ is homeomorphic to E^{n+m}. Since there exist wild arcs in E^n for $n \geq 3$ such that their complement is not simply connected, the result of Kwun gives us a class of "cartesian elements" or roots of E^{2n}, for $n \geq 3$, which are not open topological manifolds. Another class of cartesian elements of E^{2n} was constructed by A.J. Boals [3] in 1970. This class includes the famous dog bone space of Bing [1, 2]. K.W. Kwun and F. Raymond constructed nontrivial "cartesian elements" of the cube $[0, 1]^{2n}$, for $n \geq 2$ [12]. In 1978, an analogous result for the Hilbert cube was published by Cerin [6]. H. Torunczyk [16] proved a number of very general results which imply, for example, that if A and B are cell-like finite dimensional continua in the Hilbert cube Q, the $Q/A \times Q/B$ is homeomorphic to Q.

It is not known whether there exist nonhomeomorphic, 3-dimensional, compact manifolds A and B such that A^2 and B^2 are homeomorphic. It is not known whether there exist 3-dimensional nonhomeomorphic polyhedra A and B so that A^2 is homeomorphic to B^2.

There are some cases for which part (a) has a positive answer. R.H. Fox [7] showed that the answer is yes if A and B are 2-dimensional compact manifolds with or without boundary. Recently, W. Rosicki, in a dissertation [15] upon which this commentary is based, gave an affirmative answer in case A and B are compact 2-dimensional polyhedra. Related problems were considered by Borsuk [4], H. Patkowska [14], Furdzik [8], and Cauty [5].

Part (b) was solved in the negative by G. Fournier [9]. However, it is open whether there is an affirmative solution to (b) in case A and B are complete metric spaces. In fact, part (b) is open in the case where A and B are assumed to be compact.

1. R.H. Bing, A decomposition of E^3 into points and tame arcs such that the decomposition space is topologically different from E^3, *Ann. of Math.* 65 (1957), 484–500.
2. _____, The Cartesian product of a certain nonmanifold and a line is E^4, *Ann. of Math.* 70 (1959), 399–412.
3. A.J. Boals, Non-manifold factors of Euclidean space, *Fund. Math.* 68 (1970), 159–177.
4. K. Borsuk, Sur la décomposition des polyèdres en produits cartesiens, *Fund. Math.* 31 (1938), 137–148.
5. R. Cauty, Sur les homéomorphismes de certain produits de courbes, *Bull. Acad. Polon. Sci. Sér. Sci. Math.,* 27 (1979), 413-416.
6. Z. Čerin, Hilbert cube modulo an arc, *Fund. Math.* 101 (1978), 111–119.
7. R.H. Fox, On a problem of S. Ulam concerning Cartesian products, *Fund. Math.* 34 (1947), 27S–287.
8. Z. Furdzik, On the uniqueness of decomposition into Cartesian product of two curves, *Bull. Acad. Polon. Sci.,* 14 (1966), 57–61.
9. G. Fournier, On a problem of S. Ulam, *Proc. Amer. Math. Soc.,* 29 (1971), 622.
10. J. Glimm, Two Cartesian products which are Euclidean spaces, *Bull. Soc. Math. France,* 88 (1960),131–135.
11. K.W. Kwun, Product of Euclidean spaces modulo an arc, *Ann. of Math.,* 79 (1964), 104–108.
12. K.W. Kwun and F. Raymond, Factors of cubes, *Amer. J. Math.,* 84 (1962), 433–440.
13. D.R. McMillan, Jr., Cartesian products of contractable open manifolds, *Bull. Amer. Math. Soc.,* 67 (1961), 510–514.
14. H. Patkowska, On the uniqueness of the decomposition of finite dimensional ANR's into Cartesian products of at most 1-dimensional spaces, *Fund. Math.,* 58 (1966), 89–110.
15. W. Rosicki, O Problemie Ulama Dotyczacym Kartezjanskich Kwadratou Wieloschianow 2-wymiarowych, *Rozprowo Doktorska,* Gdansk, 1979; On a problem of S. Ulam concerning Cartesian squares of 2-dimensional polyhedra., *Fund. Math.* 127 (1987), 101-125.
16. H. Toruńczyk, On CE-images of the Hilbert cube and characterization of Q-manifolds, *Fund. Math.,* 106 (1980), 31-40.
17. J.H.C. Whitehead, A certain open manifold whose group is unity, *Quart. J. Math.,* 6 (1935), 268–279.

R. Daniel Mauldin

Commentary—Part (c)

For abelian (i.e., commutative) groups this was one of the three "test problems" in the first edition of [4]. I did not know of its appearance in the Scottish Book. My idea was to show how Ulam's theorem for countable torsion groups could really be used to answer explicit questions.

Jónsson [3] gave a negative answer with A and B torsion-free of rank two. Crawley [2] gave an example for torsion groups (of course uncountable). Corner [1] exhibited a countable torsion-free abelian group G which is isomorphic to $G \oplus G \oplus G$ but not to $G \oplus G$—this is even more spectacular.

1. A.L.S. Corner, On a conjecture of Pierce concerning direct decompositions of torsion-free abelian groups, *Proc. of Coll. on Abelian groups,* Budapest, 1964, 43–48.
2. P. Crawley, Solution of Kaplansky's test problems for primary abelian groups, *J. Alg.* 2 (1965), 413–431.
3. B. J'onsson, On direct decompositions of torsion-free abelian groups, *Math. Scand.,* 5 (1957), 230–235.
4. I. Kaplansky, *Infinite Abelian groups,* U. of Michigan Press, 1954, 1969.

I. Kaplansky

PROBLEM 78: STEINHAUS

August 2, 1935

Find all the surfaces with following property: Through every point of the surface there lie two curves congruent, respectively, to two given curves A and B. Compare Problem 61. (Such a surface is, for example, a cylinder: the curves A and B are here a circle and a straight line.)

PROBLEM 79: MAZUR, ORLICZ

A polynomial $y = U(x)$ maps in a one-to-one fashion, a space X of type (B) onto a space Y of type (B); the inverse of this mapping $x = U^{-1}(y)$ is also a polynomial. Is the polynomial $y = U(x)$ of first degree? Not decided even in the case when X and Y are a Euclidean plane; in this case, the question is given for a one-to-one mapping $t' = \phi(t,s)$, $s' = \psi(t,s)$ of a plane onto itself where $\phi(t,s)$, $\psi(t,s)$ are polynomials; the inverse mapping has also the form $t = \Phi(t',s')$, $s = \Psi(t',s')$ where $\Phi(t',s')$, $\psi(t',s')$ are polynomials. Is the mapping affine; that is to say, of the form $t' = a_1 t + b_1 s + c_1$, $s' = a_2 t + b_2 s + c_2$ where $a_1 b_2 - a_2 b_1 \neq 0$?

Addendum. Trivial. In the Euclidean space:

$$y_1 = x_1 + k$$

$$y_2 = x_2 + \phi_2(x_1)$$

$$y_3 = x_3 + \phi_3(x_1, x_2)$$

$$\vdots$$

$$y_n = x_n + \phi_n(x_1, \ldots, x_{n-1})$$

where k is an arbitrary constant and $\phi_2, \ldots, \phi_n$ are arbitrary polynomials in their variables. The inverse mapping is obvious at once.

PROBLEM 80: MAZUR

Let E be a complete metric space; we denote by E^∞ a complete metric space formed by the set of all sequences $\{e_n\}$ of elements of E. By a distance between two such sequences $\{e'_n\}$, $\{e''_n\}$ we understand the number

$$\sum_{n=1}^\infty 2^{-n} \frac{(e'_n, e''_n)}{1 + (e'_n, e''_n)}$$

[For $e', e'' \in E$ we denote by (e', e'') the distance between the elements e', e'']. If R is a given set contained in E, then we denote by R_δ the set of all sequences $\{r_n\}$ of elements of R, and by R_σ the set of all sequences $\{r_n\}$ of elements of R such that $r_n = r_0$ almost always; r_0 is a fixed element of R. Is it true that: If the set R is an F_σ set but not closed, then R_δ is an $F_{\sigma\delta}$ set but not an F_σ; if the set R is an $F_{\sigma\delta}$ but not an F_σ, then R_σ is and $F_{\sigma\delta\sigma}$ but not an $F_{\sigma\delta}$; more generally, if R is an $F_{2\xi+1}$ but not an $F_{2\xi}$, then R_δ is an $F_{2\xi+2}$, but not an $F_{2\xi+1}$, and if R is an $F_{2\xi+2}$ but not an $F_{2\xi+1}$, then R_σ is an $F_{2\xi+3}$ but not an $F_{2\xi+2}$ ($F_0 = F$, $F_1 = F_\sigma$, $F_2 = F_{\sigma\delta}$, $F_3 = F_{\sigma\delta\sigma}$, ...)? Investigate in particular the case when the space E is compact or of type (B) or of type (F).

Commentary

If X is a metric space, let $\mathscr{F}_0(X)$ be the family of all closed subsets of X. For each ordinal $\alpha > 0$, let

$$\mathscr{F}_\alpha(X) = \{K : K = \bigcup F_n, \text{ where each } F_n \in \bigcup \{\mathscr{F}_\beta(X) : \beta < \alpha\}\},$$

if α is odd, and let

$$\mathscr{F}_\alpha(X) = \{K : K = \bigcap F_n, \text{ where each } F_n \in \bigcup \{\mathscr{F}_\beta(X) : \beta < \alpha\}\},$$

if α is even. Limit ordinals are considered even.

Thus, in the terminology of this problem it is true that if $R \in \mathscr{F}_{2\alpha+1}(E)$, then $R_\delta \in \mathscr{F}_{2\alpha+2}(E^\infty)$ and if $R \in \mathscr{F}_{2\alpha}(E)$, then $R_\sigma \in \mathscr{F}_{2\alpha+1}(E^\infty)$, for $0 \le \alpha < \omega_1$. The problem is whether these estimates are sharp. Specifically,

(a) if $R \in \mathscr{F}_1(E) - \mathscr{F}_0(E)$, does $R_\delta \in \mathscr{F}_2(E^\infty) - \mathscr{F}_1(E^\infty)$?
(b) if $R \in \mathscr{F}_2(E) - \mathscr{F}_1(E)$, does $R_\sigma \in \mathscr{F}_3(E^\infty) - \mathscr{F}_2(E^\infty)$?
(c) if $R \in \mathscr{F}_{2\alpha+1}(E) - \mathscr{F}_{2\alpha}(E)$, does $R_\delta \in \mathscr{F}_{2\alpha+2}(E^\infty) - \mathscr{F}_{2\alpha+1}(E^\infty)$?
(d) if $R \in \mathscr{F}_{2\alpha+2}(E) - \mathscr{F}_{2\alpha+1}(E)$, does $R_\sigma \in \mathscr{F}_{2\alpha+3}(E^\infty) - \mathscr{F}_{2\alpha+2}(E^\infty)$?

If all of these questions have affirmative answers, then one would have an elegant method of generating Borel sets of arbitrarily high class. One could simply take a subset of E whose structure is known and iterate the described procedure through

$E^{\infty}, (E^{\infty})^{\infty}, \ldots$. However, the solutions to these general problems are unknown. There are some positive solutions in specific cases.

One can show that if E is compact, then the answer to (a) is yes as follows:

Let $R = \bigcup K_n$, where each K_n is compact and suppose $R_\delta = \bigcup T_n$, where each T_n is closed in E^{∞}. Suppose R is not compact. Then for each n, $R \neq \pi_n(T_n)$, since T_n is compact and $\pi_n(T_n)$ is compact where π_n is the projection of E^{∞} onto the nth coordinate. For each n, let $r_n \in R - \pi_n(T_n)$. Then $\{r_n\}_{n=1}^{\infty} \in R_\delta - \bigcup T_n$. This contradiction establishes (a) in case E is compact.

Using a variation of the procedure described by Mazur and a wonderful application of Brouwer's fixed point theorem, Sikorski [2] and Engelking, Holsztynski, and Sikorski [1] gave a positive solution to the iterative process in a special case. Sikorski [3] also used the Brouwer fixed point theorem to give a specific example of an analytic set which is not a Borel set. In [1], Sikorski raised a problem which is closely associated with Problem 80.

1. R. Engelking, W. Holsztynski, and R. Sikorski, Some examples of Borel sets, *Coll. Math.*, 15 (1966), 271–274.
2. R. Sikorski, Some examples of Borel sets, *Coll. Math.*, 5 (1958), 170–171.
3. _____, On an Analytic Set, *Bull. Acad. Sci. Pol.*, 14 (1966), 15–16.

R. Daniel Mauldin

PROBLEM 81: STEINHAUS
August 6, 1935

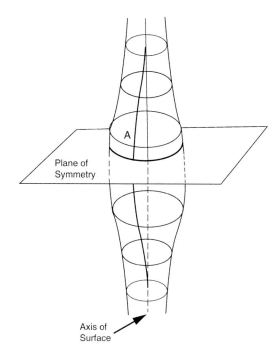

A hyperbolic paraboloid and a plane are composed, in *two* ways, of curves which are imbedded in the surface $(AA;BB)$, straight lines and parabolas. Do there exist other surfaces of this kind? Are they composed of (AB), (CD)? Is it true that such surfaces, namely, all surfaces having at each point two intersecting curves congruent to A and B, respectively (*exceptis excipiendis*), are necessarily of the form $z = f(x) + g(y)$? (The plane, sphere, and circular cylinder are considered trivial.) (Compare Problems 44 and 61)

Commentary

This problem is not clearly stated. We will make an attempt to interpret it. Let us disregard the plane since the plane as an example of a composed surface is usually trivial. With this exception, the first sentence says that a hyperbolic paraboloid is composed in two ways of straight lines and parabolas. (That the hyperbolic paraboloid and the hyperboloid of one sheet are the only doubly ruled surfaces is well known. See, for example, M. Spivak, *A Comprehensive Introduction to Differential Geometry,* Vol. 3, 2nd edition, page 345, problem 14, Publish or Perish, Inc., 1979.)

The notation $(AA;BB)$ might have the following interpretation. Suppose one can find on a surface four families of curves (A_1), (A_2), (B_1), and (B_2) such that

a. All curves of the family (A_1) are congruent to one another; all curves of the family (A_2) are congruent to one another and to the curves (A_1);
b. Likewise for (B_1) and (B_2);
c. Through each point of the surface there pass four curves, one from each family.

One then needs to show that the hyperbolic paraboloid has this property where (A_1) and (A_2) consist of straight lines and (B_1) and (B_2) consist of parabolas. Of course, with this interpretation, any surface of revolution with a plane (Fig. 81.1) of symmetry perpendicular to the axis of revolution has these properties. Such a surface is fibered by circles parallel to the plane of symmetry and if we take any curve A at all on the circle which meets each circle in one point, we see that the surface is generated by congruent copies of A. And, provided A is not itself symmetric in the plane of symmetry of the surface, there will be a second system of curves on the surface congruent to A (viz. the reflections of the first system in the plane of symmetry). In fact, we see that the surface contains a noncountable infinity of pairs of curve-systems (A_1, A_2), $(B_1, B_2, \ldots$ having the properties (a) and (b) above with one curve of every system passing through any chosen point of the surface. It may be a reasonable conjecture that only the hyperbolic paraboloid, other than surfaces of revolution, has properties (a), (b), and (c).

In the third sentence, Steinhaus asks if there are surfaces composed of (AB), (CD). He might mean that $(AB) \equiv (AA;BB)$ and then is asking if the structure with the structure $(AA;BB)$ can have also a different structure $(CC;DD)$.

A Reviewer [Ed. who requested to remain anonymous.]

PROBLEM 82: STEINHAUS

August 6, 1935

$f(t)$ is independent (in the sense of correlation) from $y_1(t), y_2(t), \ldots, y_n(t)$ $(0 \le t \le 1)$, if, for every function of n-variables $F(y_1, y_2, \ldots, y_n)$ and for every 4-tuple of numbers $(\alpha_1, \beta_1, \alpha_2, \beta_2)$ the sets which are defined as follows: $A = E_t(\alpha_1 \le f(t) \le B_1)$, $B = E_t(\alpha_2 \le F(y_1(t), \ldots, y_n(t)) \le \beta_2)$ have the property that $|AB| = |A| \cdot |B|$.]

Problems

(1) Is a set of functions mutually independent (that is to say, each independent of all the other n) at most countable?

(2) Does a system like that have to be complete and orthogonal, or only complete?

Remark: The notion of independence introduced above is what natural scientists call "complete lack of correlation." (Their definitions are, however, not too precise.)

Addendum. Under the assumption that the functions which are independent are integrable, together with their ℓth power, we have the following relation:

$$\int_0^1 y_{k_1}(t) y_{k_2}(t) \cdots y_{k_\ell}(t) dt = \prod_{i=1}^{\ell} \int_0^1 y_{k_i}(t) dt.$$

It follows immediately that the system $\{\phi_i(t)\}$ where $\phi_i(t) = y_i(t) - \int_0^1 y_i(t)$ is orthogonal. If we assume that $y_i(t) \in L$, then the system is "lacunary" (and therefore cannot be complete). Lacunarity follows in this case from the relation

$$\int_0^1 \left| \sum_{i=1}^n \phi_i(t) \right| dt \ge \sqrt{M \sum_{i=1}^n \int_0^1 \phi_i^2(t) dt},$$

where M does not depend on n nor does it depend on the sequence $\int_0^1 \phi_i^2(t) dt$.

October 12, 1935

Commentary

The solution to part (1) is as follows. Let I be a set of mutually independent functions; then all but countably many members of I are constant almost everywhere. To see this, let $I' = \{\arctan \phi : \phi \in I\}$. Then $I' \subset L^2([0,1])$. Let $I'' = \{\phi \in I' : \text{variance}(\phi) > 0\}$. If $\arctan \phi \in I' - I''$, then ϕ is constant almost everywhere. Since $B = \{X - E(X)/\sigma(X) : X \in I''\}$, is an orthonormal subset of $L^2([0,1])$, B, and therefore I'', is countable.

One answer to (2) is given by Steinhaus at the end of the problem. The following may be a more elementary argument for the fact that a sequence of independent random variables $X_i \in L^2([0,1])$ cannot span L^2.

We can assume without loss of generality that X_i is the constant 1 and that $\sigma(X_i) > 0$, for $i > 1$. For each $i > 1$, set $Y_i = X_i - E(X_i)/\sigma(X_i)$. Then $1, Y_1, Y_2, \dots$ is an independent, orthonormal sequence. Also, $Y_2 \cdot Y_3$ is orthogonal to each of these functions and, by independence, $E(Y_2^2 Y_3^2) = E(Y_2^2)E(Y_3^2) = 1$. But, if the X_i's spanned L^2, then we would have $Y_2 \cdot Y_3 = 0$, which is inconsistent.

D. Stroock

PROBLEM 83: AUERBACH

We assume that a continuous function $f(x)$ satisfies at every point the condition

$$\varlimsup_{h \to 0} \left| \frac{f(x+h) - f(x)}{h^a} \right| < M$$

(M a constant, $0 < a < 1$, a constant). Does the function $f(x)$ satisfy a Hölder condition? (It is easy to prove that in every interval of a certain dense set of intervals, Hölder condition holds with the exponent a and with the same constant.)

Addendum. The answer is negative. We define the function $f(x)$ as a triangular one in intervals $1/n - x_n, 1/n + x_n$ with the height $1/n$.

Marcinkiewicz

PROBLEM 84: AUERBACH

One assumes that for a convex surface in the 3-dimensional space all its plane sections by planes going through a fixed point 0 inside the surface are projectively equivalent. Is this surface an ellipsoid?

PROBLEM 85: BANACH

Does there exist a sequence of measurable functions $\{\phi_n(t)\}$ $(0 \le t \le 1)$, belonging to L^2, orthogonal, normed, complete, and such that the development of every polynomial is divergent almost everywhere?

(b)   The same question if, instead of polynomials, we consider analytic functions for $0 \le t \le 1$.

One can prove that the answer to Question (a) is an affirmative one if we admit only polynomials of degree less than n (n arbitrary given ahead of time).

Second Edition Commentary

Constructions of divergent Fourier series go back to A.N.Kolmogorov and D.E.Menshov. In particular, Menshov (1923) discovered that there is an orthonormal system of functions $\{\varphi_n\}$ such that some series

$$\sum c(n)\varphi_n \ , \{c(n)\} \in l^2$$

diverges almost everywhere (a.e.)

The subject was developed substantially in the beginning of 1960s. This progress was inspired by Kolmogorov-Zahorski's theorem which says that a trigonometric L^2- series may diverge a.e. after a suitable rearrangement of its terms.

See [O-75] for a presentation of results obtained in the area.

In particular, one can find there (p.115) the following result of A. Krantsberg [K-74]: one can construct a complete orthonormal system in $L^2(I)$ such that Fourier expansion of every (nontrivial) continuous function with respect to this system diverges a.e.

This result answers Banach's Problem 85 and also its stronger version - Problem 186.

References

[K-74] A.S.Krantsberg, On divergent orthogonal Fourier series, Mat. Sb. 93 (1974), 540–553 (in Russian).

[O-75] A.M.Olevskii, Fourier series with Respect to General Orthogonal Systems. Springer, 1975.

A. Olevskii

PROBLEM 86: BANACH

Given a sequence of functions $\{\phi_n(t)\}$ orthogonal, normed, measurable, and uniformly bounded; can one always complete it, using functions with the same bound, to a sequence which is orthogonal, normed, and complete? Consider the case when infinitely many functions are necessary for completion.

Second Edition Commentary

K. S. Kazarian, in his article *On a problem of S. Banach from the Scottish Book*, Proc. Amer. Math. Society 110 (1990), 881-887 showed that the answer is no. Working with the Haar system of functions, it is shown that there is an ONS $\{\Phi_n\}$ on the interval $[0,1]$ such that the dimension of the orthogonal complement is infinite and there is no uniformly bounded orthonormal basis for the orthogonal complement. In fact, stronger theorem are proven in the paper. Kazarian mentions that Banach probably knew how to construct counterexamples in case the complement is finite dimensional. Kazarian mentions a letter from B. S. Kašin. in which it is related that he and A. M. Olevskii had obtained similar results. Finally, Kazarian mentions a letter from Z. Cielsielski to Kašin where it is stated that S. Kaczmarz solved the problem in 1936 with the help of the Banach method. Kazarian was unable to find any published result of Kaczmarz along these lines. I also made queries about Kaczmarz's result, but no one was able to provide a reference. Perhaps it is lost.

R. Daniel Mauldin

PROBLEM 87: BANACH

Let $y = U(x)$ be an operation which is continuous and satisfies a Lipschitz condition. The operation is defined in L^β ($\beta \geq 1$) and its image is also contained in L^β. We assume that for a certain $\alpha > \beta$ there exists a constant M_α such that if $x \in L^\alpha$ then $U(x) \in L^\alpha$ and $\|U(x)\|_\alpha \leq M_\alpha \|x\|_\alpha$. Show that for every γ such that $\beta < \gamma < \alpha$ there exists an M_γ with the property: If $x \in L^\gamma$, then $U(x) \in L^\gamma$ and $\|U(x)\|_\gamma \leq M_\gamma \|x\|_\gamma$. This theorem is true under the additional assumption that U is a linear operation (follows from a theorem of M. Riesz). Banach showed that the theorem is true if $\alpha = \infty$.

$$[\ \|x(t)\|_\gamma = (\textstyle\int_0^1 |x(t)|^\gamma dt)^{1/\gamma}.]$$

Commentary

The fundamental objects of linear and nonlinear functional analysis are operators which map one Banach space (or more generally, topological vector space) into another. One of the methods of studying operators is an interpolation theory, called the theory of interpolation spaces. This theory has been applied to other branches of analysis (e.g., Fourier series, Schauder basis, partial differential equations, numerical analysis, approximation theory), but it is also of considerable interest in itself. In order to explain the essence of these questions let us define linear interpolation spaces.

Let A, B, A_i, B_i, $i = 0, 1$, be complex Banach spaces (or quasi-Banach spaces, i.e., topological vector spaces which are complete and locally bounded) and let A, A_i, $i = 0, 1$ be continuously embedded into some fixed Hausdorff topological vector space $\mathscr{A}$, likewise B, B_i, $i = 0, 1$ into $\mathscr{B}$; A is intermediate between A_0 and A_1, and B is intermediate between B_0 and B_1 in the sense that

$$A_0 \cap A_1 \subset A \subset A_0 + A_1, B_0 \cap B_1 \subset B \subset B_0 + B_1,$$

with the continuous inclusions ([1], p. 24–25).

The pair of spaces (A, B) is called a linear interpolation pair with a constant $C > 0$ between the pairs of spaces (A_0, B_0) and (A_1, B_1) if for any linear continuous operator L from A_i into B_i, $i = 0, 1$, L (or its appropriate unique extension $\hat{L}$) is a linear continuous operator from A into B and $\|L\|_{A \to B}$ (or $\|\hat{L}\|_{A \to B}$) is majorized by $C \max(\|L\|_{A_0 \to B_0}, \|L\|_{A_1 \to B_1})$. If $A_i = B_i$, $i = 0, 1$ and $A = B$, we shall say shortly that the space A is a linear interpolation space with constant $C > 0$ between the spaces A_0 and A_1.

The first linear interpolation theorem was the M. Riesz theorem [22] of 1926, formulated as inequalities for bilinear forms. This theorem was improved and presented in an operator form by G.O. Thorin [26], who used the classical method of analytic functions (the three-line theorem). Thorin's theorem states that (L^p, L^q) is a linear interpolation pair with constant 1 between the pairs (L^{p_0}, L^{q_0}) and (L^{p_1}, L^{q_1}), where $1/p = (1 - \theta)/p_0 + \theta/p_1$, $1/q = (1 - \theta)/q_0 + \theta/q_1$, $0 < \theta < 1$, and is now generally known as the Riesz-Thorin theorem. A further important generalization was the theorem of J. Marcinkiewicz. Its proof, obtained by the real method, was published by A. Zygmund [28]. Next, E.M. Stein and G. Weiss [24] have proved an important generalization of the Riesz-Thorin and Marcinkiewicz theorems. All those theorems, however, concern L^p spaces or Lorentz L^{pq} spaces. Another theorem which should be mentioned ! here, although not known in the literature, is the Orlicz interpolation theorem [18] of 1954 exceeding L^p spaces and concerning the interpolation of continuous function spaces and Lipschitz function spaces.

The theory of linear interpolation was created and has been developed by many authors in the last twenty years, among them J.L. Lions, E. Gagliardo, A.P. Calderon, S.G. Krein, and J. Peetre. Almost all information on this problem, its applications and bibliography can be found in the following monographs: P.L. Butzer and H. Berens [3], J.L. Lions and F. Magenes [7], J. Bergh and J. Löfström [1], H. Triebel [27], S.G. Krein, Ju. I. Petunin, and E.M. Semenov [5].

Problem 87 is the first problem concerning nonlinear interpolation. Before giving a positive solution of the problem we shall introduce some definitions which will simplify the formulation of the problem and the theorems connected with this problem.

The pair of spaces (A, B) is called a semi-Lipschitz (Lipschitz) interpolation pair with constant $C > 0$ between the pairs (A_0, B_0) and (A_1, B_1) if for any operator U (possibly nonlinear) which maps A_i into B_i, $i = 0, 1$ and satisfies the conditions

$$\|Ua - Ub\|_{B_0} \le M_0 \|a - b\|_{A_0} \qquad a, b \in A_0$$

$$\|Ua\|_{B_1} \le M_1 \|a\|_{A_1} \qquad a \in A_1$$

$$[\, \|Ua - Ub\|_{B_1} \le M_i \|a - b\|_{A_i} \qquad a, b \in A_i, \ i = 0, 1 \,]$$

then U (or its appropriate unique extension, denoted as U instead of $\hat{U}$) maps A into B and

$$\|Ua\|_B \le C \max(M_0, M_1) \|a\|_A \qquad a \in A$$

$$[\, \|Ua - Ub\|_B \le C \max(M_0, M_1) \|a - b\|_A \qquad a, b \in A \,].$$

If $A_i = B_i$, $i = 0, 1$ and $A = B$, we can say shortly that the space A is a semi-Lipschitz (respectively Lipschitz) interpolation space with constant $C > 0$ between A_0 and A_1.

Using these formulations, Banach's problem can be presented as follows:

Is $L^\gamma[0, 1]$ a semi-Lipschitz interpolation space

with constant $C > 0$ between $L^\beta[0,1]$ and $L^\alpha[0,1]$ for   (B)

any γ, $1 \le \beta < \gamma < \alpha \le \infty$?

Banach notes that he proved this theorem for $\alpha = \infty$; it is not known whether this proof was published.

In [11] the problem of the Riesz-Thorin theorem for Lipschitz operators is considered. The problem in this particular case has the form:

Is $L^\gamma[0,1]$ a Lipschitz interpolation space with

constant $C > 0$ between $L^\beta[0,1]$ and $L^\alpha[0,1]$ (M)

for any γ, $1 \le \beta < \gamma < \alpha \le \infty$?

W. Orlicz [19] proved that if $0 < z < \infty$ then any Orlicz space $L^\Phi(0,z)$ with the Luxemburg norm

$$\|f\|_{L^\Phi} = \inf\left\{ r > 0 : \int_0^z \Phi\left(\frac{|f(t)|}{r}\right) dt \le 1 \right\}$$

is a semi-Lipschitz and Lipschitz interpolation space with constant $C > 0$ between $L^1(0,z)$ and $L^\infty(0,z)$. *This contains an answer for problems B and M if $\beta = 1$ and $\alpha = \infty$.* Next, G.G. Lorentz and T. Shimogaki in [10] have given a generalization of Orlicz' theorem, replacing $\beta = 1$ by $\beta \ge 1$. Following G.G. Lorentz, T. Shimogaki and W. Orlicz, we shall prove the following theorem.

Theorem 1. Let $\Phi(u) = \int_0^u (u-t)^\beta d\xi(t)$, $u > 0$, where $1 \le \beta < \infty$ and ξ is a positive nondecreasing left continuous function with $\xi(0) = 0$. Then

(i) $L^\Phi(0,z)$ is a semi-Lipschitz interpolation space with constant 1 between $L^\beta(0,z)$ and $L^\infty(0,z)$;

(ii) $L^\Phi(0,z)$ is a Lipschitz interpolation space with constant 1 between $L^\beta(0,z)$ and $L^\infty(0,z)$ if $0 < z < \infty$;

(iii) $L^\Phi(0,\infty)$ is a semi-Lipschitz interpolation space with constant 1 between $L^\beta(0,\infty)$ and $L^\infty(0,\infty)$ if Φ satisfies Δ_2-condition;

Proof. (i) (Lorentz-Shimogaki [10]). We may assume that $M_\beta = M_\infty = 1$. Then by Fubini's theorem,

$$\int_0^z \Phi(|Uf(t)|)dt = \int_0^{|Uf(t)|} \left\{ \int_{E_s} (|Uf(t)| - s)^\beta dt \right\} d\xi(s)$$

$$= \int_0^\infty \left\{ \int_{E_s} (|Uf(t)| - s)^\beta dt \right\} d\xi(s),$$

where $E_s = \{ t \in (0,z) : |Uf(t)| > s \}$ for $s > 0$. We consider an s-truncation $f^{(s)}$ of the function f:

$$f^{(s)} = \begin{cases} f(t), & \text{if } |f(t)| \leq s, \\ s\frac{f(t)}{|f(t)|}, & \text{if } |f(t)| > s. \end{cases}$$

Since $|Uf| - |Uf|^{(s)} = |Uf - (Uf)^{(s)}|$, then

$$\int_0^z \Phi(|Uf(t)|)dt = \int_0^\infty \left\{ \int_{E_s} |Uf(t) - (Uf)^{(s)}(t)|^\beta dt \right\} d\xi(s)$$

$$= \int_0^\infty \left\{ \int_0^z |Uf(t) - (Uf)^{(s)}(t)|^\beta dt \right\} d\xi(s)$$

$$= \int_0^\infty \|Uf - (Uf)^{(s)}\|_{L^\beta}^\beta d\xi(s).$$

From the assumption we have $\|U(f^{(s)})\|_{L_\infty} \leq \|f^{(s)}\|_{L_\infty} \leq s$, hence the inequality $|Uf - (Uf)^{(s)}| \leq |Uf - U(f^{(s)})|$ a.e. (see [23, 14]) follows. Then

$$\int_0^z \Phi(|Uf(t)|)dt \leq \int_0^\infty \|Uf - U(f^{(s)})\|_{L^\beta}^\beta d\xi(s)$$

$$\leq \int_0^\infty \|f - f^{(s)}\|_{L^\beta}^\beta d\xi(s) = \int_0^z \Phi(|f(t)|)dt$$

or

$$\|Uf\|_{L^\Phi} \leq \|f\|_{L^\Phi}, \qquad f \in L^\Phi.$$

Taking $U_0f = (Uf)/\max(M_\beta, M_\infty)$, we obtain the theorem for the general case.

Proof. (ii) and (iii) ([19], see also [12] and [15]). We take any fixed $f_0 \in L^\beta(0,z) \cap L^\infty(0,z)$ and define an operator T by

$$Tf = \frac{U(f+f_0) - Uf_0}{\max(M_\beta, M_\infty)}, \quad f \in L^\beta(0,z) \cup L^\infty(0,z).$$

Now T satisfies assumptions (i) for $f \in L^\beta(0,z)$. Hence

$$\|Tf\|_{L^\Phi} \leq \|f\|_{L^\Phi}, \quad f \in L^\Phi(0,z) \cap L^\beta(0,z).$$

This means that

$$\|Uf - Uf_0\|_{L^\Phi} = \max(M_\beta, M_\infty)\|T(f - f_0)\|_{L^\Phi}$$

$$\leq \max(M_\beta, M_\infty)\|f - f_0\|_{L^\Phi}$$

for $f \in L^\Phi(0,z) \cap L^\beta(0,z)$. Let $z < \infty$.

For arbitrary $f, g \in L^{\Phi}(0, z)$, we consider the truncations $f^{(n)}$, $g^{(n)}$. Then $U(f^{(n)})$ and $U(g^{(n)})$ converge to Uf and Ug, respectively, in the $L^{\beta}(0, z)$-norm. Therefore, for a properly chosen sequence n_i, $U(f^{(n_i)})$ and $U(g^{(n_i)})$ converge almost everywhere to Uf and Ug. Since $f^{(n)} \in L^{\infty}(0, z)$ and $|f^{(n)} - g^{(n)}| \leq |f - g|$, by the Fatou property of the Luxemburg norm we obtain

$$\|Uf - Ug\|_{L^{\Phi}} \leq \varliminf_{i \to \infty} \|U(f^{(n_i)}) - U(g^{(n_i)})\|_{L^{\Phi}}$$

$$\leq \max(M_{\beta}, M_{\infty}) \|f - g\|_{L^{\Phi}}.$$

If $z = \infty$ and Φ satisfies Δ_2-condition, then $L^{\beta}(0, \infty) \cap L^{\infty}(0, \infty)$ is dense in $L^{\Phi}(0, \infty)$ and U can be uniquely extended to $L^{\Phi}(0, \infty)$.

Since $\Phi(u) = u^{\gamma}$ for $\beta \leq \gamma < \infty$ has a representation as Φ in theorem 1, we obtain a positive answer to problem B and M for $\alpha = \infty$.

In [9, 12, and [13]], there are generalizations of the Orlicz and Lorentz-Shimogaki theorems. Those generalizations replace Orlicz spaces with rearrangement invariant spaces. Further considerations on the interpolation of Lipschitz operators of weak type can be found in [23] and [14].

Now we shall deal with any Banach space. For $0 < \omega < 1$ and $1 \leq p \leq \infty$, let

$$(A_0, A_1)_{\theta, p; K} = \{a \in A_0 + A_1 :$$

$$\|a\|_{\theta, p; K} \equiv \int_0^{\infty} (t^{-\theta} K(t, a))^p dt/t)^{1/p} < \infty\},$$

where

$$K(t, a) \equiv K(t, a; A_0, A_1)$$

$$= \inf\{ \|a_0\|_{A_0} + t\|a_1\|_{A_1} : a_0 \in A_0, a_1 \in A_1, a = a_0 + a_1 \}.$$

The real method presented above is called the K-method. The properties of this interpolation space are characterized in [1, 3, 27]. There also exist other real methods such as the trace method and the mean method, which are equivalent to the K-method. The method which, in general, is not equivalent to the K-method is the complex method introduced independently by J.L Lions and A.P. Calderon (see [1]). We shall prove that the K-method can be applied to nonlinear interpolation.

Theorem 2 (Lions [6], Peetre [20], see also [15]). *Let $A_1 \subset A_0$ and $B_1 \subset B_0$. For all θ, $0 < \theta < 1$ and all p, $1 \leq p \leq \infty$ the pair of Banach spaces $((A_0, A_1)_{\theta, p; K}, (B_0, B_1)_{\theta, p; K})$ is a semi-Lipschitz interpolation pair with constant 1 between the pairs (A_0, B_0) and (A_1, B_1).*

The proof is immediate and is obtained by showing that $K(t, Ua; B_0, B_1) \leq \max(M_0, M_1) K(t, a; A_0, A_1)$, for $t > 0$. J.L. Lions [6] proved this theorem using the trace method with the additional assumptions: B_0 is reflexive and $U : A_1 \to B_1$ is a continuous operator. Theorem 2 (stated as above) was proved by J. Peetre [20] and myself [15]. Further generalizations of Theorem 2, together with its applications, can be found in [20, 25, and [15]].

Using s-truncation, it can be shown that $(L^\beta, L^\alpha)_{\theta,\gamma;K} = L^\gamma$ with equivalent norms, where $0 < \beta < \alpha < \infty$, $1/\gamma = (1 - \theta)/\beta + \theta/\alpha$ and $0 < \theta < 1$ (see [1], th. 5.2.1). *This fact and theorem 2 gives us a positive answer to problem B.*

J.L. Lions [6, 21] put forward the following problem:

> Does Theorem 2 hold for the complex
>
> interpolation method?   (L)

A negative solution with one operator and one Banach space, where $A_i = B_i$, $i = 0, 1$, was obtained by M. Cwikel [4]; a negative solution for a family of operators and a family of Banach spaces can be found in [15]. Arriving at this solution was difficult, since for many important and well-studied Banach pairs, the complex interpolation spaces coincide with suitably chosen real interpolation spaces. In [15] it is shown that with an additional assumption that the operator $U : A_0 \to B_0$ is differentiable in the sense of Fréchet, Lions' problem has a positive solution.

The theory of nonlinear interpolation has not been studied thoroughly. Hence, some problems, such as the two given below, emerge.

Problem 1. Does the assumption on the continuity of the operator $U : A_1 \to B_1$ in Theorem 2 imply continuity of the operator $U : (A_0, A_1)_{\theta,p;K} \to (B_0, B_1)_{\theta,p;K}$?

Problem 2. Does Lions' problem have a positive solution with the assumption of:

(a) differentiability of the operator $U : A_0 \to B_0$ in the sense of Gateaux;
(b) Lipschitz condition on the operator $U : A_1 \to B_1$?

If the spaces A_i, B_i, $i = 0, 1$ satisfy some additional conditions, Problem 1 has a positive solution (see [15, th. 7.5]).

Theorem 3 ([15, th. 5.3]). Let $A_1 \subset A_0$ and $B_1 \subset B_0$. For any θ, $0 < \theta < 1$ and any p, $1 \le p < \infty$, the pair of Banach spaces $((A_0, A_1)_{\theta,p;K}, (B_0, B_1)_{\theta,p;K})$ is a Lipschitz interpolation pair with constant 1 between the pairs (A_0, B_0) and (A_1, B_1).

Proof. (The method is based on the paper of W. Orlicz [19]). For any fixed $a_1 \in A_1$ we set for $a \in A$, $Ta = U(a + a_1) - Ua_1$. Then

$$\|Ta - Tb\|_{B_0} = \|U(a + a_1) - U(b + a_1)\|_{B_0}$$

$$\le M_0 \|a - b\|_{A_0}$$

for $a, b \in A_0$, and

$$\|Ta\|_{B_1} = \|U(a + a1) - Ua_1\|_{B_1}$$
$$\leq M_1 \|a\|_{A_1}$$

for $a \in A_1$. From Theorem 2 we get

$$\|Ta\|_{\theta,p;K} \leq \max(M_0, M_1)\|a\|_{\theta,p;K}$$

for $a \in (A_0, A_1)_{\theta,p;K}$. Hence,

$$\|Ua - Ua_1\|_{\theta,p;K} = \|T(a - a_1)\|_{\theta,p;K}$$
$$\leq \max(M_0, M_1)\|a - a_1\|_{\theta,p;K}.$$

Because $p < \infty$, A_1 is dense in $(A_0, A_1)_{\theta,p;K}$ (see [1, th. 3.4.2(b)]). Taking any $a, b \in (A_0, A_1)_{\theta,p;K}$ there are sequences (a_n), (b_n) in A_1 convergent to a and b, respectively, in the $\| \cdot \|_{\theta,p;K}$ norm. Hence,

$$\|Ua - Ub\|_{\theta,p;K} \leq \|Ua - Ua_n\|_{\theta,p;K} + \|Ua_n - Ub_n\|_{\theta,p;K}$$
$$+ \|Ub_n - Ub\|_{\theta,p;K}$$
$$\leq \max(M_0, M_1)(\|a - a_n\|_{\theta,p;K} + \|a_n - b_n\|_{\theta,p;K}$$
$$+ \|b_n - b\|_{\theta,p;K})$$
$$\leq \max(M_0, M_1)(2\|a - a_n\|_{\theta,p;K} + \|a - b\|_{\theta,p;K}$$
$$+ 2\|b_n - b\|_{\theta,p;K}).$$

Taking $n \to \infty$ we get $\|Ua - Ub\|_{\theta,p;K} \leq \max(M_0, M_1)\|a - b\|_{\theta,p;K}$, $a, b \in (A_0, A_1)_{\theta,p;K}$. From Theorem 3 we obtain a positive solution of problem M.

The papers in [8, 2, and [15]] contain some general considerations on the subject of Lipschitz operator interpolation in rearrangement invariant spaces or arbitrary Banach spaces.

1. J. Bergh, J. Löfström, *Interpolation spaces. An introduction,* Springer, Berlin, 1976.
2. F. Browder, Remarks on nonlinear interpolation in Banach spaces, *J. Functional Anal.,* 4 (1969), 390–403.
3. P.L. Butzer, H. Berens, *Semi-groups of operators and approximation,* Grundlehren Math. Wiss. 145, Berlin, 1967.
4. M. Cwikel, A counterexample in nonlinear interpolation, *Proc. Amer. Math. Soc.,* 62 (1977), 62–66.
5. S.G. Krein, Ju.I. Petunin, E.M. Semenov, *Interpolation of linear operators* (in Russian), Nauka, Moscow, 1978.
6. J.L. Lions, Some remarks on variational inequalities, *Proc. Internat. Conf. Functional Analysis and Related Topics* (Tokyo 1969), Univ. of Tokyo Press, Tokyo, 1970, 269–282.

7. J.L. Lions, F. Magenes, *Problèmes aux limites non homogènes et applications* (in Russian), Mir, Moscow, 1971.

8. G.G. Lorentz, T. Shimogaki, Interpolation theorems for operators in function spaces, *J. Functional Anal.,* 2 1968), 31–51.

9. _____, Majorants for interpolation theorems, *Publ. Ramanujan Inst.,* 1 (1969), 115–122.

10. _____, Interpolation theorems for the pairs of spaces (L^p, L^∞) and (L^1, L^q), *Trans. Amer. Math. Soc.,* 159 (1971), 207–221.

11. L. Maligranda, Riesz Thorin theorem for Lipschitz operators, *V Session of Functional Analysis,* Poznan 1978 (abstract will appear in *Functiones et Approximatio* X).

12. _____, Interpolation of Lipschitz operators for the pairs of spaces (L^1, L^∞) and (l^1, c_0), *Comment. Math. Prace Mat.,* 21 (1979), 327-339.

13. _____, Interpolation of Lipschitz operators for the pairs of spaces (L^p, L^∞) and (l^p, c_0), $0 < p < \infty$, *Functiones et Approximatio* 9 (1980), 107-115.

14. _____, A generalization of the Shimogaki theorem, *Studia Math.,* 71 (1981), 69-83.

15. _____, Interpolation of nonlinear operators in Banach spaces, Thesis, University of A. Mickiewicz Poznan 1979 (in Polish).

16. J. Marcinkiewicz, Sur l'interpolation d'opérations, *C. R. Acad. Sci. Paris,* 208 (1939), 1272–1273.

17. W. Orlicz, Ein Satz über die Erweiterung von linearen Operationen, *Studia Math.,* 5 (1934), 127–140.

18. _____, On a class of operations over the space of continuous vector-valued functions, *Studia Math.,* 14 (1954), 285–297.

19. _____, On a class of operations over the space of integrable functions, *Studia Math.,* 14 (1954), 302–309.

20. J. Peetre, Interpolation of Lipschitz operators and metric spaces, *Mathematica (Cluj),* 12 (1970), 325–334.

21. Problems in interpolation of operators and applications (Problem list of the Special Session on Interpolation of Operators and Applications), *Notices of the AMS,* 22 (1975), 124–126.

22. M. Riesz, Sur les maxima des formes bilinéares et sur les fonctionelles linéares, *Acta Math.,* 49 (1926), 465–497.

23. T. Shimogaki, An interpolation theorem on Banach function spaces, *Studia Math.,* 31 (1968), 233–240.

24. E.M. Stein, G. Weiss, An extension of a theorem of Marcinkiewicz and some of its applications, *J. Math. Mech.,* 8 (1959), 263–284.

25. L. Tartar, Interpolation non linéaire et régularité, *J. Functional Anal.,* 9 (1972), 469–489.

26. G.O. Thorin, An extension of a convexity theorem due to M. Riesz, *Kungl. Fysiogr. Sällsk. i Lund Förh.,* 8 (1938), 166–170.

27. H. Triebel, *Interpolation theory, function spaces, differential operators,* North-Holland, Amsterdam, 1978.

28. A. Zygmund, On a theorem of Marcinkiewicz concerning interpolation of operations, *J. Math. Pures Appl.,* 35 (1956), 223–248.

Lech Maligranda

The Mathematical Institute of the Polish Academy of

Sciences, Poznań, Poland

PROBLEM 88: MAZUR

Given is a sequence of real numbers (a_n) with the property that for every bounded sequence (x_n) the series $|a_1x_1 + a_2x_2 + \ldots + a_nx_n + \ldots| + |a_2x_1 + a_3x_2 + \ldots + a_{n+1}x_n + \ldots| + \ldots + |a_mx_1 + a_{m+1}x_2 + \ldots + a_{m+n-1}x_n + \ldots| + \ldots$ converges. Is the series

$$\sum_{n=1}^{\infty} n|a_n|$$

convergent?

Remark: If sequences of numbers $(a_{1n}), (a_{2n}), \ldots, (a_{mn})$ are given with the property that for every bounded sequence of numbers (x_n) the series $|a_{11}x_1 + a_{12}x_2 + \ldots| + |a_{21}x_1 + a_{22}x_2 + \ldots + a_{2n}x_n + \ldots| + \ldots + |a_{m1}x_1 + a_{m2}x_2 + \ldots + a_{mn}x_n + \ldots| + \ldots$ converges, then, according to a remark by Mr. Banach, the series

$$\sum_{m=1}^{\infty} (|a_{m1}| + |a_{m2}| + \ldots + |a_{mn}| + \ldots)$$

can diverge.

Second Edition Commentary

This problem was solved in negative by S. Kwapień and A. Pełczyński, *The main triangle projection in matrix spaces and its applications*, Studia Math. **34** (1970) 43–68. It is closely related to Problem 8. Via duality arguments its negative solution provides a negative solution to Problem 8. For more details we refer to the papers from References in Commentary to Problem 8.

Stanisław Kwapień

PROBLEM 89: MAZUR

Let W be a convex body located in the space (L^2), and such that its boundary W_b does not contain any interval; let $x_n \in W$, $(n = 1, 2, \ldots)$, $x_0 \in W_b$, and in addition let the sequence (x_n) converge weakly to x_0. Does then the sequence (x_n) converge strongly to x_0? It is known that this statement is true in the case where W is a sphere. Examine this problem for the case of other spaces.

Second Edition Commentary

The first part of this problem can be answered in negative by the following example : let $W = \{(\alpha_n) \in \ell_2 : \sum_{n=1}^{\infty} \alpha_n^2/n \leq \sum_{n=1}^{\infty} \alpha_n/n\}$ and let $x_n = e_n, n \in N$ be the standard base sequence in $\ell_2, x_0 = 0$. The answer to the second question in Problem 89 is: whenever in Banach space there is weakly convergent sequence which is not norm convergent we can produce a similar example without the property.

Stanisław Kwapień

PROBLEM 90: ULAM, AUERBACH

It is known that every semisimple Lie group (e.g., the projective group in n variables) contains four elements generating a dense subgroup. Can one lower the number 4?

Commentary

Every connected semisimple Lie group G has a free subgroup with two free generators which is dense in G (see [1]). Related results and references are given in [2].

1. M. Kuranishi, On everywhere dense imbeddings of free groups in Lie topological groups, *Nagoya Math. J.* 2 (1951), 63–71.
2. J. Mycielski, Almost every function is independent, *Fund. Math.*, 81 (1973), 43–48.

J. Mycielski

PROBLEM 91: MAZUR

A convex body W with a center is given, in the n-dimensional Euclidean space. it is affine to its conjugate body. Is W then an ellipsoid? The answer is negative in the case when n is an even number; for odd n the problem is not solved. It is equivalent to this: If a space of type (B) of n dimensions is isometric to its conjugate space, is it then isometric to the Euclidean space?

Commentary

The answer is negative for odd dimensions as well. If the dimension is $n = 3$, a simple example due to K. Leichtweiss (Zur expliziten Bestimmung der Norm der selbstadjungieren Minkowksi-Räume, *Resultate der Mathematik,*1 (1978), 61–87) is obtained by taking as W the convex polyhedron given in a cartesian system of coordinates as the convex hull of the eight points $\pm(1,1,1), \pm(1,1,-1), \pm(1,0,0), \pm(0,1,0)$. The conjugate (dual, polar) convex polyhedron W^* has as vertices the points $\pm(-1,1,-1), \pm(1,-1,-1), \pm(0,1,0), \pm(1,0,0)$. Thus W is isometric to W^*, although it clearly is not an ellipsoid. Analogous examples can be constructed in all odd dimensions ≥ 3.

Branko Grünbaum

PROBLEM 92: MAZUR

Given is a bounded sequence of numbers (s_n). There exist sequences of numbers (ℓ_n) with the property that:

(1) $\ell_n > 0$ $(n = 1, 2, \ldots)$;
(2) $\ell_1 + \ell_2 + \ldots = \infty$;
(3) The sequence $(\ell_1 s_1 + \ldots + \ell_n s_n)/(\ell_1 + \ldots + \ell_n)$ converges.

Do there exist sequences (ℓ_n) which, in addition to properties (1), (2), and (3), satisfy the condition:

(4a) The sequence (ℓ_n) is fully monotonic; that is, all the differences $\Delta_n^1 = \ell_n - \ell_{n+1}, \Delta_n^2 = \Delta_n^1 - \Delta_{n+1}^1, \ldots$ are nonnegative;

   or only the condition:

(4b) The sequence (ℓ_n) is nonincreasing.

If two sequences are given (ℓ_n'), (ℓ_n'') which satisfy the conditions (1), (2), (3), (4a), or merely (1), (2), (3), (4b), then can the limits

$$\lim_{n \to \infty} \frac{\ell_1' + \ldots + \ell_n' s_n}{\ell_1' + \ldots + \ell_n'}$$

and

$$\lim_{n \to \infty} \frac{\ell_1'' + \ldots + \ell_n'' s_n}{\ell_1'' + \ldots + \ell_n''}$$

be different?

Addendum. There exist sequences (ℓ_n'), (ℓ_n'') satisfying conditions (1), (2), (3), (4b) such that for a certain bounded sequence s_n composed of 0s and 1s, the two limits

$$\lim_{n \to \infty} \frac{\ell_1' + \ldots + \ell_n' s_n}{\ell_1' + \ldots + \ell_n'}$$

and

$$\lim_{n \to \infty} \frac{\ell_1'' + \ldots + \ell_n'' s_n}{\ell_1'' + \ldots + \ell_n''}$$

exist but are different.

Mazur

August 10, 1935

PROBLEM 93: MAZUR

Let R be a planar set. The system of functions $x = f(t)$, $y = g(t)$ $(0 \le t \le 1)$ is called a parametric description of the set R, if the set of points $(f(t), g(t))$ is identical with R. Assume that for a given set R there exists a parametric description $x = f_1(t)$, $y = g_1(t)$ for which the functions $f_1(t)$, $g_1(t)$ are continuous and there also exists another parametric description $x = f_2(t)$, $y = g_2(t)$ where the functions $f_2(t)$, $g_2(t)$ are of bounded variation; does there exist a parametric description of R: $x = f(t)$, $y = g(t)$ so that the functions $f(t)$, $g(t)$ are simultaneously continuous and of bounded variation? Assume that for a given set R there exist parametric descriptions $x = f(t)$, $y = g(t)$ for which the functions are of bounded variation and continuous—for every such description, we determine the length $d(f(t), g(t))$ of the set R and we take the lower bound of the numbers $d(f(t), g(t))$ denoted by d; does there exist a parametric description of R: $x = f_0(t)$, $y = g_0(t)$ also with functions of bounded variation and continuous and such that $d(f_0(t), g_0(t)) = d$? The same problem in the case of the n-dimensional Euclidean space.

Addendum.[*]   The theorem is true; we can represent R by functions $x = f_\xi(t)$, $y = g_\xi(t)$, continuous and of bounded variation, in such a way that the length of the curve (by Jordan's definitions) is at most twice the Carathéodory measure of R

A.J. Ward

March 23, 1937

[*]Original manuscript in English.

Commentary

For references to the early work of Gołab and Ważewski, where the solution announced here is proved, as well as related work, see V. Faber, J. Mycielski and P. Pedersen, On the shortest curve which meets all the lines which meet a circle, *Ann. Polon. Math.* 44 (1984), 249–266.

Jan Mycielski

PROBLEM 94: Z. LOMNICKI, ULAM

Let $\lim_{n \to \infty} k_n / n = f < 1$, where always $k_n < n$. Prove that

$$\lim_{n \to \infty} \int_0^P \int_K x_1 \cdots x_{k_n} (1 - x_{k_n+1}) \cdots (1 - x_n) dx_1 \cdots dx_n dp = \begin{cases} 0, & P < f \\ 1, & P \ge f \end{cases}$$

where

$$K = \{ (x_1, \ldots, x_n) : x_1 + \ldots + x_n = nP, 0 \le x_i \le 1, i = 1, \ldots, n \}.$$

Compare Problem 17.

Addendum. This conjecture was proved by S. Bochner in April, 1936—he even gave the order of convergence. A paper on this topic will appear in *Annals of Math.*

S. Ulam 1936

Commentary

The problem is better formulated as follows:

Let (ξ_i, X_i), $i = 1, 2, \ldots$ be independent random vectors, where each ξ_i is (*a priori*) uniformly distributed in $(0, 1)$ and each X_i takes the value 1 with probability ξ_i and the value 0 with probability $1 - \xi_i$. Thus the two components of each vector are dependent. It is easy to compute the conditional expectations below:

$$E\{\xi \mid X_i = 1\} = \frac{\int_0^1 \xi \xi \, d\xi}{\int_0^1 \xi \, d\xi} = \frac{2}{3};$$

(1)

$$E\{\xi \mid X_i = 0\} = \frac{\int_0^1 \xi (1 - \xi) \, d\xi}{\int_0^1 \xi \, d\xi} = \frac{1}{3}.$$

In fact, owing to the stated independence, if $\mathscr{F}_i$ denotes the σ-field of all X_j for $j \geq 1$ except X_i, the conditional expectations above are not affected if we adjoin $\mathscr{F}_i$ to the two conditions shown. Therefore,

$$\frac{1}{N} \sum_{i=1}^{N} E\{\xi \mid X_1 + \ldots + X_N = m\} = \frac{2}{3} \frac{m}{N} + \frac{1}{3}\left(\frac{N - m}{N}\right).$$

This is the curious discovery of Bochner (see [1], Formula (10)).

Given all X_i, $i \geq 1$, the random variables $\{\xi, i \geq 1\}$ are independent with *a posteriori* expectations determined by the X_i's as shown in (1). Hence, if $(m/N) \to t$, then

$$P\left\{\frac{1}{N} \sum_{i=1}^{N} \xi_i \to \frac{1}{3}(1 + t) \,\middle|\, \frac{1}{N} \sum_{i=1}^{N} X_i \to t\right\} = 1 \qquad (2)$$

by the classical law of large numbers (Borel's form is sufficient). Bochner gave the following explicit analytic formula (cf. his (3) which is really the same):

$$P\left\{\frac{1}{N} \sum_{i=1}^{N} \xi_i \leq p \,\middle|\, \sum_{i=1}^{N} X_i = m\right\}$$

$$= 2^N \int_{\substack{\sum\limits_{1 \leq i \leq N} \xi_i \leq pN \\ 0 \leq \xi_i \leq 1}} \xi_1 \cdots \xi_m (1 - \xi_{m+1}) \cdots (1 - \xi_N) \, d\xi_1 \cdots d\xi_N.$$

Therefore, if $(m/N) \to t$, the latter integral converges to 0 if $p \le (1/3)(1+t)$ and to 1 if $p > (1/3)(1+t)$. (Note: The equality case is included because the integral has the same value if the constraint $\sum_{i=1}^{N} \xi_i \le pN$ is changed to $\sum_{i=1}^{N} \xi_i < pN$.) This result proved by Bochner by the method of Fourier transforms. He pointed out that, contrary to what might be facilely conjectured, the critical value of p is $(1/3)(1+t)$ and not t. (Only when $t = 1/2$ do these two values coincide, and the original problem concerns only this case.) The latter is indeed the critical value if the ξ_i's take the values 1 or 0 with probability 1/2 each. This is seen by reevaluating the conditional expectations in (1) under the new *a priori* distributions for the ξ_is, as Bochner did. (When the ξ_i's are constants the result reduces to Problem 17 which can be done by bounded convergence.) His improvement of (2) in obtaining the speed of convergence and a central limit theorem can also be derived, presumably, by applying (by now) well-known limit theorems to the sequence of independent ξ_i's, but using their *a posteriori* moments. Surely it was a remarkable achievement in 1936.

K.L. Chung

1. S. Bochner, A converse of Poisson's theorem in the theory of probability *Annals of Math.*, 37 (1936), 816–822.

PROBLEM 95: SCHREIER, ULAM

Is the group $\mathbb{R}$ of real numbers (under addition) isomorphically contained in the group S_∞ of all permutations of the sequence of natural integers?

Addendum. The answer is affirmative.

Schreier, Ulam *November, 1935*

Commentary

The rationals $\mathbb{Q}$ can be embedded into S_∞ by letting $\mathbb{Q}$ act on itself by left translations. The direct product of countably many copies of S_∞ can be embedded into S_∞ since the integers may be decomposed into countably many disjoint infinite subsets. Hence, a countable direct product of $\mathbb{Q}$'s may be embedded into S_∞. This last direct product, however, is isomorphic to the reals under addition since both are vector spaces of the same dimension over $\mathbb{Q}$. Schreier and Ulam noted this argument; Schreier and Ulam also asked if every Lie group can be embedded (as an abstract subgroup) into S_∞. In particular, can $SO(3)$ be embedded into S_∞? It is easy to check that this question has an affirmative answer if and only if $SO(3)$ has a subgroup of countable index. It is unknown at this time whether $SO(3)$ has such a subgroup. It is simple to check that the analogous question for second countable connected locally compact groups reduces to the connected Lie group case, for any second countable connected locally compact group is a projective limit of a sequence of Lie groups. It is also simple to check that any totally disconnected second countable locally compact group can be embedded as an abstract group into S_∞, for any such group has a neighborhood basis of the identity consisting of compact open subgroups. The embedding may be constructed by letting the group act on a suitably chosen sequence of (countable discrete) quotient spaces.

Robert R. Kallman

Second Edition Commentary

The obvious generalization of Problem 95 was stated by Ulam ([5], p. 58) as follows: "…is every Lie group isomorphic (as an abstract group) to a subgroup of the group S_∞?" In case one regards any discrete group to be a zero-dimensional Lie group, one should perhaps refine Ulam's question to ask if every Polish (i.e., complete metrizable separable) Lie group can be abstractly embedded into S_∞? As noted in the original commentary, Schreier and Ulam proved that the additive group of the reals can be embedded into S_∞.

A certain amount of progress has been made on this problem. In fact if F is any field whose cardinality is less than or equal to $2^{\aleph_0}$, then $GL(n, F)$ is isomorphic to a subgroup of S_∞ ([3]). Therefore any connected matrix Lie group can be abstractly embedded into S_∞. In particular, any semisimple complex analytic group can be abstractly embedded into S_∞ since any such group has a faithful finite dimensional representation ([1], p. 200, Theorem 3.2). The same is true for any simply connected solvable analytic group ([1], p. 219, Theorem 3.1) or, more generally, any semidirect product $G = B \rtimes_\eta H$, where H is a reductive analytic group and B is a simply connected solvable analytic group, normal in G ([1], p. 223, Theorem 4.3).

However, not every Lie group is isomorphic to a matrix Lie group. For example, the simply connected cover $\widetilde{SL(2, \mathbb{R})}$ of $SL(2, \mathbb{R})$ has center isomorphic to $(\mathbb{Z}, +)$ but any semisimple Lie group with faithful continuous matrix representation has finite center ([1], Proposition 4.1, p. 221). More generally, any nontrivial covering group of $SL(2, \mathbb{R})$ has no continuous faithful finite dimensional representation ([1], Exercise 1, p. 210) and the same is true for a group covered by the Heisenberg group ([1], Exercise 1, p. 225). Therefore the most general case of Ulam's extrapolation of Problem 95 does not follow from the matrix result and the general Lie group question remains unanswered.

Consider the following two questions:
Question #1: Let G be a group with a finite central cyclic subgroup Z such that G/Z can be injected into S_∞. Can G be injected into S_∞? In particular, is this true if $Z = Z_2$?
Question #2: Let $G \subset S_\infty$ be a subgroup and let Z be a central cyclic subgroup of G. Can G/Z be injected into S_∞?
If one is optimistic and believes that these two questions have affirmative answers, then Ulam's general Lie group question has an affirmative answer. To see this, recall that the fundamental group of every connected Lie group is a finitely generated abelian group. The fundamental theorem of finitely generated abelian

groups implies that such a group is a finite direct sum of cyclic groups. Therefore every connected Lie group is of the form G/Z, where G is a simply connected Lie group and Z is a finitely generated central subgroup of G. On the other hand, Ado's theorem ([2]) implies that every finite dimensional Lie algebra has a faithful finite dimensional representation. Therefore basic facts about Lie groups imply that every simply connected Lie group covers a matrix Lie group, which, of course, can be injected into S_∞. In this same circle of ideas, recall that if G is a matrix group and $Z \subset G$ is a finite central subgroup, then G/Z has a faithful finite dimensional representation ([1], Lemma 3.1, p. 199) and therefore can be injected into S_∞. On the other hand, and more generally, it is an easy exercise to prove that if $G \subset S_\infty$ is a subgroup and $N \subset G$ is a finite normal subgroup, then G/N can be injected into S_∞.

Again, if one is optimistic and believes that Ulam's general Lie group question has an affirmative answer, one might well ask if still more might be true. For example, can every locally compact Polish group be algebraically injected into S_∞? This has a ring of plausibility since every locally compact group has an open subgroup which is a projective limit of Lie groups ([4], Yamabe's Main Approximation Theorem, p. 175). The left regular representation of a locally compact group G is a homeomorphism onto its (closed) range in the unitary group of the Hilbert space $L^2(G)$. Therefore one might further ask if the unitary group of a separable Hilbert space can be algebraically injected into S_∞? Unfortunately this is asking too much, for it is known that any algebraic injection of the unitary group of an infinite dimensional separable Hilbert space into a separable topological group is continuous. It is an easy consequence of the spectral theorem that the unitary group is connected in the strong operator topology. But there is no continuous injection of the connected unitary group into the totally disconnected group S_∞. The homeomorphism group of $[0,1]$ cannot be injected into S_∞ for similar reasons. A fortiori the same is true for the universal Polish groups the homeomorphism group of the Hilbert cube and the isometry group of the universal Urysohn space.

1. Gerhard Hochschild, The Structure of Lie Groups, Holden-Day, San Francisco, 1965.
2. Nathan Jacobson, Lie Algbras, Dover, New York, 1979, ISBN 0-486-63832-4.
3. Robert R. Kallman, Every reasonably sized matrix group is a subgroup of S_∞, Fundamenta Mathematicae, volume 164, Issues 1–2, 2000, pp. 35 – 40.
4. Dean Montgomery and Leo Zippin, Topological Transformation Groups, Robert E. Krieger, New York, 1974.
5. S.M. Ulam, Problems in Modern Mathematics, Wiley, New York, 1964 (first published under the title A Collection of Mathematical Problems, Wiley, New York, 1960).

Robert R. Kallman

PROBLEM 96: ULAM

Can the group S_∞ of all permutations of integers be so metrized that the group operation (composition of permutations) is a continuous function and the set S_∞ becomes, under this metric, a compact space? (locally compact?)

Addendum. One cannot metrize in a compact way.

Schreier, Ulam *November, 1935*

Commentary

Two solutions of generalizations of this problem have appeared in the literature. Gaughan [1] showed that there is no nontrivial, locally bounded, Hausdorff topological group structure on S_∞. Kallman [2] showed that S_∞ has a unique topology (the usual one) under which it is a complete separable metric group.

1. E.D. Gaughan, Topological Group Structures of Infinite Symmetric Groups, *Proceedings of the National Academy of Sciences U.S.A.,* 58 (1967), 907–910.
2. R.R. Kallman, A Uniqueness Result for the Infinite Symmetric Group, Studies in Analysis, Advances in Mathematics Supplemental Studies, 4 (1979), 321–322.

Robert R. Kallman

PROBLEM 97: KURATOWSKI, ULAM

Two sets (spaces) A AND B are called quasihomeomorphic if, for every ε, there exists a continuous mapping f_ε of the space A onto the space B such that the counterimages are smaller than ε (that is to say, from $|x' - x''| > \varepsilon$ it follows that $f(x') \neq f(x'')$) and, conversely, a continuous mapping g_ε with counterimages smaller than ε of the space B onto the space A. Problem: Are two manifolds (topological spaces such that every point has a neighborhood homeomorphic to the n-dimensional Euclidean sphere) which are quasihomeomorphic, of necessity homeomorphic?

Commentary

For dimensions strictly greater than 4, the problem has an affirmative solution. In fact, the following much stronger statement is true:

Theorem [5] If M_n, $n \geq 5$, is a compact manifold without boundary, then there is an $\varepsilon > 0$ such that if $f : M_n \to N_n$ is a map with $\mathrm{diam} f^{-1}(x) < \varepsilon$ for each $x \in N_n$, then f is homotopic to a homeomorphism.

The proof uses techniques of Siebenmann [11] and Chapman and Ferry [3] which arise out of the general Kirby-Siebenmann [8] approach to the study of topological manifolds. Further developments in this general area are due to Chapman and to Quinn [10].

In dimensions ≤ 2 the problem is easily seen to be true. In dimension 3, work of Waldhausen [12], Armentrout [1], and Siebenmann [11] is certainly relevant; see also Hamilton's paper [7]. While the details have not been worked out, this work, together with the approach of [3] and [5] may well add up to a proof of Problem 97 modulo the Poincaré conjecture in dimension 3. Nothing seems to be known in dimension 4.

More interestingly, no counterexample seems to be known to the following: If X and Y are compact *ANR*s and for each $\varepsilon > 0$ there are surjective ε-maps $f : X \to Y$ and $g : X \to Y$, then X and Y are homeomorphic. Perhaps a reasonable first step would be to show that X and Y are homotopy equivalent. [2, 4, 6], and [9] are relevant. Compare the commentary to Problem 21.

1. S. Armentrout, Cellular decompositions of 3-manifolds that yield 3-manifolds, *Bull. Am. Math. Soc.* 75 (1969), 453–456.
2. I. Bernstein and T. Ganea, Remark on spaces dominated by manifolds, *Fund. Math.,* 47 (1959), 45–56.
3. T.A. Chapman and S. Ferry, Approximating homotopy equivalences by homeomorphisms, *American J. of Math.* 101 (1979), 583–607.
4. S. Eilenberg, Sur les transformations á petites tranches, *Fund. Math.,* 30 (1938), 92–95.
5. S. Ferry, Homotoping ε-maps to homeomorphisms, *Amer. J. of Math.,* 101 (1979), 567–582.
6. T. Ganea, On ε-maps onto manifolds, *Fund. Math.,* 47 (1959), 35–44.
7. A.J.S. Hamilton, The triangulation of 3-manifolds, *Q. J. Math. Oxford,* 27 (1976), 63–70.
8. R.C. Kirby and L.C. Siebenmann, On the triangulation of manifolds and the Hauptvermutung, *Bull. Amer. Math. Soc.* 75 (1969), 742–749.
9. S. Mardešic and J. Segal, ε-mapping onto polyhedra, *Trans. Amer. Math. Soc.* 109 (1963), 146–164.
10. F. Quinn, Ends of maps. I, *Ann. Math.* 110 (1979), 275–331.
11. L.C. Siebenmann, Approximating cellular maps by homeomorphisms, *Topology* 11 (1972), 271–294.
12. F. Waldhausen, On irreducible 3-manifolds which are sufficiently large, *Ann. Math.,* 87 (1968), 56–88.

Steve Ferry, The Institute for Advanced Study

and The University of Kentucky

PROBLEM 98: SCHREIER, ULAM

Do there exist a finite number of analytic transformations of the n-dimensional sphere into itself, $f_1, \ldots, f_n$, such that by composing these transformations a finite number of times, one can approximate arbitrarily any continuous transformation of the sphere into itself? How is it for one-to-one transformations? (Analytic here means differentiable any number of times.)

Remark

The first problem is still open. The second problem about homeomorphisms has a positive answer—see J. Schreier and S. Ulam, Über topologische Abbildungen der euklidischen Sphären, *Fund. Math.,* 23 (1934), 102–118. Further comments are in Stanislaw Ulam, *Sets, Numbers, and Universes; Selected Works,* edited by W.A. Beyer, J. Mycielski, and Gian-Carlo Rota, in the series Mathematicians of Our Time, MIT Press, Cambridge, Mass., 1974.

PROBLEM 99: ULAM

By a product set in the unit square, we understand the set of all pairs (x, y) where x belongs to a given set A, y to a given set B. Do there exist sets which cannot be obtained through the operations of forming countable sums and differences of sets starting from product sets? Do there exist nonprojective sets with respect to product subsets?

Commentary

If the continuum hypothesis or Martin's axiom holds, then the answer to the first question if no [2, 6]. In fact, under either of these assumptions, we have $\mathscr{R}_{\sigma\delta} = \mathscr{P}(I^2)$, where $\mathscr{R} = \{A \times B : A, B \subseteq [0, 1] = I\}$.

If every subset of I^2 is generated from $\mathscr{R}$ by the operations of forming countable unions and differences, then all sets are generated by some countable stage [1]. A. Miller [5] has shown that the stage at which all sets are generated can be any ordinal α, $2 \le \alpha < \omega_1$. This problem has some interesting connections to other problems of set theory, for example, the existence of Q-sets [4] and whether the continuum is real-valued measurable [3]. In connection with this it is unknown whether a universal analytic set is in $\mathscr{B}(\mathscr{R})$, the Borel field generated by $\mathscr{R}$, if the continuum is real-valued measurable.

The situation regarding the second question does not seem to be clear. It also seems to be unknown whether every subset of I^2 can be analytic with respect to $\mathscr{R}$ and yet $\mathscr{B}(R) \neq \mathscr{P}(I^2)$.

1. R.H. Bing, W.W. Blesdoe, R.D. Mauldin, Sets generated by rectangles, *Pac. J. Math.,* 51 (1974), 27–36.
2. K. Kunen, *Inaccessibility properties of cardinals,* Ph.D. Thesis, Department of Mathematics, Stanford University, August 1968.
3. R.D. Mauldin, Countably generated families, *Proc. Amer. Math. Soc.* 54 (1976), 291–297.
4. _____, On rectangles and countably generated families, *Fund. Math.* 95 (1977), 129–139.
5. A. Miller, On the length of Borel heirarchies, *Ann. Math. Log.* 16 (1979), 233–267.
6. B.V. Rao, On discrete Borel spaces and projective sets, *Bull. Amer. Math. Soc.,* 75 (1969), 614–617.

R. Daniel Mauldin

Second edition Commentary

Let $\mathscr{S}(\mathscr{R})$ be the family of sets obtained by applying the Souslin operation to sets in $\mathscr{B}(\mathscr{R})$). Miller [3] Theorem 6: It is relatively consistent with ZFC that $\mathscr{S}(\mathscr{R}) = \mathscr{P}(I^2)$ and $\mathscr{B}(\mathscr{R}) \neq \mathscr{P}(I^2)$. This answers the last question of the commentary above.

In regard to the second question: "Do there exist nonprojective sets with respect to product subsets?" We might define the projective sets with respect to product subsets to be the smallest family of sets containing

$$\{A_1 \times A_2 \times \cdots \times A_n \ : \ n < \omega \text{ and } A_i \subseteq \mathscr{R}\}$$

and closed under complementation, countable unions, and projection. In his thesis [2] Kunen showed that in the random real model or Cohen real model that $\mathscr{B}(\mathscr{R}) \neq \mathscr{P}(I^2)$ by showing that no well-ordering of the reals is in $\mathscr{B}(\mathscr{R})$. For a different argument for why $\mathscr{B}(\mathscr{R}) \neq \mathscr{P}(I^2)$ in the Cohen real model, see Miller [3] Remark 4 page 180. It is also true that by combining results of Rothberger [4] and Bing, Bledsoe, Mauldin [1], that certain cardinal arithmetic, for example, $2^\omega = \aleph_2 + 2^{\omega_1} = \aleph_{\omega_1}$, implies that $\mathscr{B}(\mathscr{R}) \neq \mathscr{P}(I^2)$, see Miller [3] Remark 5 page 180. I am not sure if any of these arguments can be generalized to show that it is consistent that not every subset of the plane is a projective set with respect to the product subsets.

1. Bing, R. H.; Bledsoe, W. W.; Mauldin, R. D.; Sets generated by rectangles. Pacific J. Math. 51 (1974), 27–36.
2. Kunen, Kenneth; Inaccessibility Properties of Cardinals. Thesis (Ph.D.) Stanford University. 1968. 124 pp, ProQuest LLC
3. Miller, Arnold W.; Generic Souslin sets. Pacific J. Math. 97 (1981), no. 1, 171–181.
4. Rothberger, Fritz; A remark on the existence of a denumerable base for a family of functions. Canadian J. Math. 4, (1952). 117–119.

                                                            Arnold W. Miller, January 2015

PROBLEM 100: ULAM, BANACH

Let Z be a closed set contained in the surface of the n-dimensional sphere. Does there exist a sequence of homeomorphic mappings of the surface of the sphere onto itself, converging to a mapping of the surface onto Z?

Addendum. For $n = 2$, affirmative answer by Borsuk.

Solution

First, we notice that the answer is yes if Z is a singleton, e.g., for the case of the circle S^1, the iterates of the map $e^{2\pi it} \mapsto e^{2\pi it}$ have the required property for $Z = \{1\}$. For S^n we take an appropriate fibration into circles and do a similar thing.

If Z has more than one point, then by choosing an appropriate coordinatization of S^n we can assume that the north pole and the south pole are in Z. Then let h move each point p down toward the south pole along a great circle by the angular distance $(1/2)\text{dist}(p, Z)$. Then it is easy to check that h is continuous, that $h(p) = p$ for each $p \in Z$, and that $h^m(p) \to Z$ as $m \to \infty$ for each $p \in S^n$. To check that h is one-to-one, we note that the great circles of our movement meet only at the poles. Also, if p

and q are on the same great circle, and p is above q, then clearly $h(p) \neq h(q)$, if the arc from p to q meets Z. If the arc from p to p does *not* meet Z, then $\text{dist}(p,Z) < 2\text{dist}(p,q) + \text{dist}(q,Z)$. Thus h is one-to-one.

Jan Mycielski

PROBLEM 101: ULAM

A group U of permutations of the sequence of integers is called infinitely transitive if it has the following property: If A and B are two sets of integers, both infinite and such that their complements to the set of all integers are also infinite, then there exists in the group U an element (permutation) such that $f(A) = B$. Is the group U identical with the group S_∞ of all permutations?

Solution

The answer is no. Let G be the group of all permutations of $N = \{1,2,3,\ldots\}$ such that for each $\pi \in G$ there is a finite partition P of N such that π is order preserving on each set in P. It is clear that G has the desired property, since there exists a $\pi \in G$ which maps A onto B preserving order and $N - A$ onto $N - B$ preserving order. On the other hand, G is a proper subgroup of S_∞ since a permutation which reverses the order in all the blocks $n^2, n^2 + 1, \ldots, (n+1)^2 - 1$ does not belong to G.

A. Ehrenfeucht

Note

S. Ulam informs us that C. Chevalley had found the solution and a number of interesting related results shortly after the end of the second world war.

PROBLEM 102: ULAM

(a) Let ε be a positive number; p and q two points of the unit square. In the first case let the point p be fixed and q wander at random. In the other case, assume that both points move at random. Is the probability of approach of the two points p and q within a distance $\leq \varepsilon$ of each other, after n steps, greater in the first case than in the second?

(b) Let a,b denote two rotations of a circle of radius 1 through angles a,b. Let ε be a positive number. We define a set of pairs $E_\varepsilon^1(a,b)$ as follows: Two rotations a,b belong to it if $(na - b) \bmod 2\pi$ is smaller than ε earlier than $(na - nb) \bmod 2\pi$; that is to say, for smaller n than is the case for $(na - nb) \bmod 2\pi$. We denote by $E_\varepsilon^2(a,b)$ the complement of the set of pairs $E_\varepsilon^1(a,b)$ with respect to the set E of all pairs. Which of the two sets $E\varepsilon^1(a,b)$, $E_\varepsilon^2(a,b)$ has greater measure? (Show that asymptotically these sets have equal measures.)

PROBLEM 103: SCHREIER, ULAM

Does there exist a separable group S, universal for all locally compact groups? (That is, a group such that every locally compact group should be continuously isomorphic with a subgroup of it?) The authors deduced from J. von Neumann's representation of compact groups the existence of a compact group, universal for all compact groups.

Commentary

This problem, if taken literally, is false by cardinality considerations. A better formulation of the question might be the following: Does there exist a Polish group S such that every second countable locally compact group G is continuously isomorphic to a subgroup of S? The answer to this question is yes. The existence of Haar measure on such a G shows that G is continuously isomorphic to a subgroup of $U(H)$, the unitary group on the complex separable infinite dimensional Hilbert space H, by considering the left regular representation of G on $L^2(G)$. It will follow from results given later that S cannot be locally compact and second countable.

This problem has a number of variants and perturbations, most of which are not quite so easy to settle; some of them are considered below. Unless otherwise noted, G will denote a typical locally compact group with a countable basis for its topology. The prerequisites for the following discussion may be found in Montgomery and Zippin's book on topological groups, Hochschild's book on Lie groups, and Kaplansky's book on infinite abelian groups.

Is there a locally compact group S such that every compact group is isomorphic to a subgroup of S? This is false by cardinality considerations. However, there is a compact metric group S such that every compact metric group is continuously isomorphic to a subgroup of S. Take S to be the product of countably many copies of $U(n)$, for each positive integer n, where $U(n)$ is the unitary group on a complex n-dimensional Hilbert space. This is the Peter-Weyl theorem.

Is there a locally compact group S such that S/S^0 is compact and such that every connected G is isomorphic as an abstract group to a subgroup of S? The answer is no. In fact there is no such S for all G which are noncompact centerless simple Lie groups. Let K be the maximal compact normal subgroup of S. S/K is a Lie group whose connected component of the identity is of finite index. For any G, either G intersects K only in the identity or G is contained in K. In the latter case, let π be any finite dimensional unitary representation of K. Then $\pi(G)$ is the identity or is isomorphic with G. Choose a π such that the latter holds. Now any solvable subgroup of a compact connected Lie group has an abelian subgroup of finite index. But any noncompact simple Lie group G has a solvable subgroup with no abelian subgroup of finite index. Hence, G intersects K only in the identity for all noncompact simple Lie groups. Hence,! if S exists, we may assume that S is a Lie group such that S/S^0 is finite. Since each G is simple and S/S^0 is finite, each G is contained in S^0. Hence, we may assume that S is a connected Lie group. Each

G intersects the radical of S in the identity, so we may further assume that S is a centerless semisimple Lie group. But this is impossible, for there is a finite upper bound on the length of a maximal solvable series of subgroups of S, but the length of a maximal solvable series of subgroups of an arbitrary noncompact simple G has no upper bound.

Note that there is a countable abelian group S such that every countable abelian G is isomorphic to a subgroup of S. Take S to be the direct sum of countably many copies of Q and the $Z_{p,\infty}$'s. However, there is no countable group S such that every countable group G is isomorphic to a subgroup of S, for the number of two-element subsets of such an S is countable, but the number of nonisomorphic countable groups with two generators is uncountable.

Finally, is there a Polish group S such that every Polish group G may be injected continuously into S? Is there a Polish group S such that every Polish group G is abstractly isomorphic to a subgroup of S? Is there a second countable locally compact group S so that every countable G is isomorphic to a subgroup of S? For this last question such an S cannot be connected since the group of all finite permutations S_f has a simple subgroup of index 2, either S_f is contained in a compact group or $S_f \cap K = (e)$. In the latter case, there is an injection of S_f to the Lie group S/K. Now in both cases there is therefore a faithful matrix representation of S_f. But this is a contradiction, for any nontrivial representation of S_n occurs on a vector space of dimension of least n, and $S_f = \bigcup_{n \geq 1} S_n$. A slight modification of this argument also shows that S cannot have the property that S/S^0 is compact.

Robert R. Kallman

Second Edition Commentary

It is now known ([1]) that every Polish group is topologically isomorphic to a closed subgroup of the homeomorphism group of the Hilbert cube. It is also known that every Polish group is topologically isomorphic to an isometry group of some Polish space and that every isometry group of a Polish space is topologically isomorphic to a closed subgroup of the isometry group of the universal Urysohn space ([2]).

1. Vladimir Vladimirovich Uspenskij, A universal topological group with a countable base, Functional Analysis and Its Applications, volume 20, 1986, pp. 160–161.
2. Vladimir Vladimirovich Uspenskij, On the group of isometries of the Urysohn universal metric space, Commentationes Mathematicae Universitatis Carolinae, volume 31, number 1, 1990, pp. 181–182.

Robert R. Kallman

PROBLEM 104: SCHAUDER

Let $f(x,y,z,p,q)$ denote a function of five variables possessing a sufficient number of derivatives and satisfying the inequalities: $f > M(|p|^{z+a} + |q|^{z+a})$; M constant, $a > 0$.

One has to find a minimum of the integral where $z = z(x,y)$, $p = z_x$, and $q = z_y$:

$$\iint_\Omega f(x,y,z,p,q)dx\,dy \tag{1}$$

(the region Ω should be sufficiently regular), among all z which possess all the first, possibly also the second continuous derivatives, and which assume the same values on the boundary. One may assume that the given boundary value has a given number of derivatives with respect to the arc length of the boundary curve. Expression (1) is assumed regular. A similar condition for free boundary conditions. Prove the existence of a function, minimizing in a given class. (Regular problem: $f_{pp}f_{qq} - 4f_{pq} > 0$.)

PROBLEM 105: SCHAUDER

The question is to find a system of functions $x(u,v)$, $y(u,v)$, $z(u,v)$ minimizing the parametric variational problem

$$\int \cdots \int f(x,y,z,X,Y,Z)du\,dv, \quad X = \begin{vmatrix} x_u & y_u \\ x_v & y_v \end{vmatrix}, \text{ etc.} \tag{2}$$

corresponding to Problem 104. It is allowed to change the class of admissible functions; these could be, for example, functions which are absolutely continuous in the sense of Tonelli. If not, then the problem is not solved. Mazur and Schauder solved Eq. (2) in the case when f does not contain x,y,z explicitly (even without any conditions analogous to those in Problem 104) but only within the class of functions absolutely continuous in the sense of Tonelli. Even this case (x,y,x does not appear) was not solved for functions $x(u,v),\ldots,z(u,v)$ sufficiently regular.

PROBLEM 106: BANACH

Prize: One bottle of wine, S. Banach

Let

$$\sum_{i=1}^\infty x_i$$

be a series [x_i are elements of a space of type (B)] with the property that under a certain ordering of its terms the sum $= y_0$, under some other ordering, equals y_1. Prove that for every real number ℓ there exists an ordering of the given series such that the sum of it will be: $\ell y_0 + (1 - \ell)y_1$. In particular, consider the case where x_i are continuous functions defined on the interval $(0,1)$. The convergence according to norm means uniform convergence.

Addendum. This does not hold in the space L^2 and also not in C. We define, for every n, 2^n functions $f_{n,i}(x)$ as follows:

$$f_{n,i}(x) = \quad 1, \quad \frac{i-1}{2^n} < x < \frac{i}{2^n} \quad \text{if } i = 2^n$$

$$f_{n,i}(x) = -1, \quad \frac{i-1}{2^n} < x < \frac{i}{2^n} \quad \text{if } i \neq 2^n$$

$$f_{n,i}(x) = 0 \quad \text{otherwise.}$$

Consider the orderings:

$$0 \equiv f_{1,1} + f_{1,3} + f_{1,2} + f_{1,4} + f_{2,1} + f_{2,2^2+1} + f_{2,2}$$

$$+ f_{2,2^2+2} + f_{2,3} + f_{2,2^2+3} + f_{2,4} + f_{2,2^2+4} + \cdots$$

$$1 \equiv f_{1,1} + f_{1,2} + f_{1,3} + f_{2,2^2+1} + f_{2,2^2+2} + f_{1,4}$$

$$+ f_{2,2^2+3} + f_{2,2^2+4} + \cdots$$

Since the $f_{i,k}$ assume integer values, one cannot order the series in such a way that it converges in L^2 to $0 < \ell < 1$.

Marcinkiewicz

Commentary

The addendum does not completely make sense as it stands. In particular, the definition of $f_{n,i}$ for $i > 2^n$ seems not to apply for $(0,1)$, if this is the interval the writer of the addendum is considering. Problem 106 inquires about the generalization to Banach spaces of the theorem of Steinitz [5] which asserts that if a series of vectors in R^m, $\sum v_i$, is convergent, then its sums (allowing all orderings of the terms) form a flat in R^m. For early discussions of the generalization of Steinitz's theorem to abstract spaces, see Wald [6], Hadwiger [2, 3], and Pracher [4]. As is stated by Damsteeg and Halperin [1], Steinitz's theorem follows from the theorem that for $c_i \geq 0$, $i = 1, 2$, and positive integer m there is a finite constant $K_m(c_1, c_2)$ with the property: whenever, for any n, the m-dimensional vectors $u_1, v_1, v_2 \ldots, v_n, u_2$ satisfy $|u_1| \leq c_1$, $|u_2| \leq c_2$, $|v_i| \leq 1$ for all i and $u_1 + \sum_{i=1}^{n} v_i + u_2 = 0$, then by reordering the v_i it is possible to satisfy $|u_1 + \sum_{i=1}^{h} v_i| \leq K_m(c_1, c_2)$ for $h = 1, 2, \ldots, n$. Assume now that $K_m(c_1, c_2)$ denotes the least possible such constant. Obviously, $K_m(\bar{c}_1, \bar{c}_2) \geq K_m(c_1, c_2)$ if $\bar{c}_1 \geq c_1$ and $\bar{c}_2 \geq c_2$. Damsteeg and Halperin [1] have proved that $K_m(0, 0) \geq (1/2)\sqrt{m+3}$ and thus that this method of proof of Steinitz's theorem cannot be used to generalize Steinitz's theorem to Hilbert space.

One can adjust the definition given in the addendum to provide a counterexample. However, perhaps the following geometric description, due to Israel Halperin is more transparent.

For every half-open interval $J = [a,b)$, let $J_L = [a,(a+b)/2)$ and $J_R = [(a+b)/2,b)$. Also, let $\chi(J)$ denote the characteristic function of J. Now, let $I = [0,1)$. Then $I_L, I_R, I_{LL}, I_{LR}, \ldots$ are determined by the above definitions.

Consider two sequences (different orderings of the same elements):

$$(s_1)\ \chi(I), -\chi(I), \chi(I_L), -\chi(I_L), \chi(I_R), -\chi(I_R), \chi(I_{LL}),$$
$$-\chi(I_{LL}), \chi(I_{LR}), -\chi(I_{LR}), \chi(I_{RL}), -\chi(I_{RL}), \ldots$$

and

$$(s_2)\ \chi(I), -\chi(I), \chi(I_L), \chi(I_R), -\chi(I_L), \chi(I_{LL}), \chi(I_{LR}),$$
$$-\chi(I_R), \chi(I_{RL}), \chi(I_{RR}), -\chi(I_{LL}), \chi(I_{LLL}), \chi(I_{LLR}), \ldots$$

The first sequence sums to 0 in every L^p, $0 < p < \infty$. This is easily seen by grouping each odd term with its successor. The second sequence sums to 1 in every L^p. This can be seen be grouping the terms in s_2 after the first term into sets of three consecutive terms:

$$-\chi(I), \chi(I_L), \chi(I_R), \qquad -\chi(I_L), \chi(I_{LL}), \chi(I_{LR})$$

and noticing that the sum of the three terms in each set is zero.

The final statement in the addendum states a reason why no rearrangement of this series converges to any constant function ℓ, $0 < \ell < 1$. Finally, since L^2 can be embedded in C, this gives an example in C.

1. I. Damsteeg and I. Halperin, The Steinitz-Gross theorem on sums of vectors, *Trans. Royal Soc. Canada Sect. III.* 44, series 3 (1950), 31–35.
2. H. Hadwiger, Über das Umordnungsproblem im Hilbertschen Raum, *Math. Zeit.*, 46 (1940), 70–79.
3. _____, Über das Umordnungsstärke und eine Erweiterung des Steinitzschen Satzes, *Math. Ann.*, 118 (1943), 702–717.
4. K. Prachar, Über bedingt konvergente Vektorreihen im Banach'schen Raum, *Monatsh. Math.*, 54 (1950), 284–307.
5. E. Steinitz, Bedingt konvergente Reihen und konvexe Systeme, *Journal Reine u. Angew. Math.*, 143 (1913), 128–175.
6. A. Wald, Reihen in topologischen Gruppen, *Ergebnisse eines Math. Koll. Wien* (1933), 59.

R. Daniel Mauldin

W.A. Beyer

Second Edition Commentary

The issue of Marcinkiewicz's *addendum* to Problem 106 was resolved by the Soviet probabilist Kornilov who was able to reconstruct what Marcinkiewicz was talking about. The example is similar to that constructed by Israel Halperin and included in the **Commentary** of Mauldin and Beyer. This example was used by V. M. Kadec (Functional Anal. Appl 20 (1986)) to solve Banach's problem completely when he was able to surgically implant a suitable sequence of finite dimensional versions of this example in any infinite dimensional Banach space. Kadec's main tool (other than his personal cleverness) was the celebrated Dvoretzky Spherical Sections theorem which says that any infinite dimensional Banach space contains subspaces of arbitrarily large dimension that are nearly isometric to Euclidean spaces of the same dimension. Kadec's theorem was generalized by W. Banaszczyk (J. Angew. Math. 403 (1990), 187–200 and Studia Math. 107 (1993), 213–222); the upshot of Banaszcyk's work is that the Levy-Steinitz theorem holds in a (locally convex) Fréchet space if and only if that space is nuclear. Remarkably, the key to Banaszczyk's proof is a sharpening of Dvoretzky's result due to V. Milman (Proc. AMS 94 (1985), 445–449). Alas, a companion to Banaszcyk's Fréchet space theorem was uncovered by J. Bonet and A. Defant (Israel J. math. 117 (2000), 131-156), who investigated what the situation is in spaces dual to nuclear Fréchet spaces.

A beautiful exposition of the original Levy-Steinitz theorem is due to Peter Rosenthal (Amer. Math. Monthly 94 (1987), 342–351) while the little book of M. I. Kadets and V. M. Kadets (*Series in Banach Spaces*, Birkhauser Verlag, 1987) is chock full of interesting examples and the full proof of V. M. Kadets' result as well as a proof of Dvoretzky's result. Regarding the Dvoretzky result no better source can be found than the Springer Lecture Notes in Mathematics volume of V. D. Milman and G. Schechtman *Asymptotic Theory of Finite Dimensional Normed Spaces.*

Joe Diestel

Kent, Ohio

PROBLEM 107: STERNBACH

Does there exist a fixed point for every continuous mapping of a bounded plane continuum E, which does not cut the plane, into part of itself? The same for homeomorphic mappings of E into all of itself.

Commentary

This problem was well known when Sternbach recorded it in the Scottish Book. In a recent conversation Professor Kuratowski mentioned that he, Mazurkiewicz, and Knaster first considered this problem in the late 1920s. Ayres [1] proved in 1930 that every locally connected nonseparating plane continuum has the fixed-point property for homeomorphisms. In 1932 Borsuk [1] introduced the concept of a retract to prove that every locally connected nonseparating plane continuum has the fixed-point property (for all continuous functions). In 1938 Hamilton [17] showed that this problem is related to the notion of an indecomposable continuum. Let G be a bounded simply connected plane domain whose closure does not separate the plane and whose boundary is hereditarily decomposable. Hamilton proved that the closure of G has the fixed-point property for homeomorphisms. Bell [2] in 1967 and Sieklucki [26] in 1968 proved that every nonseparating plane continuum that has a hereditarily decomposable boundary has the fixed-point property. In 1972 Hagopian [20] extended his theorem to every nonseparating plane continuum with the property that every pair of its points can be joined by a hereditarily decomposable subcontinuum.

In 1951 Cartwright and Littlewood [16] proved that every homeomorphism of a nonseparating plane continuum onto itself that can be extended to an orientation-preserving homeomorphism of the plane has a fixed point. Recently Bell [3] proved that every homeomorphism of a nonseparating plane continuum onto itself that can be extended to the plane has a fixed point.

The question whether a nonseparating plane continuum has the fixed point property for mappings is one of the most famous unsolved problems in plane topology. There was a rekindling of interest in this problem when Bellamy [5] constructed an example of a locally planar tree-like continuum without the property. However, in spite of concerted efforts, we have been unable to convert Bellamy's example to a nonseparating plane continuum without the fixed point property for mappings.

R. H. Bing

Second Edition Commentary

The fixed point problem for a planar continuum, nonseparating the plane, is one of the most persistent unsolved problems in topology [7]. K. Borsuk proved that there is no generalization to higher dimensions: there exists a cellular continuum in $\mathbb{R}^3$ without the fixed point property [12].

The planar problem seems to have been considered already by L.E.J. Brouwer in [14]; E.R. Reifenberg [25] reconstructs Brouwer's proof of the Cartwright-Littlewood theorem in [25]. In 1912, Brouwer [15] proved his famous "Translationssatz" from which it follows that an orientation preserving homeomorphism of the plane such that the orbit of some point is bounded must have a fixed point. In 1977, Morton Brown [13] used the Brouwer Translation Theorem to give a very concise proof of the Cartwright-Littlewood theorem, but not Bell's theorem for arbitrary homeomorphisms of the plane, without the assumption of preserving orientation. The Brouwer Translation Theorem does not generalize to orientation reversing homeomorphisms of the plane. In 1981, S.M. Boyles [9] constructed such homeomorphism without a fixed point and the orbit of every point bounded. It is still not known whether a nonseparating plane continuum has the fixed point property even with respect to homeomorphisms of the continuum.

In 1982, Bell announced a version of the Cartwright-Littlewood theorem for analytic maps. In 2013, A. Blokh, A., R. Fokkink, J. Mayer, L. Oversteegen, and E. Tymchatyn [8] extended the theorem to compositions of open and monotone maps.

Brouwer also considered a locally connected "circular continuum" invariant under an orientation reversing homeomorphism of the plane showing that the homeomorphism possesses two fixed points in the continuum. This theorem has been extended to any continuum in the plane with two invariant complementary domains by K. Kuperberg [21], and further by J.P. Boroński [10], who proved that if h is a planar orientation reversing homeomorphism of the plane, the continuum X is invariant under h, and there are at least $n + 1$ complementary domains, then the set of fixed point of h in X has at least $n + 1$ components.

A very significant result obtained independently by Bell [3], K. Sieklucki [26], and S.D. Iliadis [22] is that a nonseparating plane continuum without the fixed point property must contain an indecomposable continuum in the boundary. In 1971, C. Hagopian [19] proved that an arcwise connected, nonseparating plane continuum possesses the fixed point property. His 1972 paper generalizes this result and also the Bell-Iliadis-Sieklucki theorem.

O.J. Hamilton [18] proved in 1951 that chainable (arc-like) continua have the fixed point property with respect to continuous maps. In 1990, P. Minc extended the theorem to weakly chainable planar continua [23]. Having interest in showing that there is just one pseudo-arc (a homogeneous chainable continuum) Bing [6] showed that all chainable continua are planar. The tree-like continuum without the fixed point property constructed by Bellamy is not embeddable in the plane. In 1999, Minc gave an example of a non-planar weakly chainable tree-like continuum without the fixed point property [24]. The fixed point problem for planar, tree-like continua is still open.

1. W.L. Ayres, Some generalisations of Scherrer fixed-point theorem, *Fund. Math.,* 16 (1930), 332–336.
2. H. Bell, On fixed point properties of plane continua, *Trans. Amer. Math. Soc.,* 128 (1967), 539–548.
3. _____, A fixed point theorem for plane homeomorphism, *Bull. Amer. Math. Soc.,* 82 (1976), 778–780.
4. H. Bell, A fixed point theorem for plane homeomorphisms, *Fund. Math.,* 100 (1978), 119–128.
5. D.P. Bellamy, A tree-like continuum without the fixed point property, *Houston J. of Math.,* 6 (1980), 1–13.
6. R. H. Bing, Snake-like continua, *Duke Math. J.,* 18 (1951), 653–663.
7. R. H. Bing, The elusive fixed point property, *Amer. Math. Monthly,* 76 (1969), 119–132.
8. A. Blokh, A., R. Fokkink, J. Mayer, L. Oversteegen, and E. Tymchatyn, Fixed point theorems for plane continua with applications, *Mem. Amer. Math. Soc.,* 224 (2013), no. 1053.
9. S.M. Boyles, A counterexample to the bounded orbit conjecture, *Trans. Amer. Math. Soc.,* 266 (1981), 415–422.
10. J. P. Boroński, Fixed points and periodic points of orientation-reversing planar homeomorphisms, *Proc. Amer. Math. Soc.,* 138 (2010), 3717–3722.
11. K. Borsuk, Einige Satze über stetige Streckenbilder, *Fund. Math.,* 18 (1932), 198–213.

12. K. Borsuk, Sur un continu acyclique qui se laisse transformer topologiquement en lui même sans points invariants, *Fund. Math.*, 24 (1935), 51–58.
13. M. Brown, A short proof of the Cartwright-Littlewood theorem, *Proc. Amer. Math. Soc.*, 65 (1977), 372.
14. Brouwer, L. E. J., Über eineindeutige, stetige Transformationen von Flächen in sich, *Math. Ann.*, 69 (1910), 176–180.
15. Brouwer, L. E. J., Beweis des ebenen Translationssatzes, *Math. Ann.*, 72 (1912), 37–54.
16. M.L. Cartwright and J.E. Littlewood, Some fixed point theorems, *Ann. of Math.*, 54 (1951), 1–37.
17. O.H. Hamilton, Fixed points under transformations of continua which are not connected im kleinen, *Trans. Amer. Math. Soc.*, 44 (1938), 18–24.
18. Hamilton, O. H., A fixed point theorem for pseudo-arcs and certain other metric continua, Proc. Amer. Math. Soc., 2 (1951), 173–174.
19. C.L. Hagopian, A fixed point theorem for plane continua, *Bull. Amer. Math. Soc.*, 77 (1971), 351–354.
20. _____, Another fixed point theorem for plane continua, *Proc. Amer. Math. Soc.*, 31 (1972), 627–628.
21. K. Kuperberg, Fixed points of orientation reversing homeomorphisms of the plane, *Proc. Amer. Math. Soc.*, 112 (1991), 223-229.
22. S.D. Iliadis, Positions of continua on the plane, and fixed points, *Vestnik Moskov. Univ. Ser. I Mat. Meh.*, 25 (1970), 66–70.
23. P. Minc, A fixed point theorem for weakly chainable plane continua, *Trans. Amer. Math. Soc.*, 317 (1990), 303–312.
24. P. Minc, A weakly chainable tree-like continuum without the fixed point property, *Trans. Amer. Math. Soc.*, 351 (1999), 1109–1121.
25. E.R. Reifenberg, On the Cartwright-Littlewood fixed point theorem, *Ann. of Math.*, 61 (1955), 137–139.
26. K. Sieklucki, On a class of plane acyclic continua with the fixed point property, *Fund. Math.*, 63 (1968), 257–278.

K. Kuperberg

Lex Oversteegen

PROBLEM 108: BANACH, MAZUR, ULAM

Let E be a space of type (B) which has a basis and H a set everywhere dense in E.

(1) Does there exist a basis whose terms belong to H?
(2) The same question under the additional assumption that the set H is linear.

Addendum. Affirmative answer.

Krein

PROBLEM 109: MAZUR, ULAM

October 16, 1935

Given are n functions of a real variable: $f_1, \ldots, f_n$. Denote by $R(f_1, \ldots, f_n)$ the set of all functions obtained from the given functions through rational operations (expressions of the form

$$\frac{\sum a_{k_1 \cdots k_n} f_1^{k_1} \cdots f_n^{k_n}}{\sum b_{k_1 \cdots k_n} f_1^{k_1} \cdots f_n^{k_n}} \Bigg).$$

Must there always exist, in the set R, a function f such that its indefinite integral does not belong to the set R?

An analogous question in the case where we include in the set R all the functions obtained by *composing* functions belonging to R.

Addendum. An affirmative answer for the first question was found by Docents, Dr. S. Kaczmarz and Dr. A. Turowicz.

March, 1938

Commentary

The solution to the first question appears as S. Kaczmarz and A. Turowicz, "Sur l'irrationalité des intégrales indefinies," *Studia Mathematica* 8 (1939), 129–134. Their solution is more general and is the following: Let $f_i(x)$, $1 \leq i < \infty$, be an infinite sequence of functions of a real variable that are assumed to be finite and summable in an interval (a,b) $(-\infty \leq a < b \leq \infty)$. Denote by Z the functions $g(x) = R(f_1(x), f_2(x), \ldots, f_n(x))$ where $R(y_1, \ldots, y_n)$ is an arbitrary rational function with real coefficients and n is arbitrary. A function $g(x)$ in Z is defined for all x in (a,b) for which the denominator is not zero. Then for each interval (α, β) such that $a < \alpha < \beta < b$ there exists a function $g_0(x)$ in Z such that (1) $g_0(x)$ is finite and summable in (α, β), (2) the function $F(x) = \int_{-\infty}^x g_0(t) dt$ is not in Z. If a and b are both infinite, one can take $\alpha = a$ and $\beta = b$.

The title of the paper is well chosen. The theorem states that if one starts with a countable collection F of functions and denotes by Z the closure of F under the operation of forming rational functions of functions in F, then there are functions in Z whose integrals are not in Z. That is, integrals of functions in Z may be "irrational."

The proof depends on two lemmas.

Lemma 1. If $r_j(x_i; 1 \leq i \leq k)$, $1 \leq j \leq k+1$, are rational functions of k variables, then there exists a polynomial $G(y_i; 1 \leq i \leq k+1)$ of $k+1$ variables which is not identically zero such that $G(r_i(x_j; 1 \leq j \leq k); 1 \leq i \leq k+1) \equiv 0$.

Lemma 2. Suppose $Q(\log|x - c_1|, \log|x - c_2|, \ldots, \log|x - c_n|) \equiv 0$ for all x in (α, β), Q being a polynomial in n variables and c_i being outside (α, β). Then $Q(y_1, \ldots, y_n) \equiv 0$.

The result of Kaczmarz and Turowicz seems to be unrelated to the theory of Liouville-Ritt-Risch of integration in finite terms. Also, the theorem appears to be nonconstructive in the sense that a function $g_0(x)$ is not exhibited. The reviewer is not aware of any work on the second question.

W.A. Beyer

Second Edition Commentary

The phrasing of Problem 109 is somewhat vague, and is subject to interpretation. Also, it is possible that the original question formulated by Mazur and Ulam was different. Indeed, in [1] Kaczmarz and Turowicz quote the first part of Problem 109 as follows.

Let $f_1(x), \ldots, f_n(x)$ be a set of n real functions continuous in an interval (a,b), and let Z be the set of the functions of the form $R(f_1(x), \ldots, f_n(x))$, where R denotes an arbitrary rational function of n variables. Does there exist in Z a function which is summable in (a,b) and whose indefinite integral does not belong to Z?

That is, according to Kaczmarz and Turowicz, in the original problem the functions $f_1, \ldots, f_n$ were continuous. In order to formulate the possible interpretations of the problem, let us start with fixing the ideas.

Let $\mathscr{R}$ denote the set of all real-valued functions defined on subsets of the real line. We shall denote by $\mathscr{F}_c$ the family of all functions $f : I \to \mathbb{R}$ such that I is a subinterval of $\mathbb{R}$ and f is continuous on I. The domain of $f \in \mathscr{R}$ is denoted by $D(f)$. We define addition, multiplication, division, and composition on $\mathscr{R}$ as usual with the obvious conventions $D(f + g) = D(f \cdot g) = D(f) \cap D(g)$, $D(f/g) = D(f) \cap \{x \in D(g) : g(x) \neq 0\}$ and $D(f \circ g) = \{x \in D(g) : g(x) \in D(f)\}$.

For every family $\mathscr{F} \subset \mathscr{R}$ we denote by $E(\mathscr{F})$ the set of all functions obtained from $\mathscr{F}$ and from the constant functions through addition, multiplication, division, and composition.

The second question of Problem 109 asks if for every finite system $\mathscr{F} \subset \mathscr{R}$ (or $\mathscr{F} \subset \mathscr{F}_c$) there exists an $f \in E(\mathscr{F})$ such that the indefinite integral of f does not belong to $E(\mathscr{F})$. It is not specified what we should mean by the indefinite integral of f. In the Kaczmarz-Turowicz variant the function f is summable on (a,b), so it is very likely that by the indefinite integral of f Mazur and Ulam meant the integral function $x \mapsto \int_{x_0}^x f \, dx$. But we could also interpret the indefinite integral as the primitive (antiderivative) of f. In this case we have to assume that f is defined on an interval. Summing up, we interpret the problem as follows.

Is it true that for every finite system $\mathscr{F} \in \mathscr{R}$ (or $\mathscr{F} \in \mathscr{F}_c$) there exists an $f \in E(\mathscr{F})$ such that f is defined a.e. on an interval I, f is summable on I, and its integral function on I does not belong to $E(\mathscr{F})$ (or, f is defined on an interval I, f has an antiderivative F on I and F does not belong to $E(\mathscr{F})$)?

In the sequel we answer three of these questions and give a partial answer to the fourth one.

I. First we show that if there are no restrictions on the functions f_i, then the answer is negative, for both interpretations of the indefinite integral. We prove that *there are functions $f, g \in \mathscr{R}$ such that $\mathscr{F}_c \subset E(f, g)$.* Since the integral functions and the antiderivatives belong to $\mathscr{F}_c$, this will solve the problem in the negative.

Let A and B be subsets of $\mathbb{R}$ such $|A| = |B| = 2^\omega$ and the map $(x, y) \mapsto x + y$ $(x \in A, \ y \in B)$ is injective. (Let, e.g., A be the set of all numbers in $[0,1]$ whose decimal expansion only contains the digits 0 and 1, and let $B = 2A$.) Let g be an injective map from $\mathbb{R}$ into A.

The cardinality of the family $\mathscr{F}_c$ is of the continuum, and thus there is a bijection ϕ from B onto $\mathscr{F}_c$. We put $X = \{(x,y) \subset \mathbb{R}^2 : y \in B,\ x \in D(\phi(y))\}$, and define

$$f(g(x) + y) = \phi(y)(x) \qquad ((x,y) \in X).$$

Since the map $(x,y) \mapsto g(x) + y$ is injective on X, it follows that f is well defined on the set $\{g(x) + y : (x,y) \in X\}$.

Let $h \in \mathscr{F}_c$ be arbitrary. If $h = \phi(y)$, then we have $h(x) = f(g(x) + y)$ for every $x \in D(h)$. This proves that $h \in E(f,g)$. $\square$

II. We show that if the functions $f_1,\dots,f_n$ are supposed to be continuous, and if we interpret the indefinite integral as the antiderivative of a function defined on an interval, then the answer to the question is again negative. We shall prove the following.

Theorem 1. *There are continuous functions $f_1,\dots,f_6 \in C[0,1]$ and $f_7, f_8 : \mathbb{R} \to \mathbb{R}$ such that whenever $f \in E(f_1,\dots,f_8)$ is defined on an interval, then the indefinite integral of f belongs to $E(f_1,\dots,f_8)$.*

Note that if $f \in E(f_1,\dots,f_8)$ is defined on an interval I, then f is continuous on I, and thus the integral function of f is the same as the antiderivative of f.

The proof of Theorem 1 is based on the following observation.

Theorem 2. *There are continuous functions $g, \phi_1,\dots,\phi_5 \in C[0,1]$ with the following property. For every Lipschitz function $f : [0,1] \to \mathbb{R}$ there are constants α, β and $\gamma_1,\dots,\gamma_5$ such that*

$$f(x) = \alpha \cdot \sum_{i=1}^{5} g\left(\beta \phi_i(x) + \gamma_i\right)$$

for every $x \in [0,1]$.

Proof. If $f \in \mathscr{R}$, $A \subset D(f)$ and f is Lipschitz on A, then we shall denote by $Lip(f;A)$ the infimum of the numbers K such that $|f(x) - f(y)| \le K \cdot |x - y|$ for every $x,y \in A$. The set

$$K = \{f \in C[0,1] : |f(0)| \le 1 \quad \text{and} \quad Lip(f;[0,1]) \le 1\}$$

is a compact, convex subset of the Banach space $C[0,1]$. Then the intersection of K with any open ball is also convex, hence connected. This shows that K is a compact and locally connected subset of $C[0,1]$. Therefore, by the Hahn-Mazurkiewicz-Sierpiński theorem [3, Theorem 2, §50, II, p. 256], K is the continuous image of $[0,1]$. Let ψ be a continuous map from $[0,1]$ onto K. We define

$$\Psi(x,y) = \psi(y)(x) \qquad (x,y \in [0,1]);$$

then Ψ is a continuous function defined on the unit square.

Now we apply a variant of Kolmogorov's theorem representing continuous functions of several variables using functions of one variable and addition [2, Statement (5), p. 231]. We obtain that there are functions $g, \phi_1, \dots, \phi_5 \in C[0,1]$ and positive constants λ_1, λ_2 such that

$$\Psi(x,y) = \sum_{i=1}^{5} g\left(\lambda_1 \phi_i(x) + \lambda_2 \phi_i(y)\right) \tag{1}$$

for every $x, y \in [0,1]$.

Let $f \colon [0,1] \to \mathbb{R}$ be Lipschitz. Then there is a constant $\varepsilon > 0$ such that $\varepsilon \cdot f \in K$. By the choice of the function Ψ, there is a $y \in [0,1]$ such that $\varepsilon \cdot f(x) = \Psi(x,y)$ for every $x \in [0,1]$. Writing c_i for $\lambda_2 \phi_i(y)$ ($i = 1, \dots, 5$), (1) gives

$$\varepsilon \cdot f(x) = \sum_{i=1}^{5} g\left(\lambda_1 \phi_i(x) + c_i\right) \qquad (x \in [0,1]),$$

which completes the proof. $\qquad\qquad\qquad\qquad\qquad\qquad\qquad\qquad\qquad\qquad\qquad \square$

Proof of Theorem 1. Let $g, \phi_1, \dots, \phi_5$ as in Theorem 2. We put $f_1 = g$ and $f_i = \phi_{i-1}$ for every $i = 2, \dots, 6$. In addition, we define $f_7(x) = x$ and $f_8(x) = \arctan x$ for every $x \in \mathbb{R}$. We prove that the functions $f_1, \dots, f_8$ satisfy the requirements. We put $E_0 = E(f_1, \dots, f_8)$.

Suppose that $f \in E_0$ is defined on an interval I. Then f is continuous on I. It is clear that the integral function of f is Lipschitz on every closed and bounded subinterval of I. Therefore, it is enough to prove the following: if F is defined on an interval I and if F is Lipschitz on every closed and bounded subinterval of I, then $F \in E_0$.

First suppose that $I = [a,b]$ is a closed, bounded interval. Then F is Lipschitz on $[a,b]$, and thus the function $x \mapsto F((b-a)x+a)$ ($x \in [0,1]$) is Lipschitz on $[0,1]$. By Theorem 1 we have

$$F((b-a)x+a) = \alpha \cdot \sum_{i=1}^{5} f_1\left(\beta f_{i+1}(x) + \gamma_i\right)$$

for every $x \in [0,1]$. Thus

$$F(x) = \alpha \cdot \sum_{i=1}^{5} f_1\left(\beta f_{i+1}\left(\frac{1}{b-a} \cdot (x-a)\right) + \gamma_i\right)$$

for every $x \in [a,b]$. Therefore, we have

$$F = \alpha \cdot \sum_{i=1}^{5} f_1\left(\beta f_{i+1}(\delta f_7 + \lambda) + \gamma_i\right)$$

on $[a,b]$, where $\delta = 1/(b-a)$ and $\lambda = -a/(b-a)$. This proves $F \in E_0$.

Next suppose that $I = (a,b)$ is a bounded open interval. Then F is Lipschitz on the interval $[a+2^{-n}, b-2^{-n}]$ for every $n \geq n_0$, where n_0 is a positive integer with $2^{-n_0} < (b-a)/2$. Let

$$M_n = \max\{1 + |F(x)| : x \in [a+2^{-n}, b-2^{-n}]\} \quad \text{and}$$

$$K_n = 1 + Lip(F; [a+2^{-n}, b-2^{-n}])$$

for every $n \geq n_0$. It is easy to construct a Lipschitz function $h \in C[a,b]$ such that $h(a) = h(b) = 0$, $h(x) > 0$ for every $a < x < b$,

$$|h(x)| \leq 1/K_n \quad \text{if} \quad x \in [a+2^{-n}, a+2^{-n+1}] \cup [b-2^{-n+1}, b-2^{-n}] \quad \text{and}$$

$$Lip(h; [a+2^{-n}, a+2^{-n+1}]) = Lip(h; [b-2^{-n+1}, b-2^{-n}]) < 1/M_n$$

for every $n \geq n_0$. Let

$$k(x) = \begin{cases} F(x) \cdot h(x) & \text{if } x \in (a,b), \\ 0 & \text{if } x = a \text{ or } x = b. \end{cases}$$

Then h and k are both Lipschitz functions on $[a,b]$, and thus, as we proved above, they belong to E_0. Since $F = k/h$, we obtain $F \in E_0$.

If $I = (a,b]$ or $I = [a,b)$, then an obvious one sided modification of the previous argument shows that $F \in E_0$.

Next suppose that $I = \mathbb{R}$. Then the function $G(x) = F(\tan x)$ is Lipschitz on every closed subinterval of $(-\pi/2, \pi/2)$. Therefore, we have $G \in E_0$. Thus we have $F = G \circ f_8 \in E_0$.

A similar argument applies if $I = [0,\infty)$ (or $I = (0,\infty)$). Put $J = [0,\pi/2)$ (or $J = (0,\pi/2)$) and $H(x) = F(\tan x)$ for every $x \in J$. Then H is Lipschitz on every closed subinterval of J. Therefore, we have $H \in E_0$. Thus we have $F = H \circ f_8 \in E_0$.

Using suitable linear substitutions, one can check the statement in the cases when I is one of $[a,\infty)$, $(-\infty, b]$, (a,∞) or $(-\infty, b)$. $\qquad\square$

III. Now we turn to the case when the functions f_i are supposed to be continuous, and we interpret the indefinite integral as the integral function of a summable function defined a.e. on an interval. This case seems to be the most difficult among the variants, and we only can give a partial solution. We show that the answer to the corresponding question is *affirmative*, even if we start with a countable system of continuous functions, assuming that some elementary functions are among them. The precise statement is the following.

Theorem 3. *For every countable set of continuous functions $\mathscr{F} \subset \mathscr{F}_c$ there is a function $f \in E(\mathscr{F} \cup \{e^x, \log x, \sin x, \arcsin x\})$ such that f is defined on $\mathbb{R}$ except one point, f is summable on $\mathbb{R}$, and the integral function of f does not belong to $E(\mathscr{F} \cup \{e^x, \log x, \sin x, \arcsin x\})$.*

Proof. Let $\mathscr{F} \subset \mathscr{F}_c$ be countable, and put $\mathscr{F}_1 = \mathscr{F} \cup \{e^x, \log x, \sin x, \arcsin x\}$ and $E_1 = E(\mathscr{F}_1)$. One can prove that there is a sequence of functions $\omega_i : [0, \infty) \to \mathbb{R}$ ($i = 1, 2, \ldots$) such that $\lim_{\delta \to 0+} \omega_i(\delta) = 0$ for every i, and whenever $f \in E_1$ is defined on $[0, 1]$, then the modulus of continuity of $f|_{[0,1]}$ is dominated by one of the functions ω_i. That is,

$$\max\{|f(x) - f(y)| : x, y \in [0, 1], \ |x - y| \le \delta\} \le \omega_i(\delta)$$

for every $\delta > 0$. (See the proof of [4, Theorem 6.4].) Let $x_i \in [0, 1]$ be a strictly decreasing sequence such that $x_i \to 0$ and $0 < x_i < \omega_i(2^{-i})$ for every i. If F is defined on $[0, 1]$, $F(0) = 0$ and $F(x_i) > 2^{-i}$ for every i, then the modulus of continuity of F cannot be dominated by any one of the functions ω_i, and thus $F \notin E_1$.

We shall construct a function $f_0 \in E_1$ such that f_0 is defined on $\mathbb{R}$, $f(x) = x^{-2} \cdot f_0(1/x)$ is summable on $\mathbb{R}$, and the integral function $F(x) = \int_o^x f \, dx$ satisfies $F(x_i) > 2^{-i}$ for every i. By the choice of the numbers x_i this will imply $F \notin E_1$, and will complete the proof.

We shall construct f_0 such that $|f_0(x)| \le 1/x^2$ holds for $x \le -1$. This will ensure that f is summable on the intervals $(-\infty, -1]$, $[-1, 0]$ and $[1, \infty)$. We have

$$\int_{x_{i+1}}^{x_i} f \, dx = \int_{1/x_i}^{1/x_{i+1}} f_0 \, dx$$

for every i. We shall construct f_0 such that the value of this integral will be between 2^{-i} and 2^{1-i} for every $i = 2, 3, \ldots$. This will ensure that f is summable on $[0, 1]$, and $F(x_i) > 2^{-i}$ for every i.

In order to construct f_0, we apply Theorem 6.4 of [4]. We find that for every pair of continuous functions $g : \mathbb{R} \to \mathbb{R}$ and $\varepsilon : \mathbb{R} \to (0, \infty)$ there is a function $f_0 \in E_1$ such that $|f_0(x) - g(x)| < \varepsilon(x)$ for every $x \in \mathbb{R}$. It is clear that with a suitable pair of functions g and ε, f_0 will have the required properties.   □

1. S. Kaczmarz and A. Turowicz, Sur l'irrationalité des intégrales indéfinies, *Studia Math.* **8** (1939), 129-134.
2. Jean-Pierre Kahane, Sur le théorème de superposition de Kolmogorov, *J. Approximation Theory* **13** (1975), 229-234.
3. K. Kuratowski, Topology, vol 2, Academic Press, 1968.
4. M. Laczkovich and I. Z. Rusza, Elementary and integral-elementary functions, *Illinois J. Math.* (44) (2000), 161-182.

M. Laczkovich

PROBLEM 110: ULAM

October 1, 1935; Prize: One bottle of wine, S. Ulam

Let M be a given manifold. Does there exist a numerical constant K such that every continuous mapping f of the manifold M into part of itself which satisfies the condition $|f^n x - x| < K$ for $n = 1, 2, \ldots$ [where f^n denotes the nth iteration of the image $f(x)$] possesses a fixed point: $f(x_0) = x_0$? (By a manifold, we mean a set such that the neighborhood of every point is homeomorphic to the n-dimensional Euclidean sphere.) The same under more general assumptions about M (general continuum?).

Addendum. An affirmative answer in the case where M is a locally contractible 2-dimensional continuum.

March, 1936

J. von Neumann observed that from the n-dimensional theorem an affirmative answer would follow for Hilbert's problem concerning the introduction of analytic parameters in n-parameter groups.

March, 1936

Updated Commentary

The second part of the problem has been answered in the negative by W. Kuperberg, who gave an example of a 1-dimensional metric continuum, which for every $K > 0$ admits a fixed-point free K-involution. Subsequently, W. Kuperberg and P. Minc proved that the Cartesian product of the Hilbert cube Q and the circle S^1 has the property: For every $K > 0$ there exists a dynamical system Φ on $Q \times S^1$ such that for each $x \in Q \times S^1$ the trajectory $\Phi(t,x)$ is of diameter smaller than K, and $\Phi(n,x) \neq x$ for each nonzero integer n. Thus, by defining $f(x) = \Phi(1,x)$, the authors have found a fixed-point-free homeomorphism f satisfying the above property $|f^n(x) - x| < K$, for $n = 1, 2, \ldots$, and defined on an absolute neighborhood retract.

The problem for a manifold has been answered in the negative by K. Kuperberg and C. Reed [9] who gave an example of a C^∞ dynamical system Φ on the 3-dimensional Euclidean space $\mathbb{R}^3$ with all trajectories bounded by the given constant K and no rest points (in fact, $\Phi(n,x) \neq x$ for any $x \in \mathbb{R}^3$). Again, by taking $f(x) = \Phi(1,x)$, the authors obtain an example of a fixed-point free homeomorphism of $\mathbb{R}^3$ onto $\mathbb{R}^3$ such that $|f^n(x) - x| < K$ for any iteration f^n of f. An example of a dynamical system with the same properties can be constructed on a closed manifold $S^1 \times S^1 \times S^1$.

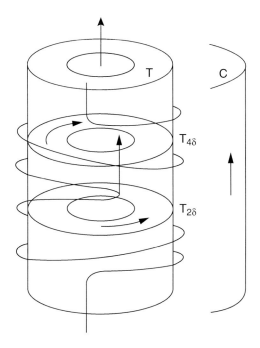

W. Kuperberg

C. Reed

The examples by W. Kuperberg and P. Minc, which were given in 1979, are described in the survey [8].

The dynamical system on $\mathbb{R}^3$ with uniformly bounded orbits in [9] is based on a construction of a plug, first constructed by F.W. Wilson in [12]. In fact, without explicitly specifying, Wilson's paper yields a C^∞ dynamical system on any smooth, connected, metrizable, boundaryless n-manifold, $n \geq 3$, non-compact or of Euler characteristic zero, with each orbit contained in an element of a given open cover, thus answering Ulam's question. The Kuperberg-Reed plug possesses invariant annuli, a property which led to the radius inequality method used by K. Kuperberg in [7] for the construction of a smooth aperiodic dynamical system on S^3 (Seifert conjecture [11]), and by G. Kuperberg and K. Kuperberg [5] for the construction of real analytic counterexamples to the Modified Seifert conjecture. The C^1 Seifert conjecture was solved in the negative earlier by P.A. Schweitzer [10]. A modification of Schweitzer's plug by G. Kuperberg [6] gave a volume preserving, aperiodic, C^1 dynamical system Φ on S^3; the total space of the line bundle tangent to Φ has a symplectic structure, and Φ is Hamiltonian with respect to this structure. The classical Hamiltonian Seifert conjecture for S^3 embedded in $\mathbb{R}^4$ was solved by V.L. Ginzburg and B.Z. Gürel [3]. It is not known whether there exists a volume preserving, C^∞ counterexample to the Seifert conjecture.

The following question posed by G. Kuperberg remains open: Does there exist a non-singular, volume preserving dynamical system on $\mathbb{R}^3$ with uniformly bounded orbits?

There are examples of non-singular dynamical systems on $\mathbb{R}^3$ with bounded (though not uniformly) orbits:

1. B. Brechner and R.D. Mauldin [2], which is based on K. Borsuk's example in [1] of an acyclic compact subset of $\mathbb{R}^3$ without the fixed point property and cannot be modified to be volume preserving.
2. G.S. Jones and J.A. Yorke [4], which can be modified to be volume preserving.

1. K. Borsuk, *Sur un continu acyclique qui se laisse transformer topologiquement en lui même sans points invariants,* Fund. Math. 24 (1935), 51–58.
2. B.L. Brechner and R.D. Mauldin, *Homeomorphisms of the plane,* Pacific J. Math. 59 (1975), 375–381.
3. V.L. Ginzburg and B.Z. Gürel, *A C^2-smooth counterexample to the Hamiltonian Seifert conjecture in $\mathbb{R}^4$,* Ann. of Math. 158 (2003), 953–976.
4. G.S. Jones and J.A. Yorke, *The existence and nonexistence of critical points in bounded flows,* J. Differential Equations 6 (1969), 238–246.
5. G. Kuperberg and K. Kuperberg, *Generalized counterexamples to the Seifert conjecture,* Ann. of Math. 144 (1996), 239–268.
6. G. Kuperberg, *A volume-preserving counterexample to the Seifert conjecture,* Comment. Math. Helv. 71 (1996), 70–97.
7. K. Kuperberg, *A smooth counterexample to the Seifert conjecture,* Ann. of Math. 140 (1994), 723–732.
8. K. Kuperberg, W. Kuperberg, P. Minc and C.S. Reed, *Examples related to Ulam's fixed point problem,* Topol. Methods in Nonlinear Analysis 1 (1993), 173–181.
9. K. Kuperberg and C. Reed, *A rest point free dynamical system on $\mathbb{R}^3$ with uniformly bounded orbits,* Fund. Math. 114 (1981), 229–234.
10. P.A. Schweitzer, *Counterexamples to the Seifert conjecture and opening closed leaves of foliations,* Ann. of Math. 100 (1974), 386–400.
11. H. Seifert, *Closed integral curves in 3-space and isotopic two-dimensional deformations,* Proc. Amer. Math. Soc. 1 (1950), 287–302.
12. F.W. Wilson, *On the minimal sets of non–singular vector fields,* Ann. of Math. 84 (1966), 529–536.

K. Kuperberg

PROBLEM 111: SCHREIER

Does there exist an uncountable group with the property that every countable sequence of elements of this group is contained in a subgroup which has a finite number of generators? In particular, do the groups S_∞ and the group of all homeomorphisms of the interval have this property?

Commentary

The answer to the first problem is yes. It can be obtained by taking the union of an appropriate chain of groups obtained by amalgamation using the fact that every countable group is a subgroup of a group with two generators.

The second question remains open.

J. Mycielski

Second edition Commentary

Fred Galvin gave an affirmative answer to the second question in his paper: Generating countable sets of permutations, *J. London Math. Soc.* 51 (1995), 230–242. Whether the homeomorphism group of the interval has this property remains open.

R. Daniel Mauldin

PROBLEM 112: SCHREIER

Is an automorphism of a group G which transforms every element into an equivalent one of necessity an inner automorphism?

Commentary

The answer is no. Burnside gave an example of a finite group with outer automorphisms mapping every conjugacy class onto itself. For further work and references, see G.E. Wall, Finite groups with class-preserving outer automorphisms, *J. London Math. Soc.* 22 (1947), 315–320.

Jan Mycielski

PROBLEM 113: SHREIER

Let C denote the space of continuous functions of a real variable (under uniform convergence in every bounded interval); let $F(f)$ denote an operation which is continuous, which has an inverse which maps C onto itself, and such that it maps the composition of two functions $f(g)$ into the composition of $F(f)$ and $F(g)$.

Is $F(f)$ of the form $F(f(t)) = hfh^{-1}(t)$, where h is a continuous function strictly monotonic in this interval $(-\infty, +\infty)$ and

$$\lim_{t \to -\infty} h(t) = -\infty, \quad \lim_{t \to +\infty} h(t) = +\infty?$$

Solution

Theorem. Let T be a continuous automorphism of the semigroup C onto C. Then there is a homeomorphism h of $\mathbb{R}$ onto $\mathbb{R}$ so that $T(f) = h^{-1}fh$.

Notice that an element f of C is constant if and only if $fg = f$, for all g in C. From this it follows that T takes constant functions to constant functions. For each $x \in \mathbb{R}$, let $w(x)$ be the number such that the constant function $\bar{x}$ is taken to $w(x)$ by T. Notice that w is a one-to-one map of $\mathbb{R}$ onto $\mathbb{R}$. Also, since T is continuous (under uniform convergence on compact sets), w is continuous. Therefore, w is a homeomorphism. Let $h = w^{-1}$. We plan to show that $T(f) = h^{-1}fh$, for all f in C.

Let H be the subset of C consisting of all homeomorphisms of $\mathbb{R}$. It can be checked that $T(H) = H$. Thus, $T|_H$ is a continuous automorphism of the group of homeomorphisms of $\mathbb{R}$. It follows that $T|_H$ is inner [1]. So, there is a homeomorphism k such that $T(f) = k^{-1}fk$ for all homeomorphisms f.

In order to see that $k = h$, fix x_0 and for each n, set $h_n(x) = (1/n)(x - x_0) + x_0$. Then $\{h_n\}_{n=1}^{\infty}$ is a sequence of homeomorphisms converging to $\bar{x}_0$. Thus, $T(h_n) = k^{-1}h_nk$ converges to $T(\bar{x}_0) = h^{-1}(x_0)$. In particular, $(Th_n)(k^{-1}(x_0)) = k^{-1}(h_n(x_0)) = k^{-1}(x_0)$ converges to $h^{-1}(x_0)$. Thus, $k = h$.

Now fix $f \in C$ and $x_0 \in \mathbb{R}$. We will show that $(T(f))(h^{-1}(f(x_0)) = h^{-1}(f(x_0))$. For each n, let h_n be a homeomorphism of $\mathbb{R}$ such that $h_n(x_0 + n) = x_0 + 1/n$. Then the sequence fh_n converges to the constant function $f(x_0)$ uniformly on every interval. Thus, $T(fh_n)$ converges to $T(f(\bar{x}_0)) = h^{-1}(\bar{f}(x_0))$. Since $T(fh_n) = Tf(h^{-1}h_nh)$ and $T(fh_n)(h^{-1}(x_0)) \to h^{-1}(f(x_0))$, it follows that $(Tf)(h^{-1}(x_0)) = h^{-1}(f(x_0))$. The theorem follows from this.

1. N.J. Fine and G.E. Schweigert, On the group of homeomorphisms of an arc, *Ann. Math.* 62 (1955), 237–253.

R. Daniel Mauldin

PROBLEM 114: AUERBACH, ULAM

The circumference of a circle can be approximated by a one-to-one continuous image of a half line p in an *essential* manner; that is to say, the Abbildungsgrad of the transformation obtained by central projecting of the line into the circumference is equal to $+\infty$ and the approximated circle is the set of points of condensation.

Is it possible to approximate analogously the surface of a sphere in the 3-dimensional space by a one-to-one continuous image of a plane?

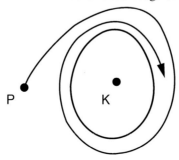

P   K

Remark

The answer is no. See A. Calder, For $n > 1$ any map $R^n \to S^n$ is uniformly homotopic to a constant, *Indag. Math.* 34 (1972), 32–36, and A. Calder, Uniformly trivial maps into spheres, *Bull. Amer. Math. Soc.* 81 (1975), 189–191.

PROBLEM 115: ULAM

Does there exist a homeomorphism h of the Euclidean space R^n with the following property? There exists a point p for which the sequence of points $h^n(p)$ is everywhere dense in the whole space. Can one even demand that all points except one should have this property? For a plane such a homeomorphism (with the desired property only for certain points) was found by Besicovitch.

Commentary

The answer to the first question is yes for $n \geq 2$ [8, 10] (and of course no for $n = 1$). In this part of the problem it does not matter whether the sequence $h^n(p)$ is taken to refer to the full orbit of p or just to the positive semiorbit (because if some point has a dense orbit, then a dense G_δ set of points have dense positive and negative semiorbits [9, p. 70]), but it makes a difference in the second part. If the sequence is taken to refer to the positive semiorbit of p, the answer to the second question is always no. By a theorem of Dowker [6, Th. K] (applied to $R^n \cup \{\infty\}$), if some point has a dense orbit, then there are points whose positive semiorbit is nowhere dense and lies outside any given sphere. More generally, Homma and Kinoshita [7, Th. 5] have shown that under any continuous mapping of R^n into itself there is a dense set of points whose positive semiorbit closure is a proper subset of R^n. See also Birkhoff [4, p. ! 202]. If the sequence $h^n(p)$ is taken to refer to the full orbit of p, the second question remains open for every $n \geq 2$. Besicovitch [2, 3] constructed a class of homeomorphisms of R^2 that are aperiodic except for a fixed point at 0. Each has a point whose positive semiorbit is dense in R^2, but also a point other than 0 whose orbit is bounded. In the case $n = 2$ the words "except one" are important— it is known that not every orbit can be dense in R^2 [5]. Indeed, by a theorem of Brouwer (see [1, Prop. 1.2]), if h is a homeomorphism of R^2 onto itself, then either $h^2(p) = p$ for some p or else $h^n(p) \to \pm\infty$ for every p. For $n \geq 3$ the second question appears to remain open even when the words "except one" are deleted.

1. S.A. Andrea, On homeomorphisms of the plane which have no fixed points, *Abh. Math. Sem. Univ. Hamburg* 30 (1967), 67–74.
2. A.S. Besicovitch, A problem on topological transformation of the plane, *Fund. Math.* 28 (1937), 61–65.
3. _____, A problem on topological transformation of the plane II, *Proc. Cambridge Philos. Soc.* 47 (1951), 38–45.
4. G.D. Birkhoff, *Dynamical systems*, Amer. Math. Soc. Colloq. Pub. Vol. 9, New York, 1927.
5. B.L. Brechner and R.D. Mauldin, Homeomorphisms of the plane, *Pacific J. Math.* 59 (1975), 375–381.

6. Y.N. Dowker, The mean and transitive points of homeomorphisms, *Ann. Math.* (2) 58 (1953), 123–133.
7. T. Homma and S. Kinoshita, On the regularity of homeomorphisms of E^n, *J. Math. Soc. Japan* 5 (1953), 365–371.
8. J.C. Oxtoby, Note on transitive transformations, *Proc. Nat. Acad. Sci. U.S.A.* 23 (1937), 443–446.
9. _____, *Measure and category*, Springer-Verlag, New York, 1971.
10. V.S. Prasad, Ergodic measure preserving homeomorphisms of R^n, *Indiana Univ. Math. J.* 28 (1979), 859–867.

John C. Oxtoby

Second Edition Commentary

The most interesting still unresolved questions from Problem 115 in the *Scottish Book* ask if there exists a homeomorphism of $\mathbb{R}^n$ or of $\mathbb{R}^n \setminus \{0\}$ with the full orbit of every point dense. Such a homeomorphism is called *minimal* because the smallest non-empty closed invariant subset is the entire space. In full generality the problem is still open, but in addition to the partial results cited in the commentary above there have been some more recent developments. The case $n = 2$ has been completely resolved. As mentioned in the commentary the Brouwer plane translation theorem (see, e.g., [1] for a modern proof) implies there can be no minimal homeomorphism of $\mathbb{R}^2$. The proof that no minimal homeomorphism exists for the punctured plane was provided by P. Le Calvez and J.-C. Yoccoz [4]. Le Calvez and Yoccoz prove, in fact, that the plane with any finite number of punctures does not admit a minimal homeomorphism. A shorter proof using the Conley index was provided by J. Franks [2]. The question has also been resolved in the orientation reversing case for $\mathbb{R}^3$. L. Hernández-Corbato, P. Le Calvez, and F. Ruiz del Portal prove in [5] that there is no minimal orientation reversing homeomorphism of $\mathbb{R}^3$. The other cases for $\mathbb{R}^n$, $n \geq 3$, with or without a deleted point, remain open.

The non-existence of forward minimal homeomorphisms for $\mathbb{R}^n$ is referenced in the commentary above. However, it is worth mentioning a more general (and earlier) result of W.H. Gottschalk (Theorem B of [3]) which asserts there is no forward minimal homeomorphism of any non-compact locally compact space.

1. J. Franks, A New Proof of the Brouwer Plane Translation Theorem, *Ergodic Theory and Dynamical Systems,* **12** (1992), 217–226.
2. J. Franks, The Conley index and non-existence of minimal homeomorphisms, *Illinois Journal of Math.*, **43** (1999) 457–64.
3. W. H. Gottschalk, Orbit-closure decompositions and almost periodic properties. *Bull. Amer. Math. Soc.* **50,** (1944). 915–919.
4. P. Le Calvez and J.-C. Yoccoz, Un théorème d'indice pour les homéomorphismes du plan au voisinage d'un point fixe, *Annals of Math.* **146** (1997), 241–293.
5. L. Hernández-Corbato, P. Le Calvez, and F. Ruiz del Portal, About the homological discrete Conley index of isolated invariant acyclic continua, *Geom. Topol.*, **17**, (2013), 2977–3026.

John Franks

PROBLEM 116: SCHREIER, ULAM

Let G be a compact group. It is known that almost every (in the sense of Haar measure) couple of elements $\phi, \psi \in G$ generates in G an everywhere dense subgroup. Let there be given a sequence $\{c_n\}$ of zeros and ones. Let us put $f_n = \phi$ if $c_n = 0$, $f_n = \psi$ if $c_n = 1$. Prove that for almost every pair ϕ, ψ and almost every sequence $\{c_n\}$ the sequence $f_1, f_1 f_2, f_1 f_2 f_3, \ldots$ is everywhere dense in G. Investigate whether this sequence is *uniformly dense;* that is, for every region $V \subset G$ we should have $\lim q_n / n = $ measure of V, if q_n denotes the number of the elements of $f_1, f_1 f_2, \ldots, f_1 f_2 \cdots f_n$ which fell into V. Investigate also whether an analogous theorem holds for similar sequences of images of a point p obtained with the aid of two transformations $\Phi(p)$ and $\Psi(p)$, which are *strongly transitive* mappings of the space S into itself preserving measure.

Commentary

The problem in its general form still seems to be open. A deep result of Veech can, however, be applied in order to construct sequences of the form $f_1, f_1 f_2, \ldots, f_1 f_2 \cdots f_n, \ldots$ that are "uniformly dense" (or, in modern terminology, "uniformly distributed") in the compact group G. Suppose $\phi, \psi \in G$ generate a dense subgroup of G, and let $y_1, y_2, \ldots$ be nonconstant sequence with $y_n = \phi$ or ψ for each n. Then, according to Veech, there exists a sequence $r_1, r_2, \ldots$ of positive integers such that, putting $f_n = y_{r_n}$ for each n, the sequence $f_1, f_1 f_2, \ldots, f_1 f_2 \cdots f_n, \ldots$ is uniformly distributed in G. The result of Veech, which refers to more general sequences $y_1, y_2, \ldots$, can be found in "Some questions of uniform distribution," *Ann. Math.* (2) 94 (1971), 125–138.

H. Niederreiter

Second Edition Commentary

We begin with a discussion about the first sentence of the problem. It is clear that some additional assumptions on the group are needed in order that almost every pairs of elements generate a dense subgroup. In particular, the group needs to be connected, since otherwise both generators may be contained in the connected component with positive probability.

In addition, the topology must have a base of small cardinality. It has been shown by Hofmann and Morris (Weight and c, J. Pure Appl. Algebra 68 (1990), no. 1–2, 181–194.) that a pair of elements generating a dense subgroup in a compact connected Hausdorff topological group exists if and only if the topology has a base of cardinality at most c. Even when there is a pair of elements generating a dense subgroup, the set of such pairs can still be non-measurable, e.g. in the case

$(\mathbb{R}/\mathbb{Z})^c$. Indeed, in this group any Borel set may depend on at most countably many coordinates.

Auerbach (Sur les groupes linéaires bornés (III), Stud. Math. 5 (1934), no. 1, 43–49.) proved that almost every pair generates a dense subgroup in a compact connected linear group. This was generalized by Schreier and Ulam (Sur le nombre des générateurs d'un groupe topologique compact et connexe, Fund. Math. 24 (1935), no. 1, 302–304) to compact connected metrizable groups with almost every in the sense of measure theory replaced by generic in the sense of Baire category. They used von Neumann's partial solution of Hilbert's 5'th problem. It seems that a very similar argument would work to show the claim for almost every pairs. We believe that this was the intended setting of the problem.

Probability measures and random walks on semigroups have been studied extensively in the past century. A comprehensive exposition of the subject can be found in the book of Högnäs and Mukherjea (Probability measures on semigroups, Springer, 2011). In particular, Theorem 3.12 (Chapter 3, p 212) gives a positive answer to the problem:

Theorem 1. *Let G be a compact metrizable group, and let μ be a Borel probability measure on it. Denote by Z_n the random walk generated by μ, that is Z_n is the product of the first n elements of a sequence of independent random elements of G with probability law μ. If the support of μ is not contained in a proper closed subgroup of G, then Z_n equidistributes in the sense that*

$$\frac{1}{N}\sum_{n=1}^{N} f(Z_n) \to \int f dm_{\text{Haar}}$$

almost surely for all continuous function $f \in C(G)$.

By the result of Schreier and Ulam, this holds in particular when $\mu = 1/2$ $(\delta_\phi + \delta_\psi)$ for almost every pair $\phi, \psi \in G$.

(The condition that suppμ is not contained in a proper closed subgroup is a standing assumption stated in the beginning of Section 3.3 (p. 198) of the book.)

For the reader's convenience we outline briefly an argument proving the above theorem. Breiman (The strong law of large numbers for a class of Markov chains, Ann. Math. Statist. 31 (1960), no.3, 801–803) proved the following "Law of large numbers" that is applicable to our situation. Let $\{X_n\}_{n\in\mathbb{N}}$ be a stationary Markov chain on a statespace G that is a compact Hausdorff topological space. Suppose that for each continuous function $f \in C(G)$, the conditional expectation $\mathbb{E}(f(X_n)|X_{n-1} = x)$ is a continuous function of $x \in G$. Suppose further that the common law of X_n is the only stationary measure with respect to the transition probabilities. Then

$$\mathbb{P}\left(\lim_{N\to\infty}\frac{1}{N}\sum_{n=1}^{N} f(X_n) = \int f dm_{\text{Haar}}\Big| X_1 = x\right) = 1$$

for every point $x \in G$.

We apply this theorem as follows. We let X_0 be a random element with law m_{Haar} that is independent of the random walk Z_n. Then we set $X_n = X_0 Z_n$ for all n and apply the theorem with $x = 1$, the unit element.

It remains to show that the Haar measure is indeed the unique stationary measure. We recall that the random walk is generated by the probability measure μ, and a measure ν is stationary if and only if $\mu * \nu = \nu$. Here and everywhere below $\mu * \nu$ denotes the convolution of μ and ν. We show that $1/N \sum_{n=1}^{N} \mu^{*n} \to m_{Haar}$ in the weak-* topology. If ν is stationary, then

$$\nu = \frac{1}{N} \sum_{n=1}^{N} \mu^{*n} * \nu \to m_{Haar} * \nu = m_{Haar}$$

and hence $\nu = m_{Haar}$ indeed.

One way to verify the claimed convergence is via spectral theory. For $g \in G$, we denote by U_g the unitary operator acting on $L^2(G)$ defined by $U_g f(h) = f(hg)$. We set $U = \int U_g d\mu(g)$. With this notation, our claim is equivalent to

$$\frac{1}{N} \sum_{n=1}^{\infty} U^n f \to \int f dm_{Haar}$$

for all $f \in C(X)$. Since $U_g f$ has the same modulus of continuity for all $g \in G$, it is enough to show the above convergence in L^2.

If $Uf = f$ for some $f \in L^2(G)$, then $\langle U_g f, f \rangle = 1$ for μ-almost all g. Since $\langle U_g f, f \rangle$ is continuous, $\langle U_g f, f \rangle = 1$ follows for all $g \in \text{supp}\mu$. Since U_g is unitary, this implies $U_g f = f$, and this holds for all $g \in G$, since $\text{supp}\mu$ is assumed to generate a dense subgroup. Thus f must be constant. Using Riesz's generalization of the von Neumann mean ergodic theorem for U, we conclude that $(1/N) \sum_{n=1}^{\infty} U^n f \to \int f dm_{Haar}$ in L^2 for all $f \in L^2(G)$. We also note that $L^2(G)$ can be decomposed as an orthogonal sum of finite dimensional U-invariant subspaces due to the Peter Weyl theorem, hence the ergodic theorem can be reduced to the finitely dimensional case in this situation.

J. Bourgain and P.P. Varjú

PROBLEM 117:FRÉCHET.

Original manuscript in French

Consider a Jordan curve which has a tangent (oriented) at every point. Does there exist at least one parametric representation of this curve where the coordinates are differentiable functions of the parameter and where the derivatives of the three coordinates do not vanish simultaneously?

Addendum.[*] In general, no; but we can represent the curve with functions of a parameter t in such a way that $dx/dt, dy/dt, dz/dt$ exist (and are not all zero), except for a set N of values of t, such that $m(N) = 0$ and also the set of points of the curve, corresponding to N, has Caratheodory measure zero (*Fund. Math.* 28).

A.J. Ward

March 23, 1937

*Original manuscript in English

PROBLEM 118: FRÉCHET.
Original manuscript in French

Let $\Delta(n)$ be the greatest of the absolute values of determinants of order n whose terms are equal to ± 1. Does there exist a simple analytic expression of $\Delta(n)$ as a function of n; or, more simply, determine an analytic asymptotic expression for $\Delta(n)$.

Commentary

This problem is misattributed. Hadamard published his famous paper [5] with a partial solution of the problem in 1893, when Fréchet was barely 15 years old. Already in 1867 Sylvester [7] studied "Hadamard" matrices, although it seems that he was not aware of the connection between these matrices and the maximal determinant problem. Hadamard [5] proved that any complex n-square matrix A, with entries not greater in absolute value than 1, satisfies

$$|\det(A)| \leq n^{n/2}. \tag{1}$$

If all the entries in A are real, then equality can hold in (1) if and only if

$$AA^T = nI_n, \tag{2}$$

which implies that all the entries of A are ± 1. A $(1. - 1)$ n-square matrix with determinant $\pm n^{n/2}$ is called a *Hadamard matrix*. It easily follows from (2) that an $n \times n$ Hadamard matrix can exist only if $n = 1, 2$, or $n \equiv 0 \bmod 4$. It has been conjectured that Hadamard matrices exist for all such n. The conjecture is unresolved although Hadamard matrices have been constructed for an infinite number of values of n. The smallest n for which the conjecture is undecided is 268. For a comprehensive listing of the orders for which Hadamard matrices are known, see [6].

Inequality (1) implies that

$$\Delta(n) \leq n^{n/2} \tag{3}$$

for all n, and if $n \equiv 0 \bmod 4$ then

$$\Delta(n) = n^{n/2},$$

provided that an $n \times n$ Hadamard matrix exists.

If $n > 2$ and $n \not\equiv 0 \bmod 4$, then $\Delta(n) < n^{n/2}$, and the bound in (1) can be improved. Barba [1] showed that if n is odd, then

$$\begin{aligned}
\Delta(n) &\leq (2n-1)^{1/2}(n-1)^{(n-1)/2} \\
&\sim (2/e)^{1/2}n^{n/2} \\
&= 0.85776n^{n/2}.
\end{aligned} \tag{4}$$

Ehrlich [4] sharpened Barba's bound for $n \equiv 3 \bmod 4$, and $n \geq 63$:

$$\begin{aligned}
\Delta(n) &\leq 2 \cdot 11^3 \cdot 7^{-7/2}(n-3)^{(n-7)/2}n^{7/2} \\
&\sim 2 \cdot 11^3 \cdot 7^{-7/2}e^{-3/2}n^{n/2} \\
&= 0.65452n^{n/2}.
\end{aligned} \tag{5}$$

For the case $n \equiv 2 \bmod 4$, Wojtas [8] proved that

$$\begin{aligned}
\Delta(n) &\leq 2(n-1)(n-2)^{(n-2)/2} \\
&\sim (2/e)n^{n/2} \\
&= 0.73576n^{n/2}.
\end{aligned} \tag{6}$$

The same result was obtained independently by Ehrlich [3]. It is also known [1, 9] that equality holds in (6) for all $n \equiv 2 \bmod 4$, $n \leq 62$, such that no prime factor of the squarefree part of $n - 1$ is congruent to 3 mod 4. The bound in (6) therefore is, in a sense, the best possible.

Comparing the bounds in (3), (4), (5), (6), and taking into consideration the known cases of equality, it appears that there is no simple analytic expression for $\Delta(n)$, nor does there exist an analytic asymptotic expression for $\Delta(n)$. Nevertheless, Clements and Lindström [2] have shown that for any n,

$$n^{(n/2)(1-C(n))} < \Delta(n) \leq n^{n/2},$$

where $C(n) = \log_2(4/3)/\log_2 n$. It follows that

$$\log \Delta(n) \sim \log n^{n/2}.$$

1. G. Barba, Intorno al teoremo di Hadamard sui determinanti a valore massimo, *Giorn. Mat. Battaglini* 71 (1933), 70–86.
2. G.F. Clements and B. Lindström, A sequence of (± 1)-determinants with large values, *Proc. Amer. Math. Soc.* 16 (1965), 548–550.

3. H. Ehrlich, Determinantenabschätzungen für binäre Matrizen, *Math. Z.* 83 (1964), 123–132.
4. _____, Determinantenabschätzung für binäre Matrizen mit $n \equiv 3 \bmod 4$, *Math. Z.* 84 (1964), 438–447.
5. J. Hadamard, Résolution d'une question relative aux déterminants, *Bull. Sci. Math.* 10 (1963), 240–246.
6. J. Seberry, A computer listing of Hadamard matrices, *Proc. International Conf. on Combinatorial Theory, Canberra,* 1978.
7. J.J. Sylvester, Thoughts on inverse orthogonal matrices, simultaneous sign-successions, and tessellated pavements in two or more colours, with applications to Newtons's rule, ornamental tile-work, and the theory of numbers, *Phil. Mag.* (4) 34 (1867), 461–475.
8. M. Wojtas, On Hadamard's inequality for the determinants of order non-divisible by 4, *Colloq. Math.* 12 (1964), 73–83.
9. C.H. Yang, On designs of maximal $(+1, -1)$-matrices of order $n \equiv 2 \pmod 4$, *Math. Comp.* 22 (1968), 174–180, and 23 (1969), 201–205.

Henryk Minc
University of California
Santa Barbara, CA 93106

PROBLEM 119: ORLICZ

Does there exist an orthogonal system composed of functions uniformly bounded and having the property possessed by the Haar system, that is to say, such that the development of every continuous function in this system is uniformly convergent?

Commentary

A.M. Olevskii [1] has shown that neither $C[0, 1]$ nor $L_1[0, 1]$ have a Schauder basis that is orthonormal and uniformly bounded. The results of Olevskii have been sharpened in some directions by S.T. Szarek [2] who has shown in particular that every normalized Schauder basis of $L_1[0, 1]$ contains a subsequence whose span is $\approx \ell$. These results were further sharpened by Szarek in [3] and by Kwapien and Szarek in [4].

1. A.M. Olevskii, Fourier series with respect to general orthogonal systems, Springer-Verlag, 1975.
2. S.T. Szarek, Bases and biorthogonal systems in the spaces C and L^1, *Arkiv Math.* 17 (1979), 255–271.

Joseph Diestel
Kent, Ohio

Second Edition Commentary

The results of Olevskii were also sharpened by S. Kwapien and S. Szarek (*Studia Math.* 66 (1970), 189–200) and S. Szarek In turn the results of Kwapein, Olevskii, and Szarek were established by J. Bourgain(*Trans. Amer. math. Soc.* 285 (1984),

133–139) for the dish algebra and L_1/H_0^1, using considerably more complicated arguments.

Joseph Diestel

Kent, Ohio

PROBLEM 120: ORLICZ

Let x^{n_i} be a sequence of powers with integer exponents on the interval (a, b) and

$$\sum_{i=1}^{\infty} \frac{1}{n_i} = +\infty.$$

Give the order of approximation of a function satisfying a Hölder condition by polynomials:

$$\sum_{i=1}^{N} a_i x^{n_i}.$$

Second Edition Commentary

This problem is now rather well-understood. Major results were obtained by M.v. Golitschek, D.J. Newman, T. Ganelius, D. Leviatan, J. Bak, J.Tzimbalario, L. Marki, G. Somorjai and J. Szabados and others.

The seminal result is as follows. Let $\Lambda = \{0 = \lambda_0 < \lambda_1 < \cdots < \lambda_N\}$ and $E_\Lambda(f) = \inf_{a_j} \|f - \sum_j a_j x^{\lambda_j}\|_p$ for $f \in L_p[0;1], 1 \le p < \infty$ or $f \in C[0;1]$ for $p = \infty$. Let $B(z) = \prod_{j=1}^{\infty} \frac{z - \lambda_j - \frac{1}{p}}{z + \lambda_j + \frac{1}{p}}$, $\varepsilon = \sup_{y \ge 0} \left| \frac{B(1+iy)}{1+iy} \right|$ and $\omega_p(f, \delta) = \sup_{|h| \le \delta} \|f(\cdot + h) - f(\cdot)\|_p$ – be the p-modulus of continuity. The main result is

$$E_\Lambda(f) \le A \omega_p(f, \varepsilon)$$

for some absolute constant A; the estimate is sharp, i.e. it can not hold for $A < 1/600$.

The history, references, and proofs can be found in G. G. Lorentz, M. v. Golitschek, Y. Makovoz "Constructive Approximation: Advanced Problems" Springer, 1996, Ch. 11.

K. S. Rjutin

PROBLEM 121: ORLICZ

Give an example of a trigonometric series

$$\sum_{n=1}^{\infty} (a_n \cos nx + b_n \sin nx)$$

everywhere divergent and such that

$$\sum_{n=1}^{\infty}(a_n^{2+\varepsilon}+b_n^{2+\varepsilon})<+\infty$$

for every $\varepsilon>0$.

Second Edition Commentary

The required example can be constructed by the following results of S.Sh. Galstyan.

Theorem 1. [2]. *Let $\{\alpha_n:n\geq 0\}$ be a nonincreasing sequence of positive numbers tending to zero and $\sum_{n=0}^{\infty}\alpha_n^2=\infty$. Then there is a series $\sum_{n=0}^{\infty}c_ne^{inx}$ of class H^1, the real and imaginary parts of which diverge everywhere, and $c_n=O(\alpha_n)$.*

Theorem 2. [3]. *Let $\{\alpha_n:n\geq 0\}$ be a nonincreasing sequence of positive numbers tending to zero and $\sum_{n=0}^{\infty}\alpha_n^2=\infty$. Then there exists a subsequence $\{\alpha_{n_k}:k\geq 0\}$ such that the series $\sum_{k=0}^{\infty}\alpha_{n_k}\cos(n_kt)$ diverges everywhere.*

Theorems 1 and 2 are sharp. We recall that, by a famous theorem of L. Carleson [4], if a_n and b_n are real numbers and $\sum_{n=1}^{\infty}(a_n^2+b_n^2)<\infty$, then the series $\sum_{n=1}^{\infty}(a_n\cos nx+b_n\sin nx)$ converges almost everywhere.

Using a result of Körner, an explicit example of such a series has been given by Akita, Gotô and Kano [1].

Everywhere divergent complex trigonometric series with rapidly decreasing coefficients are studied in [5, chapter 8, §9].

1. Akita, M.; Gotô, K; and Kano, T., A problem of Orlicz in the Scottish Book, Proc. Japan Acad. Ser. A Math. Sci. **62** (1986), 267–269.
2. Galstyan, S.Sh., Everywhere divergent trigonometric Fourier series with rapidly decreasing coefficients (Russian), Mat. Sb. (N.S.) **122 (164)** (1983), N. 2, 157–167; English translation: Mathematics of the USSR - Sbornik, **50** (1985), N. 1, 151–161.
3. Galstyan, S.Sh., Everywhere divergent trigonometric series (Russian), Mat. Zametki **37** (1985), N. 2, 186-191; English translation: Math. Notes **37** (1985), N. 1-2, 105–108.
4. Carleson, L., On convergence and growth of partial sums of Fourier series, Acta, Math. **116** (1965), N. 1–2, 135–157.
5. Kahane, J.-P., Some random series of functions, D. C. Heath and Company, Massachusetts, 1968.

Sergei Konyagin

PROBLEM 122: MAZUR, ORLICZ

Does there exist in every space of type (B) of infinitely many dimensions, a series which is unconditionally convergent but not absolutely? (A series

$$\sum_{n=1}^{\infty}x_n$$

is called unconditionally convergent if it converges under every ordering of its terms and absolutely convergent if the series

$$\sum_{n=1}^{\infty} \|x_n\|$$

converges.

Commentary

In 1950, A. Dvoretsky and C.A. Rogers (*Proc. Nat. Acad. Sci. USA* 36 (1950), 192–197) showed that if every unconditionally convergent series in a Banach space X is absolutely convergent then X must be finite dimensional; this gives an affirmative response to Problem 122. Close on the heels of the Dvoretsky-Rogers solution came a new and stunning approach to a whole circle of related problems, developed by A.Grothendieck. Central to the Grothendieck program is the idea of a p-absolutely summing operator: A bounded linear operator T between the Banach spaces X and Y is p-absolutely summing if there exists a constant $K > 0$ such that given $x_1, \ldots, x_n \in X$ the inequality

$$\sum_{k=1}^{n} \|Tx_k\|^p \leq K^p \sup \left\{ \sum_{k=1}^{n} |x^*x_k|^p : \|x^*\| \leq 1 \right\}$$

holds. A quick check in case $p = 1$ shows that the operator $T : X \to Y$ is 1-absolutely summing if and only if T takes unconditionally convergent series into absolutely convergent series; for general $p \geq 1$, T is p-absolutely summing if and only if whenever $\sum_n |x^*x_n|^p$ is finite for each $x^* \in X^*$ then $\sum_n \|Tx_n\|^p$ is finite. Through the work of Grothendieck and A. Pietsch (*Studia Math.* 28 (1967), 333–353), one can conclude that if a normed linear space X has the property that for some $p \geq 1$ the series $\sum_n \|x_n\|^p$ converges whenever $\sum_n |x^*x_n|^p$ does for each $x^* \in X^*$ then X must be finite dimensional.

Though the results of Grothendieck and Pietsch might appear to be but a marginal improvement of that of Dvoretsky and Rogers such is far from the truth. On the one hand, the theory of p-absolutely summing operators (and related classes of operators) has played a central role in the revival of Banach space theory especially as it relates to other areas of mathematical endeavor (particularly harmonic theory). On the other hand, the theory of p-absolutely summing operators is instrumental in providing a more complete answer to Problem 122, particularly for Fréchet spaces (F_0 spaces in the Polish terminology). In fact, Grothendieck (*Memoir American Mathematical Society,* volume 16 (1955)) was able to classify those Fréchet spaces in which unconditionally convergent series are absolutely convergent (such spaces are called nuclear) and showed that many of the important non-normed spaces of analysis are indeed nuclear.

The above synopsis only touches the tip of a mathematical iceberg. Improvements of the Grothendieck-Pietsch results have been obtained by B. Maurey and A. Pe}lczyński (*Studia Math.* 54 (1976), 291–300) and H. König (preprint from Bonn University). The original Dvoretsky-Rogers proof was to lead Dvoretsky to his famous "ε-spherical sections" theorem, recently given a definitive treatment by T. Figiel, J. Lindenstrauss and V. Milman (*Acta Math.* 139 (1977), 53–94). The theory of nuclear spaces has been extensively developed, principally by the Soviet school (a good report on which can be found in the articles of B. Mityagin appearing in the 1978–79 *Seminaire Functional Analyse,* Ecole Polytechnique).

In addition to these developments, the finite dimensional structure and the theory of *p*-summing operators have led to the Maurey-Rosenthal dichotomy (c.f. H.P. Rosenthal, *Studia Mathematica* 58 (1976), 21–43).

Joseph Diestel

Kent, Ohio

PROBLEM 123: STEINHAUS

Given are three sets A_1, A_2, A_3 located in the 3-dimensional Euclidean space and with finite Lebesgue measure. Does there exist a plane cutting each of the three sets A_1, A_2, A_3 into two parts of equal measure? The same for *n* sets in *n*-dimensional space.

Addendum Solution in "Z Topologii," *Mathesis Polska* 1936.

Commentary

I have not seen the solution referred to (in *Mathesis Polska* 1936); perhaps it was the one circulating orally in Princeton in 1941. The theorem for 3 sets, A_1, A_2, A_3 in R^3, was aptly named the "Ham Sandwich Theorem." The proof began by bisecting the "ham" A_3 by a continuously varying plane, and observing that the measure-differences of the parts in which it cut the two "slices of bread" A_1, A_2 provided an antipodal map from S^2 to R^2. The case $n = 2$ of what has become known as the Borsuk-Ulam theorem [2] then guarantees that some point of S^2 is mapped to the origin, giving a plane that bisects all 3 sets.

The case $n = 2$ of the Borsuk-Ulam theorem can be proved without much formal topological apparatus; hence, the device of bisecting A_3 first allows the proof to be completely elementary, as is indicated briefly in [5] and in full in [3, pp. 120–123]. However, the use of the Borsuk-Ulam theorem in full strength (an antipodal map from S^n to R^n maps onto the origin, $n = 1, 2, \ldots$) gives an easier proof that one can bisect each of *n* given sets (of finite measure) in R^n by some hyperplane, and in fact that bisection behaves like a linear condition on algebraic varieties; one can bisect 5 given sets in the plane by a conic, and so on (see [5]).

It has been pointed out that there is no need to use the same measure on all the sets. For instance, in the original case $n = 3$, one might with some realism bisect the volumes of the two slices of bread, and the surface area of the ham.

The situation for ratios other than bisection, in R^1 and R^2, was also investigated in [5]. In R^1, a necessary and sufficient condition on positive numbers α_1, α_2 that, given sets A_1, A_2 of finite Lebesgue measure, there always exists an interval (possibly infinite) in R^1 whose intersection with A_i has measure α_i times that of A_i ($i = 1,2$), is that $\alpha_1 = \alpha_2 =$ reciprocal of an integer greater than 1. In R^3, the same condition on α_1, α_2 is *necessary* in order that there always exist a circle (or straight line) cutting off α_i times the measure of A_i ($i = 1,2$). The Ham Sandwich Theorem shows that when $\alpha_1 = \alpha_2 = 1/2$ the condition is sufficient; but the case $\alpha_1 = \alpha_2 = 1/3$ remains (like the others) unsettled. To prove sufficiency here, it would suffice (by an approximation argument) to prove that given two finite subsets B_1, B_2 of the plane, B_i! having $3n_i$ points, there exists, a circle having exactly n_i points of B_i in its interior ($i = 1,2$). A recent postscript to [5] is supplied by [1], which points out that the sketch of the elementary argument in [5] is *too* sketchy; in general, one cannot bisect A_3 by a continuously varying plane taking one position in each direction. In [5] it was intended, but not stated, that one should first approximate the given sets in measure by sets having positive measure everywhere. Having bisected the new sets, one applies an obvious limiting process to bisect the given ones. (In [3] the difficulty is avoided by assuming the sets A_i are open.)

The Borsuk-Ulam theorem and some analogous theorems about continuous functions on spheres have led to very extensive developments. See [4], under the headings "Theorems about S^n of Borsuk-Ulam Type" and "Theorems of Dyson, Kakutani, Yamabe-Yujobo," pp. 1117–1125.

1. Richard Arens, On Sandwich Slicing, Topology, Vol. II (Proc. Fourth Colloq., Budapest, 1978), 57–60, Colloq. Math. Soc. János Bolyai, 23, North-Holland, Amsterdam, 1980.
2. K. Borsuk, Drei Sätze über die *n*-dimensionale euklidische Sphäre, *Fund. Math.* 20 (1933), 177–190.
3. W.G. Chinn and N.E. Steenrod, *First concepts of topology,* Random House, New York 1966.
4. N.E. Steenrod (ed.), *Reviews of Papers in Algebraic and Differential Topology, Topological Groups and Homological Algebra, Part II,* Amer. Math. Soc., 1968.
5. A.H. Stone and J.W. Tukey, Generalized "Sandwich" Theorems, *Duke Math. J.* 9 (1942), 356–359.

A.H. Stone

University of Rochester

Rochester, NY 14627

PROBLEM 124: MARCINKIEWICZ

What can one say about the uniqueness for the integral equation

$$\int_0^1 y(t) f(x-t) dt = 0, \qquad 0 \le x \le 1? \tag{1}$$

I know that if the sequence of integrals $f_k(x) = \int_0^x f_{k-1}(t)dt$, $f_0 = f$; $k = 1,2,3,\ldots$ is complete in L^2 then the only solution of Eq. 1 is $y \equiv 0$. This is the case also if f is of bounded variation and $f(0) \neq 0$. Finally, if Eq. 1 possesses even one nonzero solution, y, then every (iterated) integral of y also satisfies this equation.

I conjecture that if $f(0) \neq 0$ and f is continuous, then Eq. 1 has only the solution $y \equiv 0$.

Second Edition Commentary

The answer is affirmative. This follows from the following more general theorem of E. C. Titchmarch (Published in 1926!) (The zeros of certain integral functions, *Proc. London Math. Soc.* 25 (1926); Theorem VII.) If Φ and Ψ are integrable functions such that

$$\int_0^x \Phi(t)\Psi(x-t)dt = 0$$

almost everywhere in the interval $0 < x < K$, then $\Phi(t) = 0$ almost everywhere in $(0,\lambda)$ and $\Psi(t) = 0$ almost everywhere in $(0,\mu)$, where $\lambda + \mu \geq K$. In his book Operational Calculus (Elsevier, 1960) Jan Mikusinski proves Titchmarsh's theorem in the second chapter. He proves it following a particularly simple argument found by Ryll-Nardzewski in 1952. As Mikusinski writes, Ryll-Nardzewski never published this proof.

Z. Buczolich

M. Laczkovich

PROBLEM 125: INFELD

Originating in physics

We shall say a decent function of two variables $f(x,y)$ satisfies the condition A if there exists a function $y = \phi(x)$ such that

$$\left. \begin{array}{ll} xf_x + yf_y = 0 & (1) \\ 4f_x f_y \quad = 1 & (2) \end{array} \right\}$$

for $y = \phi(x)$. [We see that $\phi(x)$ exists for $f(x/y)$ and for $f = ax + by$, if $ab = 1/4$.] Do we have the criterion: For every function $f(x,y)$ satisfying A there exists $F(x/y)$ such that $F(x/\phi(x)) = f(x,\phi(x))$ (with the exception of the case of $f = ax + by$)?

PROBLEM 126: M. KAC

If:

$$\begin{aligned} \int_0^1 f(x)dx &= 0, \quad (1) \\ \int_0^1 f^2(x)dx &= \infty, \quad (2) \end{aligned} \Bigg\}$$

show that

$$\lim_{n\to\infty} \left[\int_0^1 \exp\left(i\frac{f(x)}{\sqrt{n}} \right) dx \right]^n = 0$$

(It is known that if $\int_0^1 f^2(x)dx = A$ then the above limit $= e^{-1/2}$).

Addendum. Solved affirmatively by A. Khintchin; it will appear in the fourth communiqué on independent functions in *Studia Math.*, Vol. 7.

PROBLEM 127: KURATOWSKI

Is it true that in every 0-dimensional metric space (in the sense of Menger-Urysohn) that every closed set is an intersection of a sequence of sets which are simultaneously closed and open? (The answer is affirmative for metric separable spaces.)

Commentary

The problem is equivalent to the question whether for every metric space X, the condition ind $X = 0$ implies that Ind $X = 0$ (see [1, p. 9] for the definitions). Indeed, if every closed subset of X is an intersection of a sequence of open and closed sets, then for every pair A, B of disjoint closed subsets of X there exists an open-and-closed set $U \subset X$ such that $A \subset U \subset X \backslash B$: the set U can be defined by the formula $U = \bigcup_{i=1}^{\infty}(U_i \backslash W_i)$, where $U_1, U_2, \dots$ and $W_1, W_2, \dots$ are decreasing sequences of open-and-closed sets satisfying $A = \bigcap_{i=1}^{\infty} U_i$ and $B = \bigcap_{i=1}^{\infty} W_i$ (cf. [1, p. 155]). It seems that the last implication, implicit in [1], was first explicitly stated in [4]. When solving, in the negative, the famous problem whether the dimensions ind and Ind coincide in metric spaces, P. Roy defined in [2] a metric space X such that ind $X! = 0$ and Ind $X = 1$ (a detailed discussion of this example is contained in [3]). Roy's space belongs among the most difficult examples in general topology.

1. J. Nagata, *Modern dimension theory*, Groningen, 1965.
2. P. Roy, Failure of equivalence of dimension concepts for metric spaces, *Bull. Amer. Math. Soc.* 68 (1962), 609–613.
3. ———————, Nonequality of dimensions for metric spaces, *Trans. Amer. Math. Soc.* 134 (1968), 117–132.
4. J. Terasawa, On the zero-dimensionality of some non-normal product spaces, *Sci. Rep. Tokyo Kyoiku Daigaku Sec. A* 11 (1972), 167–174.

R. Engelking

PROBLEM 128: NIKLIBORC

There is given, in a 3-dimensional space, a solid T which is unicoherent and homogeneous. Let $V(P) = \int_T dt_M / r_{PM}$. Assume that $V(P)$ is a polynomial in P in all of $T + S$. S is the surface of T. Show that T is an ellipsoid. It is known that if this polynomial is of second degree then the theorem is true.

PROBLEM 129: NIKLIBORC

There are given two closed spaces S and S_1, each homeomorphic to the surface of the sphere and constituting the boundary of a solid T. Suppose that $V(P) = \int_T dt_M / r_{PM}$ is a constant in T_1 (the solid boundary by S_1). Prove that S and S_1 are homothetic ellipsoids. It is known that if S and S_1 are homothetic, then they are ellipsoids.

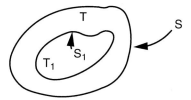

PROBLEM 130: KACZMARZ

Let $\{f_n(t)\}$ be a system of uniformly bounded, orthogonal, lacunary functions. Does there exist a constant $\gamma > 0$, such that for every finite system of numbers $c_1, c_2, \ldots, c_n$ we have:

$$\max |c_1 f_1(t) + \cdots + c_n f_n(t)| \geq \gamma \sum_{k=1}^{n} |c_k|?$$

Remark: The system is lacunary if, for every $p > 2$ there exists a constant M_p, such that

$$\sqrt[p]{\int_0^1 |c_1 f_1 + \cdots + c_n f_n|^p \, dt} \leq M_p \left(\sum_1^n c_k^2 \right)^{1/2}.$$

Second Edition Commentary

Let us formulate Kaczmarz's problem in more modern notation. Let (Ω, μ) denote a probability space and let $\{\phi_1, \phi_2, \ldots\}$ denote an orthonormal system (OS) of complex-valued functions on Ω. Such a system is said to be Sidon with constant γ if for all coefficients $\{a_j\}$ one has

$$\sup_{x \in \Omega} | \sum_{j \in \mathbb{N}} a_j \phi_j(x) | \geq \gamma \sum_{j \in \mathbb{N}} |a_j|. \tag{1}$$

An OS is said to be $\Lambda(p)$ with constant $M_p < \infty$ if for all coefficients $\{a_j\}$

$$|| \sum_{j \in \mathbb{N}} a_j \phi_j ||_{L^p(\Omega)} \leq M_p \left(\sum_{j \in \mathbb{N}} |a_j|^2 \right)^{1/2}.$$

An OS which is $\Lambda(p)$ for all $p \geq 2$ is said to be $\Lambda(\infty)$. With this notation, Kaczmarz's problem may be reformulated as: *Is a uniformly bounded $\Lambda(\infty)$ orthonormal system Sidon?*

The topic of "Sidonicity" has been extensively studied in the context of characters on groups. However, many problems regarding Sidon sets/systems can be formulated in the context of general bounded orthonormal systems. We will report on several recent results in this more general setting here. Detailed proofs will appear in [3].

Let us briefly recall the development of the theory of Sidon sets/systems in the character setting. In 1960 Rudin introduced $\Lambda(p)$ sets and constructed a subset of the integers which is $\Lambda(\infty)$ but which is not Sidon. See Section 3.2 and Theorem 4.11 of [9]. This provides a negative answer to Kaczmarz's problem, although there is no evidence there that Rudin was aware of the problem's provenance. Let us briefly discuss Rudin's construction. He first proved that a Sidon set must be $\Lambda(\infty)$ and, more restrictively, the set's $\Lambda(p)$ constants must satisfy $M_p \lesssim p^{1/2}$. From this he deduced that the size of the intersection of a Sidon set with an arithmetic progression of size n must be $\lesssim \log n$. Rudin was then able to give a combinatorial construction of a set which (1) had too large of an intersection with a sequence of arithmetic progressions to be Sidon, yet 2) was $\Lambda(p)$ for all p. He established the second property by combinatorial considerations after expanding out L^p norms in the case of even integer exponents.

On the other hand, much in the spirit of Kaczmarz's problem, Rudin asked if the stronger condition $M_p \lesssim p^{1/2}$ characterizes Sidon sets. In 1975 Rider [8] proved that the Sidon condition (1) is equivalent to the following (superficially) weaker Rademacher-Sidon condition. We say that $\{\phi_j\}$ has the Rademacher-Sidon property with constant $\tilde{\gamma}$ if the following inequality is satisfied for all coefficients $\{a_j\}$

$$\int \sup_{x \in \Omega} | \sum_{j \in \mathbb{N}} r_j(\omega) a_j \phi_j(x) | d\omega \geq \tilde{\gamma} \sum_{j \in \mathbb{N}} |a_j| \tag{2}$$

where r_n denote independent Rademacher functions. In 1978 Pisier [6] proved that Rudin's condition $M_p \lesssim p^{1/2}$ implies the Rademacher-Sidonicity property. Collectively these results show that Rudin's condition characterizes Sidonicity in the character setting. We note that both Rider's and Pisier's arguments make essential use of properties of characters.

It is well known that Rudin's condition $M_p \leq C\sqrt{p}$ is equivalent to the condition that

$$\|\sum_j a_j \phi_j\|_{\psi_2} \leq C' \left(\sum_j |a_j|^2\right)^{1/2}$$

where $\|\cdot\|_{\psi_2}$ is the Orlicz norm associated to the function $\psi_2(x) := e^{|x|^2} - 1$. We will denote this condition as $\psi_2(C')$.

Given the developments in the character setting, a natural relaxation of Kaczmarz's problem would be to ask if the ψ_2 condition implies Sidonicity in the case of general uniformly bounded orthonormal systems. Our first result is a construction of an OS that gives a negative answer to this question.

Theorem 1. *For all large n, there exists a real-valued OS $\{\phi_0, \phi_1, \ldots, \phi_n\}$ with $n+1$ elements satisfying $\|\phi_j\|_{L^\infty} \leq 7$ and satisfying the $\psi_2(C)$ condition with some universal constant C, and such that*

$$\|\sum_{i=0}^n a_j \phi_j\|_{L^\infty} \lesssim \frac{1}{\sqrt{\log n}} \sum_{j=0}^n |a_j|$$

for some choice $\{a_k\}$.

This construction makes essential use of Rudin-Shapiro polynomials.

On the other hand, the following result does provide a generalize of Pisier's theorem to the setting of general uniformly bounded orthonormal systems.

Theorem 2. *Let $\{\phi_j\}$ be a ψ_2 uniformly bounded OS. Then the OS obtained as a 5-fold tensor, $\{\phi_j \otimes \phi_j \otimes \phi_j \otimes \phi_j \otimes \phi_j\}$, is Sidon. Moreover, if the 5-fold tensor of an OS is Sidon, then the system itself is Rademacher-Sidon.*

When applied to a system of characters it follows from the homomorphism property that the system itself must be Sidon. It is an interesting problem to determine if the result holds for the two-fold tensor. A corollary of this result is that the ψ_2 condition implies the Rademacher-Sidon property. We will discuss several proofs of this fact. In fact, orthogonality beyond the ψ_2 condition is not required.

Theorem 3. *Let $\phi_1, \phi_2, \ldots$ denote a set of functions on a probability space (Ω, μ) such that $\|\phi_j\|_{L^2} = 1$ and satisfying the $\psi_2(C)$ condition. Then*

$$\int \sup_{x \in \Omega} |\sum_{j \in \mathbb{N}} r_j(\omega) a_j \phi_j(x)| d\omega \geq \tilde{\gamma} \sum_{j \in \mathbb{N}} |a_j|. \tag{3}$$

with $\tilde{\gamma} := \tilde{\gamma}(C)$.

This theorem will be a corollary of the following more general result.

Proposition 1. *Let* $\phi_1, \phi_2, \ldots, \phi_n$ *be a system of functions satisfying the* $\psi_2(C)$ *condition and* $\|\phi_j\|_{L^2} = 1$. *Let* $x_1, x_2, \ldots, x_n$ *denote (real or complex) vectors in a normed vector space satisfying* $\|x_j\| \leq 1$ *and* $\lambda_1, \lambda_2, \ldots, \lambda_n$ *scalars. Then the estimate*

$$\int \left\| \sum_{j=1}^{n} |\lambda_j| \phi_j(\omega) x_j \right\| d\omega \geq \beta \sum_{j=1}^{n} |\lambda_j|$$

implies

$$\int \left\| \sum_{j=1}^{n} \lambda_j r_j(\omega) x_j \right\| d\omega \geq \gamma \sum_{j=1}^{n} |\lambda_j|$$

for $\gamma := \gamma(\beta, C)$.

Let us explain how Proposition 1 implies Theorem 3. By truncation it suffices to prove (3) for a finite system, as long as the bounds do not depend on the size of the system. We then have that

$$\sum_{j=1}^{n} |\lambda_j| = \int \sum_{j=1}^{n} |\lambda_j| |\phi_j(x)|^2 dx \leq \int \left\| \sum_{j=1}^{n} |\lambda_j| \overline{\phi_j(x)} \phi_j(y) \right\|_{L_y^\infty} dx. \tag{4}$$

Using the $\psi_2(C)$ hypothesis, we may apply Proposition 1 to replace the functions $\{\phi_j(x)\}$ with Rademacher functions and remove the absolute values. This gives us

$$\gamma \sum_{j=1}^{n} |\lambda_j| \leq \int \left\| \sum_{j=1}^{n} \lambda_j r_j(\omega) \phi_j(y) \right\|_{L_y^\infty} d\omega$$

which is Theorem 3.

Another variant of Kaczmarz's problem would be to ask if the $\psi_2(C)$ condition implies that a system contains a large Sidon subsystem. In this direction recall the Elton-Pajor Theorem (see [4] and [5]):

Theorem 4. (Elton-Pajor) *Let* $x_1, x_2, \ldots, x_n$ *denote elements in a real or complex Banach space, such that* $\|x_i\| \leq 1$. *Furthermore, for Rademacher functions* $r_1, r_2, \ldots, r_n$ *assume that* $\gamma n \leq \int \|\sum_{i=1}^{n} r_i(\omega) x_i\| d\omega$. *Then there exists real constants* $c := c(\gamma) > 0$ *and* $\beta := \beta(\gamma) > 0$ *and a subset* $S \subseteq [n]$ *with* $|S| \geq cn$ *such that*

$$\beta \sum_{j \in S} |a_j| \leq \left\| \sum_{j \in S} a_j x_j \right\|$$

for all complex coefficients $\{a_i\}_{i \in S}$.

It immediately follows from Theorem 3 and the Elton-Pajor Theorem that a finite uniformly bounded OS satisfying the $\psi_2(C)$ condition must contain a Sidon subsystem of proportional size. More precisely:

Theorem 5. *Let $\phi_1, \phi_2, \ldots, \phi_n$ be a system of functions satisfying $\|\phi_j\|_{L^2} = 1$, $\|\phi_j\|_{L^\infty} \leq M$ and the $\psi_2(C)$ condition. Then there exists a subset $S \subseteq [n]$ of proportional size $|S| \geq \alpha(C,M)n$ such that*

$$\sup_{x \in \Omega} \left| \sum_{j \in S} a_j \phi_j(x) \right| \geq \gamma \sum_{j \in S} |a_j|.$$

where $\gamma = \gamma(C,M)$.

One interesting consequence of Proposition 1 is that one may replace the Rademacher functions in the hypothesis of the Elton-Pajor Theorem with any complex-valued functions satisfying the $\psi_2(C)$ condition.

Our approach to Theorem 2 and Proposition 1 is rather elementary. The proofs proceed by showing that one may efficiently approximate a bounded system satisfying the $\psi_2(C)$ condition by a martingale difference sequence. Once one is able to reduce to a martingale difference sequence, one may apply Riesz product-type arguments. This approach provides a new and elementary proof of Pisier's characterization of Sidon sets. It is worth noting that the first author obtained a different elementary proof of Pisier's theorem in 1983 [2]. The approach there, however, like Pisier's, relies on the homomorphism property of characters.

We have also found an alternate approach to Proposition 1 based on more sophisticated tools from the theory of stochastic processes such as Preston's theorem [7], Talagrand's majorizing measure theorem [11] and Bednorz and Latała's recent characterization of bounded Bernoulli processes. This approach yields a superior bound for the size of $\gamma(\beta, C)$ and allows for the following extension to more general norms.

Theorem 6. *Let $\phi_1, \phi_2, \ldots, \phi_n$ be a $\psi_2(C)$ system, uniformly bounded by M and let $x_1, x_2, \ldots, x_n$ be vectors in a normed space X. Then*

$$\int \|\sum_{j=1}^n \phi_j(\omega)x_j\| d\omega \lesssim CM \int \|\sum_{j=1}^n r_j(\omega)x_j\| d\omega. \tag{5}$$

In particular, one may take $\gamma(C,\beta) \gtrsim \beta \left(C \min \left(M, \sqrt{\log \frac{1}{\beta}} \right) \right)^{-1}$ in Proposition 1 for $\psi_2(C)$ systems uniformly bounded by M.

Approximating ψ_2 systems

The key ingredient in the proofs of Theorem 2 and Proposition 1 is the following:

Lemma 1. *Let $\phi_1, \phi_2, \ldots, \phi_n$ be real-valued functions on a probability space (Ω, μ) such that*

$$||\phi_j||_{L^2} = 1 \quad and \quad ||\phi_j||_{L^\infty} \leq C \tag{6}$$

$$\left\| \sum_{j=1}^n a_j \phi_j \right\|_{\psi_2} \leq C \left(\sum_{j=1}^n |a_j|^2 \right)^{1/2} \tag{7}$$

for all coefficients $\{a_j\}$. For $\varepsilon > 0$, there exists a subset $S \subseteq [n]$ such that $|S| \geq \delta(\varepsilon, C)n$ and a martingale difference sequence $\{\theta_j\}_{j \in S}$ satisfying $||\theta_j||_{L^\infty} \leq C$ such that:

$$||\phi_j - \theta_j||_{L^1} \leq \varepsilon, \tag{8}$$

and such that there exists an ordering of S, say $j_1, j_2, \ldots, j_n$, with

$$\mathbb{E}\left[\theta_{j_s} | \theta_{j_{s'}}, s' < s \right] = 0. \tag{9}$$

Moreover, one may take $\delta(\varepsilon, M) \gtrsim C^{-2} \varepsilon^2 \left(\log \frac{C}{\varepsilon} \right)^{-1}$.

For technical reasons we will require the following slightly stronger coefficient version of this result.

Lemma 2. *Let $\phi_1, \phi_2, \ldots, \phi_n$ be functions uniformly bounded by C satisfying the hypotheses of Proposition 1, and $\lambda_1, \lambda_2, \ldots, \lambda_n$ be complex coefficients such that $\sum_{j=1}^n |\lambda_j| = 1$. Then there exists a set $S \subseteq [n]$ and a martingale differences sequence $\theta_{j_1}, \theta_{j_2}, \ldots$ indexed by elements of S satisfying (8) and (9), such that $\sum_{j \in S} |\lambda_j| \gtrsim \delta(C, \varepsilon) > 0$.*

Proofs of these results may be found in [3].

Tensor systems: Theorem 2

We now describe how to obtain the first part of Theorem 2 from Lemma 2. For the sake of exposition, we prove the result for real-valued systems. The complex case can be handled in a similar manner. See [3].

Theorem 7. *Let $\{\phi_j\}$ be an OS uniformly bounded by C and satisfying the $\psi_2(C)$ condition. Then the OS obtained as a 5-fold tensor, $\{\phi_j \otimes \phi_j \otimes \phi_j \otimes \phi_j \otimes \phi_j\}$, is Sidon.*

Proof. Let $\phi_1, \phi_2, \ldots, \phi_5$ denote independent copies of the system $\{\phi_i\}$ on probability spaces $\Omega_1, \Omega_2, \ldots, \Omega_5$, respectively. Furthermore let $\tilde{\Omega} := \otimes_{s=1}^5 \Omega_s$ and let $r_i^{(1)}, r_i^{(2)}, r_i^{(3)}, r_i^{(4)}, r_i^{(5)}$ denote independent Rademacher functions on a distinct probability space $\mathbb{T}$. For a fixed set of coefficients $\{a_i\}$ and $\varepsilon > 0$, applying Lemma 1 gives a martingale difference sequence, $\{\theta_j\}$, with the following properties:

$$\sum_{i \in A} |a_i| \gtrsim C^{-2} \varepsilon^2 \left(\log \frac{C}{\varepsilon} \right)^{-1} \sum_{i=1}^{n} |a_i|, \tag{10}$$

for all $i \in [n]$

$$||\theta_i|| \le C, \tag{11}$$

$$||\phi_i - \theta_i|| \le \varepsilon. \tag{12}$$

For $0 < \delta < 1$ and $\alpha_i \in [-1, 1]$, define

$$\mu_{(\alpha,\delta)} := \int_{\mathbb{T}} \prod_{i \in A} \left(1 + \delta \alpha_i r_i^{(1)} \theta_i(x_1) \right) \prod_{i \in A} \left(1 + \delta \alpha_i r_i^{(2)} r_i^{(1)} \theta_i(x_2) \right) \prod_{i \in A} \left(1 + \delta \alpha_i r_i^{(3)} r_i^{(2)} \theta_i(x_3) \right) \times$$

$$\prod_{i \in A} \left(1 + \delta \alpha_i r_i^{(3)} r_i^{(4)} \theta_i(x_4) \right) \prod_{i \in A} \left(1 + \delta \alpha_i r_i^{(4)} \theta_i(x_5) \right) d\omega.$$

Expanding out the product, and defining $v_S(x) := \prod_{i \in S} \theta_i(x)$, we see that

$$\mu_{(\alpha,\delta)} = \sum_{S \subseteq A} \delta^{|S|} \prod_{i \in S} \alpha_i \prod_{i \in S} \theta_i(x_1) \ldots \theta_i(x_5) = \sum_{S \subseteq A} \delta^{|S|} \prod_{i \in S} \alpha_i \bigotimes_{j=1}^{5} v_S(x_j). \tag{13}$$

Assuming $\delta < C$ we clearly have that

$$||\mu_{(\alpha,\delta)}||_{L^1(\tilde{\Omega})} = 1. \tag{14}$$

To each subset $S \subseteq A$ we may associated a Walsh function on, say, the probability space $\mathbb{T}$ in the usual manner. In particular, let $r_1, r_2, \ldots, r_m$ denote a system of Rademacher functions on $\mathbb{T}$ and form the associated Walsh system element associated to S by $W_S(y) := \prod_{i \in S} r_i(y)$. Given f such that $||f||_{L_x^\infty} \le C$, observe that

$$\left| \sum_{S \subseteq A} C^{-2|S|} W_S(y) \langle v_S, f \rangle \right| \le \left\| \prod_{i \in A} \left(1 + C^{-2} r_i(y) \theta_i(x) \right) \right\|_{L_x^1} = 1$$

where we have used $|C^{-2} \theta_i(x) f(x)| \le 1$. Since the function of y defined by the expression on the left above is uniformly bounded by 1 and thus has $L^2(\mathbb{T})$ norm at most 1, Bessel's inequality gives us that

$$\sum_{S \subseteq A} C^{-4|S|} |\langle v_S, f \rangle|^2 \le 1. \tag{15}$$

Using (13), we have that

$$\left\langle \mu_{(\alpha,\delta)}, \bigotimes_{s=1}^{5} \phi_s \right\rangle = \delta \sum_{j \in A} \alpha_j |\langle \theta_j, \phi_j \rangle|^5 + \sum_{\substack{S \subseteq A \\ |S| \geq 2}} \delta^{|S|} \prod_{j \in S} \alpha_j |\langle v_S, \phi_j \rangle|^5.$$

We will estimate each of these terms separately. We start by estimating the second using (15). Provided $C^8 \delta^2 < 1$, this gives

$$\sum_{\substack{S \subseteq A \\ |S| \geq 2}} \delta^{|S|} \prod_{j \in S} \alpha_j |\langle v_S, \phi_j \rangle|^5 \leq C^8 \delta^2.$$

We now consider the first term. By orthogonality and (12) we have that

$$|\langle \theta_j, \phi_i \rangle|^5 \leq |\langle \theta_j, \phi_i \rangle|^2 \left(\langle \phi_i, \phi_i \rangle + \varepsilon \right)^3.$$

From this and (15) we have, for $j \notin A$, that

$$\left| \sum_{j \in A} |\alpha_j| \langle \theta_j, \phi_i \rangle|^5 \right| \leq \sum_{j \in A} |\langle \theta_j, \phi_i \rangle|^5 \leq C^4 \varepsilon^3.$$

For $i \in A$, using again (15), we have that

$$\left| \sum_{j \in A} \alpha_j |\langle \theta_j, \phi_i \rangle|^5 - \alpha_i |\langle \theta_i, \phi_i \rangle|^5 \right| \leq C^4 \varepsilon^3. \tag{16}$$

Finally we have

$$|\langle \phi_i, \theta_i \rangle| \geq \langle \phi_i, \phi_i \rangle - |\langle \phi_i, \phi_i - \theta_i \rangle| \geq 1 - C\varepsilon. \tag{17}$$

Setting $\alpha_j = \text{sign}(a_j)$ for $j \in A$, the preceding estimates imply

$$\left\langle \sum_{i=1}^{n} a_i \bigotimes_{s=1}^{5} \phi_i, \mu_{(\alpha,\delta)} \right\rangle$$

$$\geq \delta \sum_{i \in A} |a_i| \langle \theta_i, \phi_i \rangle^5 - \delta \left(\sum_{i \in A} |a_i| \right) \varepsilon^3 - \delta \left(\sum_{i \notin A} |a_i| \right) \varepsilon^3 - \delta^2 \sum_{i=1}^{n} |a_i|.$$

Using (17), provided $\varepsilon \lesssim C^{-1}$, we have that

$$\left\langle \sum_{i=1}^{n} a_i \bigotimes_{s=1}^{5} \phi_i, \mu_{(\alpha,\delta)} \right\rangle \geq \delta \left(\frac{1}{2} \sum_{i \in A} |a_i| - C^4 \varepsilon^3 \sum_{i=2}^{n} |a_i| - C^8 \delta \sum_{i=1}^{n} |a_i| \right).$$

Recalling (10), we have that the quantity above is

$$\geq \delta \left(\frac{1}{2} C^2 \varepsilon^2 \left(\log \frac{C}{\varepsilon} \right)^{-1} - C^4 \varepsilon^3 - C^8 \delta \right) \sum_{i=1}^{n} |a_i|.$$

The result follows by an appropriate choice of δ and ε.

We now prove the second part of Theorem 2, namely:

Proposition 2. *Let $\{\phi_i\}$ denote a real-valued OS uniformly bounded by M such that the k-fold tensored system $\{\otimes_{s=1}^{k} \phi_s\}$ is Sidon. Then $\{\phi_i\}$ has the Rademacher-Sidon property.*

Let $k \geq 2$. If $\{\otimes_{i=1}^{k} \phi_i\}$ is Sidon, we have that

$$\int \left\| \sum_{i=1}^{n} a_i g_i(\omega) \prod_{i=1}^{k} \phi_i(x_i) \right\|_{L^{\infty}(\tilde{\Omega})} d\omega \geq c \sum_{i=1}^{n} |a_i|.$$

We then claim that

$$\int \left\| \sum_{i=1}^{n} a_i g_i(\omega) \phi_i(x) \right\|_{L^{\infty}(\Omega)} d\omega \gtrsim \int \left\| \sum_{i=1}^{n} a_i g_i(\omega) \prod_{i=1}^{k} \phi_i(x_i) \right\|_{L^{\infty}(\tilde{\Omega})} d\omega.$$

Recognizing that each side can be interpreted as the expectation of the supremum of a Gaussian process, this inequality follows from Slepian's comparison lemma once one has established the following lemma.

Lemma 3. *In the notation above we have*

$$\left(\sum_{i=1}^{n} |a_i|^2 \left| \prod_{s=1}^{k} \phi_i(x_s) - \prod_{s=1}^{k} \phi_i(x_s') \right| \right)^{1/2} \leq \sqrt{k} \left(\sum_{s=1}^{k} \sum_{i=1}^{n} |a_i|^2 |\phi_i(x_s) - \phi_i(x_s')|^2 \right)^{1/2}.$$

Proof. Using the elementary inequality $|\prod_{i=1}^{n} a_i - \prod_{i=1}^{n} b_i| \leq \sum_{i=1}^{n} |a_i - b_i|$ for sequence of real numbers of modulus less than or equal to 1, we have that

$$\left(\sum_{i=1}^{n} |a_i|^2 \left| \prod_{s=1}^{k} \phi_i(x_s) - \prod_{s=1}^{k} \phi_i(x_s') \right| \right)^{1/2} \leq M^{k-1} \sum_{s=1}^{k} \left(\sum_{i=1}^{n} |a_i|^2 |\phi_i(x_s) - \phi_i(x_s')|^2 \right)^{1/2}.$$

$$\leq \sqrt{k} \left(\sum_{s=1}^{k} \sum_{i=1}^{n} |a_i|^2 \left| \phi_i(x_s) - \phi_i(x'_s) \right|^2 \right)^{1/2}.$$

To summarize, we have shown that

$$\int \left\| \sum_{i=1}^{n} a_i g_i(\omega) \phi_i \right\|_{L^\infty(\Omega)} d\omega \gtrsim_M c \sum_{i=1}^{n} |a_i|.$$

One can replace the Gaussians random variables with Rademacher functions using a truncation argument and the contraction principle. Alternatively, one may apply Proposition 1. See [3] for details. This completes the proof.

A Counterexample: Theorem 1

The purpose of this section is to prove Theorem 1. We start with the following elementary fact:

Lemma 4. *Let* $10 < n, p$ *be positive real numbers. Then*

$$\sqrt{\log n} \, n^{-1/p} \leq \sqrt{p}.$$

Proof. The claim is equivalent to $\sqrt{p} n^{1/p} \geq \sqrt{\log n}$, or $p n^{2/p} \geq \log n$. Taking logarithms, this inequality is equivalent to $\log p + \frac{2}{p} \log n \geq \log \log n$. For a fixed n, the minimum of the left-hand side occurs when $\frac{1}{p} - \frac{2}{p^2} \log n = 0$, or $p = 2 \log n$. Thus we have

$$\log p + \frac{2}{p} \log n \geq \log \log n + \log 2 + 1 \geq \log \log n,$$

which establishes the claim.

Next we estimate the $\Lambda(p)$ constant of the first n elements of the Walsh system.

Lemma 5. *Let* W_i *denote the i-th Walsh function defined on the probability space* Ω_1. *Then*

$$\frac{\sqrt{\log n}}{\sqrt{n}} \| \sum_{i=1}^{n} a_i W_i \|_p \lesssim \sqrt{p} \left(\sum_{i=1}^{n} |a_i|^2 \right)^{1/2}.$$

Proof. By the Hausdorff-Young inequality we have that

$$\frac{\sqrt{\log n}}{\sqrt{n}} \| \sum_{i=1}^{n} a_i W_i \|_p \leq \frac{\sqrt{\log n}}{\sqrt{n}} n^{1/p'-1/2} (\sum_{i=1}^{n} |a_i|^2)^{1/2}$$

$$\leq \sqrt{\log n} n^{-1/p} \left(\sum_{i=1}^{n} |a_n|^2 \right)^{1/2}.$$

Applying Lemma 4 completes the proof.

For a fixed large n, let $\sigma_i \in \{-1,+1\}$ be chosen such that

$$\| \sum_{i=1}^{n} \sigma_i W_i \|_{L^\infty(\Omega_1)} \leq 6\sqrt{n}. \tag{18}$$

In other words, $\sum_{i=1}^{n} \sigma_i W_i$ is a Walsh Rudin-Shapiro polynomial. The existence of the coefficients σ_i is guaranteed, for instance, by Spencer's "six standard deviations suffices" theorem [10]. Next let r_i denote independent Rademacher functions on Ω_2. Furthermore define

$$\Psi := \left(1 + \frac{\log n}{n^2} \left(\sum_{i=1}^{n} r_i \right)^2 \right),$$

where $\int_\Omega \Psi d\mu = \left(1 + \frac{\log n}{n} \right)$. We now define an OS $\phi_0, \phi_1, \ldots, \phi_n$ on the measure space $(\Omega, \Psi d\mu)$ where $\Omega = \Omega_1 \times \Omega_2$. For $1 \leq i \leq n$ define

$$\phi_i := \frac{1}{\sqrt{\Psi}} \left(1 + \frac{\log n}{n} \right)^{-1/2} \left(r_i - \frac{\sqrt{\log n}}{\sqrt{n}} \sigma_i W_i \right)$$

where $\|\phi_i\|_{L^\infty} \leq 1 \times 1 \times (1 + \frac{\sqrt{\log n}}{\sqrt{n}}) \leq 2$. Next define

$$\phi_0 := \frac{1}{\sqrt{\Psi}} \left(1 + \frac{\log n}{n} \right)^{-1/2} \left(\frac{\sqrt{\log n}}{n} \sum_{i=1}^{n} r_i + \frac{1}{\sqrt{n}} \sum_{i=1}^{n} \sigma_i W_i \right).$$

Using that $\left(1 + \frac{\log n}{n} \right)^{-1/2} \leq 1$, $\frac{1}{\sqrt{\Psi}} \frac{\sqrt{\log n}}{n} \sum_{i=1}^{n} r_i \leq 1$ and $\left| \frac{1}{\sqrt{n}} \sum_{i=1}^{n} \sigma_i W_i \right| \leq 1$ by (18) we then have that

$$\|\phi_i\|_{L^\infty} \leq 1 \times 1 \times \left(1 + \frac{\sqrt{\log n}}{\sqrt{n}} \right) \leq 7.$$

We now verify that this system satisfies orthonormality relations. For $1 \leq i \leq n$

$$\int_\Omega |\phi_i|^2 \Psi d\mu = \int_\Omega \frac{1}{\Psi} \left(1 + \frac{\log n}{n}\right)^{-1} \left(r_i - \frac{\sqrt{\log n}}{\sqrt{n}} \sigma_i W_i\right)^2 \Psi d\mu$$

$$= \left(1 + \frac{\log n}{n}\right)^{-1} \left(1 + \frac{\log n}{n}\right) = 1.$$

For $1 \le i, j \le n$ we have

$$\int_\Omega \phi_i \phi_j \Psi d\mu = \int_\Omega \frac{1}{\Psi} \left(1 + \frac{\log n}{n}\right)^{-1} \left(r_i - \frac{\sqrt{\log n}}{\sqrt{n}} \sigma_i W_i\right) \times \left(r_j - \frac{\sqrt{\log n}}{\sqrt{n}} \sigma_j W_j\right) \Psi d\mu$$

$$= \left(1 + \frac{\log n}{n}\right)^{-1} \int_\Omega \left(r_i - \frac{\sqrt{\log n}}{\sqrt{n}} \sigma_i W_i\right) \times \left(r_j - \frac{\sqrt{\log n}}{\sqrt{n}} \sigma_j W_j\right) d\mu = 0.$$

Next we consider ϕ_0. We have

$$\int_\Omega |\phi_0|^2 \Psi d\mu = \left(1 + \frac{\log n}{n}\right)^{-1} \int_\Omega \left(\frac{\sqrt{\log n}}{n} \sum_{i=1}^n r_i + \frac{1}{\sqrt{n}} \sum_{i=1}^n \sigma_i W_i\right)^2 d\mu$$

$$= \left(1 + \frac{\log n}{n}\right)^{-1} \left(1 + \frac{\log n}{n}\right) = 1.$$

For $1 \le i \le n$, we have

$$\int_\Omega \phi_0 \phi_i \Psi d\mu = \left(1 + \frac{\log n}{n}\right)^{-1} \int_\Omega \left(\frac{\sqrt{\log n}}{n} \sum_{i=1}^n r_i + \frac{1}{\sqrt{n}} \sum_{i=1}^n \sigma_i W_i\right) \times \left(r_i - \frac{\sqrt{\log n}}{\sqrt{n}} \sigma_i W_i\right) d\mu$$

$$= \left(1 + \frac{\log n}{n}\right)^{-1} \left(\frac{\sqrt{\log n}}{\sqrt{n}} - \frac{\sqrt{\log n}}{\sqrt{n}}\right) = 0.$$

This completes the verification that the construction gives a uniformly bounded OS. Next we verify the $\psi_2(C)$ condition.

Lemma 6. *The OS* $\phi_0, \phi_1, \ldots, \phi_n$ *satisfies the* $\psi_2(C)$ *condition for some fixed* C *independent of* n.

Proof. Let $p \ge 2$, and $\sum_{i=1}^n |a_i|^2 = 1$. We have

$$\left\|\sum_{i=1}^n a_i \phi_i\right\|_{L^p} = \left(\int_\Omega \frac{|\sum_{i=1}^n a_i \phi_i|^p}{|\Psi|^{p/2}} \Psi d\mu\right)^{1/p} \le \left(\int_\Omega \left|\sum_{i=1}^n a_i \phi_i\right|^p d\mu\right)^{1/p}$$

$$\le \|a_0 \psi_0\|_{L^p(\Omega)} + \left\|\sum_{i=1}^n a_i r_i\right\|_{L^p(\Omega_1)} + \frac{\sqrt{\log n}}{\sqrt{n}} \left\|\sum_{i=1}^n a_i \sigma_i W_i\right\|_{L^p(\Omega_2)}.$$

Estimating the first term trivially, the second term using Khintchine's inequality, and the third using Lemma 5 gives us that

$$\|\sum_{i=1}^{n} a_i \phi_i\|_{L^p(\Omega)} \lesssim \sqrt{p}.$$

This completes the proof.

Finally, we show that these systems are not uniformly Sidon in n.

Lemma 7. *There exists coefficients* $\{a_0, a_1, \ldots, a_n\}$ *with unit* ℓ^1 *norm, such that*

$$\|\sum_{i=0}^{n} a_i \phi_i\|_{L^\infty(\Omega)} \lesssim \frac{1}{\sqrt{\log n}}.$$

Proof. Set $a_0 = \frac{1}{\sqrt{\log n}}$ and $a_i = \frac{1}{n}$, for $1 \le i \le n$. Then

$$\left| -\frac{1}{\sqrt{\log n}} \psi_0 + \frac{1}{n} \sum_{i=1}^{n} a_i \phi_i \right| = \frac{1}{\sqrt{\Psi}} \left(1 + \frac{\log n}{n} \right)^{-1/2} \times$$

$$\left| -\frac{1}{n} \sum_{i=1}^{n} r_i + \frac{1}{n} \sum_{i=1}^{n} r_i - \frac{1}{\sqrt{n \log n}} \sum_{i=1}^{n} \sigma_n W_n + \frac{\log n}{n^{3/2}} \sum_{i=1}^{n} \sigma_n W_n \right|$$

$$\le \left(\frac{1}{\sqrt{\log n}} + \frac{\log n}{n} \right) \left| \frac{1}{\sqrt{n}} \sum_{i=1}^{n} \sigma_n W_n \right| \lesssim \frac{1}{\sqrt{\log n}}$$

where we have used (18).

This completes the proof of Theorem 1.

1. W. Bednorz and R. Latała, On the boundedness of Bernoulli proceses, Ann. Math. 180 (2014), 1167–1203.
2. J. Bourgain, Sidon sets and Riesz products. Ann. Inst. Fourier (Grenoble) 35 (1985), no. 1, 137–148.
3. J. Bourgain and M. Lewko, Sidoncity and variants of Kaczmarz's problem, Preprint.
4. J. Elton, Sign-embeddings of l_n^1. Trans. Amer. Math. Soc. 279 (1983), no. 1, 113–124.
5. A. Pajor, Plongement de l_k^1 complexe dans les espaces de Banach. (French) [Embedding of complex l_k^1 in Banach spaces] Seminar on the geometry of Banach spaces, Vol. I, II (Paris, 1983), 139–148, Publ. Math. Univ. Paris VII, 18, Univ. Paris VII, Paris, 1984.
6. G. Pisier, Ensembles de Sidon et processus gaussiens. (French) C. R. Acad. Sci. Paris Sér. A-B 286 (1978), no. 15, A671–A674.
7. C. Preston, Banach spaces arising from some integral inequalities. Indiana Univ. Math. J. 20 1970/1971 997–1015.
8. D. Rider, Randomly continuous functions and Sidon sets. Duke Math. J. 42 (1975), no. 4, 759–764.
9. W. Rudin, Trigonometric series with gaps. J. Math. Mech. 9 1960 203–227.

10. J. Spencer, Six standard deviations suffice. Trans. Amer. Math. Soc. 289 (1985), no. 2, 679–706.
11. M. Talagrand. Regularity of Gaussian processes. Acta Math., 159(1–2):99-149, 1987.

J. Bourgain

M. Lewko

PROBLEM 131: A. ZYGMUND

Given is a function $f(x)$, continuous (for simplicity), and such that

$$\overline{\lim_{h \to 0}} \left| \int_h^1 \frac{f(x+t) - f(x)}{t} dt \right| < \infty, \text{ for } x \in E, |E| > 0.$$

Is it true that the integral

$$\int_0^1 \frac{f(x+t) - f(x)}{t} dt$$

may not exist almost everywhere in E? Similarly, for other Dini integrals?

Remark

This problem was raised by Professor Zygmund in a lecture given in Lwów in the early thirties. The problem was solved positively by Marcinkiewicz. The solution was published in his paper "Quelques théorèms sur les series et les fonctions," *Bull. Math. Seminar at the University of Wilno,* 1938. Although the original journal is practically inaccessible now, the paper is reproduced in Marcinkiewicz's collected papers, published by the Polish Academy of Science, Warsaw, 1964.

PROBLEM 132: W. SIERPIŃSKI

February 25, 1936

Does there exist a Baire function $F(x, y)$ (of two real variables) such that for every function $f(x, y)$ there exists a function $\phi(x)$ of one real variable (depending on the function f) for which $f(x, y) = F(\phi(x), \phi(y))$ for all real x and y.

Commentary

This problem remains unsolved. Sierpiński (Sur une function universelle de deux variables réelles, *Bull. Acad. Sci. Cracovie* A (1936), 8–12) showed that assuming the continuum hypothesis there is a function $F(x, y)$ which possesses the mapping properties described in the problem.

R. Daniel Mauldin

Second Edition Commentary

A function $F : \mathscr{R} \times \mathscr{R} \to \mathscr{R}$ is universal iff for every other function $f(x,y)$ there exists a function $\phi(x)$ with

$$f(x,y) = F(\phi(x), \phi(y))$$

for all x, y.

In Larson, Miller, Steprans, Weiss [1] we show that it is relatively consistent with ZFC that there is no universal function (Borel or not), and we show that it is relatively consistent that there is a universal function but no Borel universal function. We also prove some results concerning higher arity universal functions. For example, the existence of an F such that for every G there are unary h, k, j such that for all x, y, z

$$G(x,y,z) = F(h(x), k(y), j(z))$$

is equivalent to the existence of a 2-ary universal F. However the existence of an F such that for every G there are h, k, j such that for all x, y, z

$$G(x,y,z) = F(h(x,y), k(x,z), j(y,z))$$

follows from a 2-ary universal F but is strictly weaker.

1. Larson, Paul B.; Miller, Arnold W.; Steprans, Juris; Weiss, William A. R.; Universal functions. Fund. Math. 227 (2014), no. 3, 197–246.

Arnold Miller, January 2015

PROBLEM 133: EILENBERG

There is given in a metric space E, a family of sets which are open-and-closed, covering the space E. Find a family of sets which are simultaneously open and closed and disjoint, covering the space E and smaller than the preceding family.

Remarks:

(1) A family of sets K is smaller than the family K_1 if every set of the family K is contained in a certain set of the family K_1.
(2) This problem includes Problem 127, of Prof. Kuratowski.
(3) For separable spaces, the solution is trivial.

Commentary

In general no such family can be found. Roy's space X (cf. the Commentary to Problem 127) satisfies the equality ind $X = 0$, so that (*) X has a base consisting of open-and-closed sets, and yet, since $\dim X > 0$, X has an open—and by $(^*)$ an open-and-closed—cover which does not admit a refinement by pairwise disjoint open-and-closed sets (see R. Engelking, *Dimension Theory,* Warszawa 1978, Proposition 3.2.2).

R. Engelking

PROBLEM 134: EILENBERG

Is the Cartesian product $K_1 \times K_2$ of two indecomposable continua, K_1 and K_2, of necessity an indecomposable continuum?

Commentary

Let U be a nonvoid open subset of K_1 with $\bar{U} \neq K_1$. Let x be a point of K_2. Then $(\bar{U} \times K_2) \cup (K_1 \times \{x\})$ is a proper subcontinuum of $K_1 \times K_2$ with nonvoid interior $U \times K_2$. Every proper subcontinuum of an indecomposable continuum is nowhere dense, so $K_1 \times K_2$ must be decomposable. This is essentially the argument used by F.B. Jones [*Amer. J. Math.* 70 (1948), 403–413] to show that the product of any two nondegenerate continua is aposyndetic. Jones' paper is one of the earliest published results concerning the decomposability of products of continua, though it seems unlikely that earlier workers could have been totally unaware of results along this line.

While much interesting work continues to be done on indecomposable continua, very little of it involves products since indecomposability is always lost.

Wayne Lewis
Tulane University

PROBLEM 135: EILENBERG

Is non-unicoherence of a locally connected continuum an invariant of locally homeomorphic mappings?

Second Edition Commentary

The following example due to F.Marty, Bull. Sciences Math., II série, 61 (1937), 169-172, provides a negative answer. Let S and E be a sphere and an ellipsoid in $\mathbb{R}^3$, both centered at zero, with E touching S exactly at two poles, $X = S \cup E$ and let $f : X \to Y$ glue the antipodal points of S and E, respectively. Then X is not unicoherent, f is a local homeomorphism, but $f(S)$ and $f(E)$ are projective planes with exactly one point in common, hence their union Y is unicoherent.

More details and results on this topic can be found in a paper by A.Lelek, Fund. Math. 45 (1957), 51–63.

Roman Pol

PROBLEM 136: EILENBERG

Can an interior mapping (that is to say, one such that open sets go over into open sets) increase the dimensions?

Remark: This question occupied R. Baer, who obtained some partial results.

Addendum. A. Kolmogoroff, *Annals Math.* 38 (1937), 36–38, gave an example of a continuous, interior mapping which increased the dimension from 1 to 2.

B. Knaster

PROBLEM 137: EILENBERG

Given is a continuous mapping f of a compact space X, such that $\dim X > \dim f(X) > 0$. Does there exist a closed set $Y \subset X$ such that $\dim Y < \dim f(Y)$? In particular, does there exist, for every continuous mapping of a square into the interval, a closed 0-dimensional set whose image consists of a certain interval? We assume about the set X that it has the same dimension in every one of its points.

Commentary

The particular question about mappings from a square onto an interval was settled by J.L. Kelley [2] in a paper containing results of his dissertation under G.T. Whyburn. Kelley states in the paper that he owes the theorem to S. Eilenberg and L. Zippin. Using a neat category argument be proves the following:

Theorem If f is a mapping from the unit square I^2 onto the unit interval I, then there is a closed, totally disconnected subset K of I^2 such that $f(K) = I$.

Indeed, Kelley shows that if A is the space of all subsets of I^2 (with the Hausdorff metric) which map onto I under f and $\varepsilon > 0$, then the subset of A consisting of those members of A all of whose components have diameter less than ε is a dense open set in A. Thus the set of all members of A which are closed and totally disconnected is a dense G_δ in A.

In the same paper, using a similar category argument, Kelley also proves that if f is a monotone open mapping of a compact metric space onto a finite dimensional metric space Y, then there is a closed and totally disconnected subset K of X such that $f(K) = Y$ if and only if the set of points on which f is one-to-one is a totally disconnected subset of Y.

We found only three other papers with results related to Problem 137 and we have not obtained copies of these papers. The results given below are from the reviews of these papers.

In 1955, A Kosinski [4] proved that if f is a monotone open mapping from a compact metric space X onto a finite dimensional space Y and $f^{-1}(y)$ is nondegenerate for each $y \in Y$ and $\varepsilon > 0$, then there is a finite collection $O_1, O_2, \ldots, O_n$ of open sets with disjoint closures, each of diameter less than ε and such that f maps $\bigcup_{i=1}^{n} O_i$ onto Y, a result which follows from Kelley's earlier theorem. In 1956, B. Knaster [3] showed that Kosinski's result cannot be proved without the assumption that f be open. In 1956, R.D. Anderson [1] showed that if X is a compact metric space which is either 1-dimensional or a subset of a 2-manifold and f is a monotone open map from X onto a compact metric space Y and $f^{-1}(y)$ is nondegenerate for each point $y \in Y$, then there is a closed and totally disconnected subset K of X such that $f(K) = Y$. In addition Anderson gave examples to show that his result could not be established under the assumption that X be n-dimensional for $n \geq 2$ or even that X be a manifold for $n \geq 4$.

We have a slight improvement on Kelley's theorem about mappings from I^2 onto I. We can show that if X is a locally compact, locally connected metric space with the property that each connected open subset of X is cyclically connected and f is a mapping from X onto I, then there is a Cantor set K in X such $f(K) = I$. We note that Kelley's argument applies for mappings from I^n onto I and perhaps it could be altered to obtain our result. Our proof is elementary and somewhat similar to Kelley's but is not a category argument. In view of our result and Knaster's example it seems to us that there are interesting questions which are special cases of the original problem, and which to our knowledge remain unanswered. And it seems appropriate to us to consider special cases with restrictions on the range space. We suggest two problems. If X is a continuum which is 2-dimensional at each of its points, and f maps X onto I, must there be a closed, 0-dimensional subset of X which maps onto I under f? If f maps the unit cube I^3 onto I^2 and the set of points at which f is one-to-one is totally disconnected, must there be a closed 0-dimensional subset of I which maps onto I^2 under f?

1. R.D. Anderson, Some remarks on totally disconnected sections of monotone open mappings, *Bull. Acad. Polon. Sci. Cl. III* 4 (1956), 329–330.
2. J.L. Kelley, Hyperspace of a continuum, *Trans. Am. Math. Soc.* 52 (1942), 22–36.
3. B. Knaster, Sur la fixation des decompositions, *Bull. Acad. Polon. Sci. Cl. III* 4 (1956), 193–196.
4. A. Kosinski, A theorem on monotone mappings, *Bull. Acad. Polon. Sci. Cl. III* 3 (1956), 69–72.

William S. Mahavier

North Texas State University

PROBLEM 138: EILENBERG
May 17, 1936

(a) Any compact convex set located in a linear space of type (B_0) is an absolute retract.

(b) A set compact and convex, in the sense of Wilson, is an absolute retract. [A set $Y \subset X$ is a retract with respect to X is there exists a continuous function $f \in Y^X$ such that $f(y) = y$ for $y \in Y$. We call a compact space an absolute retract if it is a retract in every space which is metric, separable, containing it.] Absolute retracts have the fixed point property: (*vide* K. Borsuk, *Fund. Math.* 17.) A set X is convex in the sense of Wilson if, for every $x, y \in X$ and $0 \le t \le 1$ there exists one and only one point $z \in X$ such that $\zeta(x, z) = t\zeta(z, y) = (1 - t)\zeta(x, y)$.

Commentary

These results have been extended, and placed in a context that does not involve any hypothesis of separability, compactness, or completeness of the metric space, nor any restriction on the convexity.

Call an (arbitrary) metric space Z an AR (Absolute Retract) if it is a retract of each metric space containing it as a closed subset; and call an (arbitrary) metric space Z an ES(M) (Extensor Space for Metric spaces) if for every (arbitrary) metric space X and each closed $A \subset X$, every continuous $f : A \to Z$ has an extension $F : X \to Z$. It was shown by Dugundji (An extension of Tietze's theorem, *Pac. J. Math.* 1 (1951), 353–367) that (a) Every convex subset (not necessarily closed) of any locally convex linear space is an ES(M), and (b) A metric space is an ES(M) if and only if it is an AR.

It is also proved in the same paper that the surface $S = \{x : \|x\| = 1\}$ of the unit ball B in a normed linear space is an ES(M) if and only if S is compact. This result answers a generalization of Problem 36; it shows that Brouwer's fixed-point theorem for the unit ball B of any normed linear space L (this theorem is equivalent to the non-retractability of B onto S) is valid if and only if L is finite-dimensional.

J. Dugundji

PROBLEM 139: ULAM

Is every one-to-one continuous mapping of the Euclidean space into itself equivalent to a mapping which brings sets of measure 0 into sets of measure 0?

Addendum. Theorems: von Neumann

(a) A compact group of transformations of the Euclidean space is equivalent to a group of transformations carrying sets of measure 0 into sets of measure 0.
(b) Let f_n be a sequence of one-to-one mappings of Euclidean space. There exists a homeomorphism h such that the mappings hf_nh^{-1} carry sets of measure 0 into sets of measure 0.

PROBLEM 140: ULAM

Two mappings (not necessarily one-to-one) f and g of a set E into a part of itself are called equivalent if there exists a one-to-one mapping h of E into itself such that $f = hgh^{-1}$. What are necessary and sufficient conditions for the existence of such an h?

Commentary

It is necessary and sufficient that the algebras $\langle E, f \rangle$ and $\langle E, g \rangle$ be isomorphic. In other words, the directed graphs over E with the arrows $x \to f(x)$ and $x \to g(x)$, respectively, should be isomorphic. The problem has been extensively studied for the case when E is the unit interval and the functions are required to be continuous (see, e.g., Jan Mycielski, On the conjugates of the function $2|x| - 1$ in $[-1, 1]$, *Bull. London Math. Soc.* 12 (1980), 4–8, where other references are given).

Jan Mycielski

PROBLEM 141: ULAM

In the group M of one-to-one measurable transformations of the circumference of a circle into itself, two transformations which are rotations through different irrational angle *are not* equivalent. An analogous theorem holds for the group of transformations of the surface of the n-dimensional sphere into itself.

PROBLEM 142: ULAM; Theorem: GARRETT, BIRKHOFF

For every abstract group G there exists a set Z and a subset X contained in the square of the set $Z : X \subset Z^2$, such that the group G is isomorphic to a group of all one-to-one transformations f of the set Z into itself, under which the mapping $(x, y) \to (f(x), f(y))$ carries the set X into itself.

Commentary

A group G is isomorphic to $\text{Aut}(P)$ for some poset P. Hence G is also isomorphic to $\text{Aut}(L)$ for the distributive lattice 2^P. This is proved in my paper: Sobre los grupos de automorfismos, *Revista de la Unión Mat. Argentina* 11 (1946), 247–256.

Garrett Birkhoff

PROBLEM 143: MAZUR

Let K denote the class of functions of two integer-valued variables x, y such that:

(1) The functions $x, y, O, x + 1, xy$ belong to K;
(2) If the functions $a(x, y), b(x, y), c(x, y)$ belong to K, then the function $f(x, y) = c(a(x, y), b(x, y))$ also belongs to K;
(3) If the function $a(x, y)$ belongs to K, then the function $f(x, y)$ for which $f(0, y) = 1, f(x + 1, y) = a(x, f(x, y))$ belongs to K.

Does the class K contain the function

$$d(x,y) = \begin{cases} 1 \; for x \neq y, \\ 0 \; for x = y \end{cases} ?$$

Commentary

Evidently, the variables should be restricted to integers ≥ 0. So far as I know, no solution to this problem has been published. An affirmative solution will be presented here. Indeed, it will be shown that all primitive recursive functions of two variables are definable. Of course, no other functions are definable.

The particular function requested by Mazur was $sgn\,|x - y|$. In the classical terminology, which we shall follow, this is the characteristic function of $x \neq y$. It would be equivalent to ask for the characteristic function of $x = y$. The use of this function is central to the further development. Was Mazur aware of this?

In [2], I made a study of restricted schemes for obtaining all primitive recursive functions. That paper starts with the standard definition of primitive recursive functions as those obtained from certain initial functions (identity, zero, and successor) by repeated substitutions and primitive recursions. It then considers various restrictions of the recursion scheme, and asks what functions should be adjoined to the initial functions in order that all primitive recursive functions can be obtained using the restricted scheme being studied. Some improvements of my results may be found in Gladstone [1].

Notice that the function defined by Mazur's recursion scheme depends on only one of the variables. That is, the scheme is essentially a definition by recursion of a function of one variable. It will not change which functions of two variables are definable if we allow functions of any number of variables, include all of the usual initial functions together with the function $F(x,y) = xy$, and allow unrestricted substitution, but allow recursion only for defining functions of one variable.

The recursion scheme for defining functions of one variable has the form

$$F0 = a, \qquad FSx = B(x, Fx).$$

Here the function F is defined in terms of a function B of two variables. To agree with Mazur's scheme, we must use only $a = 1$.

For any number c, the function c^x may be defined by $c^0 = 1$, $c^{sx} = c^x \cdot c$. We put $sgn\,x = 0^{0^x}$. Next, we define the function $Fx = |x - 1|$ by $F0 = 1$, $FSx = x$. We can use this to define the predecessor function $Px = |x - 1| \cdot sgn\,x$.

We can now define addition in an important special case. If, for every x, either $Gx = 0$ or $Hx = 0$, then

$$Gx + Hx = P(SGx \cdot SHx).$$

This enables us to define functions by piecing. For example, if

$$Fx = \begin{cases} Ax \text{ when } Cx = 0, \\ Bx \text{ when } Cx > 0, \end{cases}$$

then

$$Fx = Ax \cdot 0^{Cx} + Bx \cdot \text{sgn } Cx,$$

and this sum can be defined as above. This is very useful.

Let Rx be the distance from x to the smallest number of the form $6^n \geq x$. We see that

$$R0 = 1, \qquad RSx = \begin{cases} PRx & \text{if } Rx > 0, \\ P(5x) & \text{if } Rx = 0. \end{cases}$$

This is a legitimate definition, since piecing is allowed. But $x = y$ if and only if $2^x \cdot 3^y$ is a power of 6, and this is expressed by $R(2^x \cdot 3^y) = 0$. Thus the characteristic function of equality if $0^{R(2^x \cdot 3^y)}$, and the function sgn $|x - y|$ requested by Mazur is sgn $R(2^x \cdot 3^y)$. This completes the affirmative solution of Mazur's problem.

The above construction would not be possible if we used $a = 0$ instead of $a = 1$. Indeed, as is easily seen, we could then obtain only functions which are monotone in each variable. However, if we adjoined 0^x as well as xy to the initial functions, then all of the other functions defined above could be obtained. Indeed, if $c > 0$, then $c^x - 1$ can be defined by a recursive definition with $a = 0$, and $c^x = S(c^x - 1)$. We have $P0 = 0$, $PSx = x$, and can put $|x - 1| = Px + 0^x$. This addition is definable, as explained above. Finally, we may define $Fx = RSx$ by a recursion with $a = 0$, and then we have $Rx = FPx + 0^x$.

We want to show that all primitive recursive functions can be obtained using $a = 1$, or using $a = 0$ and adjoining 0^x as well as xy to the initial functions. Now in [2], I showed that all primitive recursive functions can be obtained if we adjoin $x + y$ and Q to the usual initial functions, where Q is the characteristic functions of squares. I used only the recursion scheme with $a = 0$. However, the use of $a = 0$ may be replaced by the use of $a = 1$. Indeed, if Fx is defined by

$$F0 = 0, \qquad FSx = B(x, Fx),$$

then $Gx = SFx$ is defined by

$$G0 = 1, \qquad GSx = SB(x, PGx),$$

and we can then obtain $Fx = PGx$. Thus it will be sufficient to define $x + y$ and Q, using $a = 0$.

We first define the function $Fx = [x^{1/2}]$ by

$$F0 = 0, \qquad FSx = \begin{cases} SFx \text{ if } Sx = (SFx)^2, \\ Fx \quad \text{otherwise.} \end{cases}$$

This is a legitimate definition, since we allow piecing and know the characteristic function of equality. But x is a square if and only if $x = [x^{1/2}]^2$, so we obtain a definition of Q.

We may define the function $Fx = [\log_2 x]$ for $x > 0$ in a quite similar way. Let

$$F0 = 0, \qquad FSx = \begin{cases} SFx \text{ if } Sx = 2^{SFx}, \\ Fx \quad \text{otherwise} \end{cases}$$

We can now define addition by $x + y = [\log_2(2^x \cdot 2^y)]$.

It follows that all primitive recursive functions can be obtained by adjoining xy to the initial functions and allowing recursion only to define functions of one variable, with the restriction that $a = 1$. The same result holds with $a = 0$, if we adjoin xy and 0^x to the initial functions. Also, we see that Mazur's class consists of all primitive recursive functions of two variables.

Gladstone [1, §4] showed that instead of adjoining $x + y$ and Q to the initial functions, as in [2], it would be sufficient to adjoin only $x + y$, provided that we did not restrict the value of a. An examination of his proof shows that only $a = 0, 1, 2$ are used, and it is easily seen that $a = 1$ alone would be sufficient. We could allow just $a = 0$ if we also adjoined 0^x to the initial functions. Thus the same results are obtained with $x + y$ as with xy.

My proof was suggested in part by the form of Gladstone's proof. He first defined the characteristic function of powers of 2, and then noted that $x = y$ if and only if $2^x + 2^y$ is a power of 2. I first defined the characteristic function of powers of 6, and then noted that $x = y$ if and only if $2^x \cdot 3^y$ is a power of 6.

1. M.D. Gladstone, Simplifications of the recursion scheme, *J. Symbolic Logic,* 36 (1971), 653–665.
2. R.M. Robinson, Primitive recursive functions, *Bull. Amer. Math. Soc.* 53 (1947), 925–942.

Raphael M. Robinson
University of California Berkeley

PROBLEM 144: MAZUR, ULAM

Let K denote a sphere in a separable space of type (B). Does there exist a one-to-one mapping of K into the interval $0 \leq x \leq 1$ under which the image of every open set in K is a set of positive measure?

Solution

The answer to the problem is yes. This may be seen as follows. Let $\{x_n\}_{n=1}^{\infty}$ be a countable dense subset of K. For each pair of positive integers $\langle n,m \rangle$, let $T(\langle n,m \rangle) = \{x \in K : \|x - x_n\| = 1/m\}$. Set $T = \bigcup \{T(\langle n,m \rangle) : n \geq 1, m \geq 1\}$ and set $R = K - T$. Then R is a dense G_δ subset of K which is 0-dimensional and dense-in-itself. Let M be a copy of the Cantor set lying in J, the set of all irrational numbers between 0 and 1, and such that the Lebesgue measure of M is zero. Of course, $J - M$ is a dense-in-itself 0-dimensional complete separable metric space. According to a theorem of Mazurkiewicz [1, p. 144], there is a homeomorphism g of R onto $J - M$. Let h be a one-to-one Borel measurable map of T onto $I - (J - M)$. Set $f(x) = g(x)$, if $x \in R$ and $f(x) = h(x)$, if $x \in K - R$. Then f is a one-to-one Borel measurable map of K onto I. Clearly, if U is an open subset of K, then $f(U)$ has a positive measure.

1. K. Kuratowski, *Topology, Volume I,* Academic Press, New York, 1966.

R. Daniel Mauldin

PROBLEM 145: ULAM

Given is a countable sequence of sets A_n. Find necessary and sufficient conditions for the possibility of introduction of a countably additive measure $m(A_n)$ such that $m(\Sigma A_n) = 1$, $m(p) = 0$; (p) denotes a set composed of a single point. [Possibly a stronger condition: $m(A_p) = 0$ for a certain given subsequence p_k.] We demand that the measure should be defined for each of the sets of a Borel ring of sets over the sequence A_n.

Commentary

A solution is given by S. Banach in Sur les suites d'ensembles excluant l'existence d'une mesure, *Coll. Math.* 1 (1948), 103–108.

Jan Mycielski

PROBLEM 146: ULAM

It is known that in sets of positive measure there exist points of density 1 [that is to say, points with the property that the ratio of the length of intervals to the measure of the part of the set contained in these intervals tends to 1 (if the length of the interval converges to 0)]. Can one determine the speed of convergence of this ratio for almost all points of the set?

Commentary

It will be more convenient to write about the natural product measure μ in the Cantor space $C = \{0,1\}^\omega$, i.e., the measure μ defined by the formula $\mu(A) =$

(the Lebesgue measure of the set $\{\sum_{i=0}^{\infty} x_i/2^{i+1} : (x_0, x_1, \ldots) \in A\}$). Let $A \subset C$ be μ-measurable. For any $\varepsilon > 0$, we put $S(n, \varepsilon, A) = \{(x_0, \ldots, x_{n-1}) \in \{0, 1\}^n : 2^n \mu(A \cap V(x_0, \ldots, x_{n-1})) > 1 - \varepsilon\}$, where $V(x_0, \ldots, x_{n-1}) = \{(y_0, y_1, \ldots) \in C : (y_0, \ldots, y_{n-1}) = (x_0, \ldots, x_{n-1})\}$. By the Lebesgue density theorem (see [2]) and an easy compactness argument we have the following.

Theorem. For every $\varepsilon > 0$

$$\lim_{n \to \infty} |S(n, \varepsilon, A)|/2^n = \mu(A).$$

On the other hand, it is not hard to prove that for every sequence $\varepsilon_0, \varepsilon_1, \ldots$ such that $1 \geq \varepsilon_0 \geq \varepsilon)1 \geq \cdots$ and $\varepsilon_n \to 0$ there exists a measurable set $A \subseteq C$ such that $\mu(A) = 1 - \varepsilon_0$ and

$$S(n, \varepsilon_n, A) = \phi$$

for $n = 0, 1, \ldots$.

For related material see [1].

1. A. Ehrenfeucht and J. Mycielski, An infinite solitaire game with a random strategy, *Colloq. Math.* 42 (1979), 115-118.
2. J.C. Oxtoby, *Measure and Category,* Springer-Verlag, New York, 1970.

Jan Mycielski

PROBLEM 147: AUERBACH, MAZUR
September 4, 1936

Suppose that a billiard ball issues at the angle $45°$ from a corner of the rectangular table with a rational ratio of the sides. After a finite number of reflections from the cushion will it come to one of the remaining three corners?

PROBLEM 148: AUERBACH

Let $P(x_1, \ldots, x_n)$ denote a polynomial with real coefficients. Consider the set of points defined by the equation $P(x_1, \ldots, x_n) = 0$. A necessary and sufficient condition for this set not to cut the Euclidean (real) space is: All the irreducible factors of the polynomial P in the real domain should be always nonnegative or always nonpositive.

Second Edition Commentary

We will interpret "not to cut" to mean that the real zero set of P has (non-empty) path-connected complement. Problem 148 admits an elementary solution via reduction to...

Lemma 1. *If $Q \in \mathbb{R}[x_1, x_2]$ and each irreducible factor of Q is always nonnegative on $\mathbb{R}^2$ or always nonpositive on $\mathbb{R}^2$, then the complement of the real zero set of Q is path-connected.*

Auerbach's Problem naturally leads to important recent advances in parts of algorithmic algebraic geometry: polynomial factorization and nonnegativity. Algorithms that are practical and efficient (as of early 2015) for multivariate polynomial factorization over $\mathbb{C}$ are detailed in [Gao03, CL07]. Algorithms that take numerical instability into account (in the coefficients and/or the final answer) include [G+04, Zen09].

For an arbitrary $f \in \mathbb{R}[x_1]$ with degree D and exactly t monomial terms, all general algorithms for factorization over $\mathbb{R}$ have complexity super-linear in D. However, such an f has at most $2t - 1$ real roots (thanks to Descartes' Rule). So counting just the degree 1 factors may admit a faster algorithm when t is fixed. Such algorithms, with complexity polynomial in $\log D$ (and the total bit size of all the coefficients when $f \in \mathbb{Z}[x_1]$), appear in [BRS09] (counting bit operations, for $t=3$) and [Sag14] (counting field operations, for any constant t). The multivariate case is touched upon in [Ave09, Gre15].

The relationship between nonnegativity and sums of squares was advanced by Hilbert and Artin, and has since evolved into a beautiful intersection of optimization, real algebraic geometry, and convexity. See, e.g., [BPT12] and the references therein.

Our key reduction to Lemma 1 hinges on the following fact:

Lemma 2. *(See [Sch00, Thm. 17, Pg. 75, Sec. 1.9]) Suppose $P \in \mathbb{R}[x_1, \ldots, x_n] \backslash \mathbb{R}[x_1]$ is irreducible in $\mathbb{R}[x_1, \ldots, x_n]$. Then there is a polynomial*

$$\Phi \in \mathbb{R}[w_1, \ldots, w_{n-1}, y_1, \ldots, y_{n-1}] \setminus \{0\}$$

with the following property: If $\alpha_2, \beta_2, \ldots, \alpha_n, \beta_n \in \mathbb{R}$ and $\Phi(\alpha_2, \ldots, \alpha_n, \beta_2, \ldots, \beta_n) \neq 0$, then $P(x_1, \alpha_2 x_2 + \beta_2, \ldots, \alpha_n x_2 + \beta_n)$ is irreducible in $\mathbb{R}[x_1, x_2]$. □

Solution to Problem 148: The case $n=1$ follows immediately upon observing that, up to real affine transformations, the only irreducible non-constant polynomials in $\mathbb{R}[x_1]$ are x_1 and $x_1^2 + 1$. So assume $n \geq 2$ and let Z be the zero set of P in $\mathbb{R}^n$.

Sufficiency: Let $u = (u_1, \ldots, u_n)$ and $v = (v_1, \ldots, v_n)$ lie in $\mathbb{R}^n \backslash Z$. Since Z is closed, u (resp. v) in fact lies in an open neighborhood U (resp. V) of points contained in the same connected component of $\mathbb{R}^n \backslash Z$ as u (resp. v). Since the complement of any real algebraic hypersurface is open, Lemma 2 implies we can find $\alpha_2, \beta_2, \ldots, \alpha_n, \beta_n$ such that the specialization $Q(x_1, x_2) := P(x_1, \alpha_2 x_2 + \beta_2, \ldots, \alpha_n x_2 + \beta_n)$ satisfies: (a) each irreducible factor P_i of P specializes to an irreducible factor Q_i of Q and (b) P_i is nonnegative on all of $\mathbb{R}^n$ (resp. nonpositive on all of $\mathbb{R}^n$) if and only if Q_i is nonnegative on all of $\mathbb{R}^2$ (resp. nonpositive on all of $\mathbb{R} r$). In particular, the condition $\Phi \neq 0$ from Lemma 2 enables us to pick $(\beta_2, \ldots, \beta_n)$ arbitrarily close to $(u_2, \ldots, u_n)$, and $(\alpha_2, \ldots, \alpha_n)$ arbitrarily close to $(v_2 - u_2, \ldots, v_n - u_n)$, so that both $Q(u_1, 0)$ and $Q(v_1, 1)$ are nonzero.

Let W denote the zero set of Q in $\mathbb{R}^2$ and note that the real 2-plane
$$H := \{(x_1, \alpha_2 x_2 + \beta_2, \ldots, \alpha_n x_2 + \beta_n)\}_{(x_1, x_2) \in \mathbb{R}^2} \subset \mathbb{R}^n$$
yields a pair $(H, H \cap Z)$ affinely equivalent to $(\mathbb{R}^2, W)$. By Lemma 1 (and Conditions (a) and (b)), $\mathbb{R}^2 \setminus W$ is path-connected, and thus $H \setminus Z$ is path-connected. Moreover, by our choice of the α_i and β_i, both U and V intersect $H \setminus Z$. So U and V, and thus u and v, are connected by a path in $\mathbb{R}^n n \setminus Z$. $\square$

(**Necessity**): Suppose now that $\mathbb{R}^n \setminus Z$ is path-connected, but P has an irreducible factor P_i attaining both positive and negative values on $\mathbb{R}^n$. Then P_i must be positive (resp. negative) at some point u_+ (resp. u_-) in $\mathbb{R}^n \setminus Z$ since $\mathbb{R}^n \setminus Z$ is open. By assumption, there is a path in $\gamma : [0,1] \longrightarrow \mathbb{R}^n \setminus Z$ connecting u_+ and u_-. In particular, $P_i(\gamma(0)) P_i(\gamma(1)) < 0$, so by the Intermediate Value Theorem, $P_i(\gamma(s)) = 0$ for some $s \in (0,1)$. In other words, $\gamma([0,1])$ intersects Z, which is a contradiction. $\square$

To prove Lemma 1 we will apply the following two facts:

Proposition 3. *If $X \subset \mathbb{R}r$ is finite, then any two points of $\mathbb{R}^2 \setminus X$ can be connected by a smooth quadric curve $\Gamma \subset \mathbb{R}^2 \setminus X$.* $\square$

Lemma 4. *Suppose $f, g \in \mathbb{C}[x_1, x_2]$ have respective degrees d and e, and no common factor of positive degree. Then $f = g = 0$ has no more than de solutions in $\mathbb{C}^2$.* $\square$

Proposition 3 follows easily by using an invertible affine map to reduce to the special case of connecting $(0,0)$ and $(1,0)$ via the graph of $cx_1(1 - x_1)$ for suitable c: The finiteness of X guarantees that all but finitely many c will work. Lemma 4 is a special case of Bézout's Theorem, but can also be easily derived from the basic properties of the univariate resultant (see, e.g., [Sch00, App. B]).

Proof of Lemma 1: Let W denote the real zero set of Q in $\mathbb{R}^2$. By Proposition 3 it suffices to prove that, under the hypotheses of Lemma 1, W is finite. It clearly suffices to restrict to the special case where Q is non-constant and irreducible in $\mathbb{R}[x_1, x_2]$. Note also that the irreducibility of Q and the assumption on the sign of Q are invariant under composition with any invertible real affine map.

Consider now any root $\zeta = (\zeta_1, \zeta_2) \in \mathbb{R}^2$ of Q. If $\delta := \left(\frac{\partial Q}{\partial x_1}(\zeta), \frac{\partial Q}{\partial x_2}(\zeta) \right) \neq 0$ then, by composing with a suitable invertible real affine map, we may assume $\delta = (1,0)$. In particular, by Taylor expansion, we see that Q changes sign in a non-empty horizontal line segment containing ζ. Therefore, *every* root of ζ of Q must satisfy $\frac{\partial Q}{\partial x_1}(\zeta) = \frac{\partial Q}{\partial x_2}(\zeta) = 0$.

Let $Q_1 \cdots Q_r$ be the factorization of Q over $\mathbb{C}[x_1, x_2]$ into factors of positive degree, irreducible in $\mathbb{C}[x_1, x_2]$. The Galois group $G := \mathrm{Gal}(\mathbb{C}/\mathbb{R})$ has order 2, is generated by complex conjugation $\overline{(\cdot)}$, and acts naturally on the Q_i. In particular, G acts trivially on Q_i if and only if $Q_i \in \mathbb{R}[x_1, x_2]$. So r must be even when $r \geq 2$, since Q is irreducible over $\mathbb{R}[x_1, x_2]$. Furthermore, $r \leq 2$ since $Q_i \bar{Q}_i$ is invariant under complex conjugation. So we either have $r = 1$ (with Q irreducible over $\mathbb{C}[x_1, x_2]$) or $r = 2$ (with $\bar{Q}_1 \neq Q_1 = \bar{Q}_2 \neq Q_2$). A simple calculation then shows that, in either case, $\frac{\partial Q}{\partial x_1}$ has no common factors with Q. So W is finite by Lemma 4. $\square$

I am grateful to Dan Mauldin and Joe Buhler for bringing Auerbach's problem to my attention. I also thank Guillaume Chèze, Michel Coste, Erich Kaltofen, Marie-Francoise Roy, and Zbigniew Szafraniec for valuable comments.

Research partially supported by NSF grant CCF-1409020.

[Ave09] Avendaño, Martín, *"The number of roots of a lacunary bivariate polynomial on a line,"* J. Symb. Comp. **44** (2009), pp. 1280–1284.

[BRS09] Bihan, Frederic; Rojas, J. Maurice; Stella, Case E., *"Faster Real Feasibility via Circuit Discriminants,"* proceedings of International Symposium on Symbolic and Algebraic Computation (ISSAC 2009, July 28–31, Seoul, Korea), pp. 39–46, ACM Press, 2009.

[BPT12] Blekherman, Grigoriy; Parrilo, Pablo A.; and Thomas, Rekha R., *"Semidefinite Optimization and Convex Algebraic Geometry,"* MPS-SIAM series on optimization, book 13, SIAM, 2012.

[CL07] Chèze, Guillaume and Lecerf, Grégoire, *"Lifting and recombination techniques for absolute factorization,"* J. Complexity, **23**, no. 3, pp. 380–420, 2007.

[Gao03] Gao, Shuhong, *"Factoring multivariate polynomials via partial differential equations,"* Math. Comp., **72**, no. 242, pp. 801–822, 2003.

[G+04] Gao, Shuhong; Kaltofen, Erich; May, John P.; Yang, Zhengfeng; and Zhi, Lihong, *"Approximate factorization of multivariate polynomials via differential equations,"* in proceedings of ISSAC 2004, pp. 167–174, ACM, New York, 2004.

[Gre15] Grenet, Bruno, *"Bounded-degree factors of lacunary multivariate polynomials,"* Math ArXiV preprint http://arxiv.org/abs/1412.3570. Submitted for publication.

[Sag14] Sagraloff, Michael, *"A near-optimal algorithm for computing real roots of sparse polynomials,"* in proceedings of ISSAC 2014, pp. 359–366, ACM, 2014.

[Sch00] Schinzel, Andrzej, *Polynomials with Special Regard to Reducibility,* Encyclopedia of Mathematics and its Applications, Book 77, Cambridge University Press, 2000.

[Zen09] Zeng, Zhonggang, *"The approximate irreducible factorization of a univariate polynomial. Revisited,"* in proceedings of ISSAC 2009, pp. 367–374, ACM, New York, 2009.

J. Maurice Rojas

Texas A&M University TAMU, College Station

PROBLEM 149: NIKLIBORC

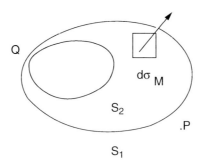

Let S_1 and S_2 denote closed and convex surfaces tangent at a point Q. Let S_2 be contained in the domain whose boundary is S_1.

Let

$$V_k(P) = \int_{S_k} \frac{d\sigma_M}{r_{PM}} \quad k = 1, 2.$$

Theorem: $V_1(Q) > V_2(Q)$.

PROBLEM 150: NIKLIBORC

Let S denote a closed surface and $f(M)$ a continuous function defined on S. Let $V(P) = \int_S f(M)(1/r_{PM})d\sigma_M$. Let us assume that the plane π has the property: If P_1 and P_2 denote two arbitrary points of space, located outside a sufficiently large sphere and symmetrically with respect to the plane π, then $V(P_1) = V(P_2)$. Prove that:

(1) The plane π is a plane of symmetry for the surface S.
(2) In points of symmetry M_1 and M_2 belonging to S we have $f(M_1) = f(M_2)$.

PROBLEM 151: WAVRE

November 6, 1936; Prize: A "fondue" in Geneva; Original manuscript in French

Does there exist a harmonic function defined in a region which contains a cube in its interior, which vanishes on all the edges of the cube? One does not consider $f \equiv 0$.

Addendum. Does there exist an algebraic function $f(z)$ homomorphic in every point of a curve traced on a surface of Riemann and such that one has

$$\int_\ell \frac{f(x)}{z-x}dz = 0, \qquad f(z) \not\equiv 0,$$

the point x being contained in a certain domain? The curve ℓ will be open. One should find $f(z)$ and ℓ.

Prize: "Fondant" in Lwów

PROBLEM 152: STEINHAUS

November 6, 1936; Prizes: For the computation of the frequency: 100 grammes of caviar. For a proof of the existence of frequency, a small beer. For counter example: A demitasse

A disk of radius 1 covers at least two points with integer coordinates (x, y) and at most 5. If we translate this disc through vectors nw ($n = 1, 2, 3, \ldots$), where w has both coordinates irrational and their ratio is irrational, then the numbers 2, 3, 4 repeat infinitely many times. What is the frequency of these events for $n \to \infty$? Does it exist?

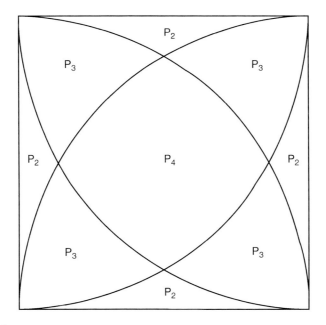

Figure 152.1

Commentary

By the 2-dimensional equipartition theorem (see J.F. Koksma, *Diophantische Approximationen,* repr. Chelsea 1936, Springer-Verlag, Berlin-New York, 1974), the frequencies exist, 2 has the frequency $4 - \sqrt{3} - (2/3)\pi$, 3 has the frequency $2\sqrt{3} - 4 + (\pi/3)$, and 4 has the frequency $1 - \sqrt{3} + (\pi/3)$. These frequencies are the areas of the three parts P_2, P_3, and P_4 of the square $[0,1] \times [0,1]$ such that if the center of a circle of radius 1 is in P_i then the circle covers i lattice points (Figure 152.1).

Jan Mycielski

PROBLEM 153: MAZUR
Prize: A live goose, November 6, 1936

   Given is a continuous function $f(x,y)$ defined for $0 \leq x,y \leq 1$ and the number $\varepsilon > 0$; do there exist numbers $a_1,\ldots,a_n$; $b_1,\ldots,b_n$; $c_1,\ldots,c_n$ with the property that

$$\left| f(x,y) - \sum_{k=1}^{n} c_k f(a_k,y) f(x,b_k) \right| \leq \varepsilon$$

in the interval $0 \leq x,y \leq 1$?

Remark: The theorem is true under the additional assumption that the function $f(x,y)$ possesses a continuous first derivative with respect to x or y.

Commentary

Grothendieck proved, in his thesis [3], that Problem 153 is equivalent to the approximation problem—a problem which was considered to be one of the central open problems of functional analysis. The statement of the approximation problem is the following: Is every compact linear operator T from a Banach space X into a Banach space Y a limit in norm of operators of finite rank? A Banach space Y is said to have the approximation property if the answer to the question is positive for every choice of X and T. Every space Y with a Schauder basis has the approximation property and thus Problem 153 is closely related to the basis problem. (Does every separable Banach space have a Schauder basis?) Since a Hilbert space has a basis (and thus the approximation property) it follows from Grothendieck's reformulation of Problem 153 that this problem (in its original formulation) has a positive answer if f satisfies a Lipschitz condition of order $1/2$.

The answer to the approximation problem (and thus also to Problem 153 as well as the basis problem) is negative. This was proved in 1972 by P. Enflo [2]. The Goose promised to the solver of 153 was given to Enflo a year or so later following his lecture on his solution in Warsaw.

By modifying Enflo's construction, Davie [1] showed that for every $\alpha < 1/2$ there is a function $f(x,y)$ satisfying a Lipschitz condition of order α but for which there is no approximation of the form required in the statement of Problem 153.

Further major progress related to the approximation problem was made by A. Szankowsky. He showed [6] that there is a natural example of Banach space which fails to have the approximation property, namely, the space of all operators from ℓ_2 into itself with the usual operator norm. He proved also [5] that unless Y is "very close" to a Hilbert space it has a subspace failing the approximation property.

A detailed treatment of questions related to the approximation property is contained in [4].

1. A.M. Davie, The Banach approximation problem, *J. Approx. Theory* 13 (1975), 392–394.
2. P. Enflo, A counterexample to the approximation problem in Banach spaces, *Acta Math.* 130 (1973), 309–317.
3. A. Grothendieck, Produits tensoriels topologiques et espaces nucléaires, *Memo. Amer. Math. Soc.,* 16 (1955).
4. J. Lindenstrauss and L. Tzafriri, *Classical Banach spaces, Vol. I, Sequence Spaces (1977), Vol. II, Function Spaces (1979),* Springer-Verlag, Berlin-New York.
5. A. Szankowski, Subspaces without the approximation property, *Israel J. Math.* 30 (1978), 123–129.
6. A. Szankowski, $B(\mathscr{H})$ does not have the approximation property, *Acta Math.* 147 (1981), 89–108.

Joram Lindenstrauss

PROBLEM 154: MAZUR

September 15, 1936

Let $\{\phi_n(t)\}$ be an orthogonal system of continuous functions and closed in C.

(a) If $f(t) \sim a_1\phi_1(t) + a_2\phi_2(t) + \cdots$ is the development of a given continuous function $f(t)$ and $n_1, n_2, \ldots$ denote the successive indices for which $a_{n_1} \neq 0, \ldots$, can one approximate $f(t)$ uniformly by the linear combinations of the functions $\phi_{n_1}(t), \phi_{n_2}(t), \ldots$?

(b) Does there exist a linear summation method M such that the development of every continuous function $f(t)$ into the system $\{\phi_n(t)\}$ is uniformly summable by the method M to $f(t)$?

Second Edition Commentary

Roughly speaking, the question is: Let $\{\varphi_n\}$ be an orthonormal system on a segment I. Suppose it satisfies the Weierstrass approximation property, that is, linear combinations are dense in $C(I)$. Is a Fejer type theorem true?

The question can be decomposed into two sub-questions:

First:

Is it true that every $f \in C(I)$ admits approximation by linear combinations which use only those $\{\varphi_n\}$, which are presented in the Fourier expansion of f with nonzero amplitudes?

(In this case one may say that spectrum-preserving approximation is possible).

And second:

Suppose that the system does satisfy the last property. Is it true that corresponding linear combinations could be obtained from the Fourier partial sums by a certain linear procedure?

Unfortunately, a counter-example can be constructed to each of the questions, see [1].

References

[1] A.M.Olevskii, On spectrum-preserving approximation, Soviet Math. Dokl. 41 (1990), no. 2, 215–218.

A. Olevskii

PROBLEM 155: MAZUR, STERNBACH

November 18, 1936

Given are two spaces X, Y of type (B), $y = U(x)$ is a one-to-one mapping of the space X onto the whole space Y with the following property: For every $x_0 \in X$ there exists an $\varepsilon > 0$ such that the mapping $y = U(x)$, considered for x belonging to the sphere with the center x_0 and radius ε, is an isometric mapping. Is the mapping $y = U(x)$ an isometric transformation?

This theorem is true if U^{-1} is continuous. This is the case, in particular, when Y has a finite number of dimensions or else the following property: If $\|y_1 + y_2\| = \|y_1\| + \|y_2\|$, $y_1 \neq 0$, then $y_2 = \lambda y_1$, $\lambda \geq 0$.

PROBLEM 156: WARD

March 23, 1937; Original manuscript in English

A surface $x = f(u,v)$, $y = g(u,v)$, $z = h(u,v)$, f, g, h being continuous functions, has at each point a tangent plane in the geometric sense; also, to each point of the surface there corresponds only one pair of values of u,v. Does there exist a representation of the surface by functions $x = f_1(u,v)$, $y = g_1(u,v)$, $z = h_1(u,v)$, in such a manner that the partial derivatives exist and the Jacobians $\partial(f_2,f_3)/\partial(u,v)$, $\partial(f_3,f_1)/\partial(u,v)$, $\partial(f_1,f_2)/\partial(u,v)$ are not all zero, except for a set N of values of u,v such that the corresponding set of points of the surface has surface measure (in Caratheodory's sense) zero? [Let (x,y,z) be a point of a surface S, and P a plane through (x,y,z). Then if, for every $\varepsilon > 0$, there exists a sphere $K(\varepsilon)$ of center (x,y,z), such that the line joining (x,y,z) to any other point of $S \cdot K(\varepsilon)$ always makes an angle of less than ε with P, we say that P is the tangent plane to S at (x,y,z).]?

PROBLEM 157: WARD

March 23, 1937; Prize: Lunch at the "Dorothy"

$f(x)$ is a real function of a real variable, which is approximately continuous. At each point x, the upper right-hand approximate derivative of $f(x)$ (that is,

$$\overline{\lim_{h \to 0^+}} \frac{f(x+h) - f(x)}{h},$$

neglecting any set of values of h which has zero density at $h = 0$) is positive. Is $f(x)$ monotone increasing?

Second Edition Remark

The answer is affirmative. This was proved by R. J. O'Malley in his paper: A density property and applications, *Trans. Amer. Math. Soc.* 199 (1974), 75–87, MR 50 13402. In fact, O'Malley's monotonicity theorem is much stronger than the statement in Problem 157.

PROBLEM 158: STOÏLOW

May 1, 1937; Original manuscript in French

Construct an analytic function $f(x)$ continuous in a domain D admitting there a perfect discontinuum set P of singularities such that $f(P)$ is a discontinuous set.

Such a function would permit one to form a "quasi-linear" function; that is to say, one which has the following properties:

(1) The function is continuous and univalent in the whole plane z.
(2) The function tends toward ∞ for $|z| \to \infty$.
(3) The function has a perfect set of singularities.

See: S. Stoïlow, Remarques sur les fonctions analytiques continues dans un domaine où elles admettant un ensemble parfait discontinu de singularités, *Bull. Math. Soc. Roum. Sci.* 38 (1936), 117–120.

PROBLEM 159: RUZIEWICZ

May 22, 1937

Let Φ denote the set of all continuous functions defined in $(0, 1), f(0) = 0, 0 \leq f(x) \leq 1$ for $0 \leq x \leq 1$. Let

$$P(x) = \sum_{n=0}^{\infty} a_n x^n$$

be a power series and let $P_k(x)$ denote the kth partial sum.

Does there exist a power series $P(x)$ with the following property; for every $\varepsilon > 0$ there exists $N(\varepsilon)$ such that for every function $f \in \Phi$ there exist $n \leq N$, so that $|f(x) - P_n(x)| < \varepsilon$?

Addenda. In this formulation the answer is negative. Since

$$\max_{0 \leq x \leq 1} |\sin^2 2^n \pi x - \sin^2 2^m \pi x| = 1 \qquad (m \neq n),$$

there does not exist a function which approximates both $\sin^2 2^n \pi x$ and $\sin^2 2^m \pi x$ with a precision $\leq 1/3$ in the interval $\langle 0, 1 \rangle$. If the requested universal $N(1/2)$ existed, then by taking

$$\sin^2 2\pi x, \sin^2 2^2 \pi x, \ldots, \sin^2 \{2^{N(1/2)+1} \pi x\}, \sin^2 \{2^{N(1/2)+2} \pi x\},$$

we would conclude on the basis of Problem 159 that for a certain $k \leq N(1/3)$, the polynomial $P_k(x)$ would approximate simultaneously $\sin^2 2^n \pi x$ and $\sin^2 2^m \pi x$ for $n \neq m$ with precision $\leq 1/3$.

Sternbach

Let Φ denote an arbitrary set of continuous functions defined in $(0, 1), f(0) = 0, |f(x)| \leq N$; in order that the set ϕ should have the requested property, it is necessary and sufficient that the functions of the set Φ be equicontinuous.

Mazur

PROBLEM 160: MAZUR

June 10, 1937

Let G denote a metric group.

(1) Let the group G be complete and have the property that for every $\varepsilon > 0$, every element $a \in G$ has a representation $a = a_1 a_2 \cdots a_n$, where $(a_k, e) < \varepsilon$. Is the group G connected in the sense of Hausdorff? (That is to say, it cannot be represented as a sum of two disjoint, closed sets $\neq \emptyset$.)
(2) If a group G is connected in the sense of Hausdorff, is it then arcwise connected?

Commentary

Part (1) is still open. It is not known whether every topological group generated by every neighborhood of the identity is connected. However, for locally compact groups, the answer is yes [1]. It is known (see [2]) that a compact group is connected if and only if each of its elements has an nth root for each integer n.

A. Gleason remarks that part (1) is open even for the completion of the infinite cyclic group Z with special metrizations which he defines as follows:

Let $p_1, p_2, \ldots$ be a sequence of positive integers and $a_1, a_2, \ldots$ be a decreasing sequence of real numbers such that

(1) $a_1 = 1 > a_2 > a_3 > \cdots$ and $a_n \to 0$;
(2) $a_n \left(\dfrac{p_{n+1}}{p_n} - 1 \right) \geq 1$;
(3) for every m there exist integers k and $n_1, \ldots, n_k$, all n_i larger than m, such that the greatest common divisor of $p_{n_1}, \ldots, p_{n_k}$ is 1.

E.g., the sequences $p_n = (n+1)! - 1$ and $a_n = (1/n)$ satisfy (1), (2), and (3) with $k = 2$.

For any $x \in Z$ we put

$$N(x) = \inf \left\{ \sum |k_i| a_i : k_i \in Z, \sum k_i p_i = x \right\}.$$

Theorem 1. N is a norm in Z, i.e., $N(x+y) \leq N(x) + N(y)$, $N(-x) = N(x)$, $N(0) = 0$ and $N(x) > 0$ for all $x \neq 0$. Moreover the metric $N(x - y)$ turns Z into a dense topological group which is generated by every neighborhood of 0.

Proof. The first two conditions for a norm are obvious from the definition of N. To prove the third we show first

(4) if $|x| \leq p_n$ and $x \neq 0$ then $N(x) \geq a_n$. By (1), if $N(x) \geq 1$ then (4) is true.

Suppose that $N(x) < 1$. Consider any representation $x = \sum k_i p_i$ such that the largest i for which $k_i \neq 0$, call it i_0, satisfies $i_0 > n$. Since $|\sum k_i p_i| = |x| \leq p_n \leq p_{i_0 - 1}$ it follows that

$$\sum |k_i| \geq \frac{p_{i_0}}{p_{i_0-1}} - 1 + |k_{i_0}|.$$

Hence, by (1) and (2), $\sum |k_i| a_i \geq 1$. Therefore there exists a representation $x = \sum k_i p_i$ in which $i \leq n$ for all i with $k_i \neq 0$. Hence, since $x \neq 0$, $N(x) \geq a_n$.

By (1) and (4) $N(p_n) = a_n \to 0$. Thus Z is dense at 0 and hence everywhere. Also by (3), Z is generated by every neighborhood of 0. Q.E.D.

Theorem 2 $N(Z)$ is dense in the interval $[0, \sup N(Z)]$.

Proof. Since Z is generated by every neighborhood of 0 for every $\varepsilon > 0$ and $x, y \in Z$ there exists a sequence $z_1, \ldots, z_n$ such that $z_1 = x$, $z_n = y$ and $N(z_i - z_{i+1}) < \varepsilon$ for $i = 1, \ldots, n-1$. Hence $N(Z)$ is dense in the interval $[N(x), N(y)]$.

Problem 1. Is $\sup N(Z) < \infty$ possible? (Notice that if

$$\frac{a_n p_{n+1}}{p_n} \to +\infty,$$

then $\sup N(Z) = \infty$, in fact $N\left(\left[\frac{p_n}{2}\right]\right) \to \infty$.)

Problem 2. Are there sequences $a_1, a_2, \ldots$ and $p_1, p_2, \ldots$ satisfying (1), (2), and (3) and such that Z can be partitioned into two non-empty sets X and Y such that for any sequences $x_1, x_2, \ldots$ and $y_1, y_2, \ldots$ where $x_i \in X$ and $y_i \in Y$, the three series

$$\sum_1^\infty N(x_i - x_{i+1}), \quad \sum_1^\infty N(y_i - y_{i+1}), \quad \sum_1^\infty N(x_i - y_i)$$

cannot all converge? (A positive solution of this problem would yield a negative answer to Mazur's problem, since the Hausdorff completion of Z would then be a disconnected complete metric abelian group generated by every neighborhood of the identity, its partition into two non-empty closed sets being given by the closures of X and Y.)

Part (2) of Problem 160 has a negative answer. The simplest example of a compact metric connected and not locally connected (in fact indecomposable) group is the subgroup of K^ω, where K is the circle group which consists of the sequences $(x_0, x_1, \ldots) \in K^\omega$ such that $x_i = x_{i+1}^2$ for $i = 0, 1, \ldots$ (such groups are called Van Danzig solenoids).

For other open problems on connected groups, see [3].

1. D. Montgomery and L. Zippen, *Topological transformation groups*, Interscience 1955.
2. J. Mycielski, Some properties of connected compact groups, *Coll. Math.* 5 (1958), 162–166.
3. J. Mycielski, On the extension of equalities in connected topological groups, *Fund. Math.* 44 (1957), 300–302.

Jan Mycielski

Remark

A negative solution to part (1) was given by T. Christine Stevens in her paper, *Connectedness of complete metric groups*, Colloq. math. 50 (1986), 233–240.

PROBLEM 161: M. KAC

Let r_n be a sequence of integers such that

$$\lim_{n \to \infty} \left(r_n - \sum_{k=1}^{n-1} r_k \right) = \infty.$$

One has then

$$\lim_{n \to \infty} \left| \underset{0 \le x \le 1}{\to} E \left\{ a < \frac{\sin 2\pi r_1 x + \cdots + \sin 2\pi r_n x}{\sqrt{n}} < b \right\} \right|$$

$$= \frac{1}{\sqrt{\pi}} \int_a^b e^{-y^2} dy.$$

(One can put, for example, $r_n = 2^{n^2}$.)

Problem: Is the theorem true for $r_n = 2^n$?

Remark

This problem is discussed in Mark Kac's conference lecture, on pages 17–26. Also, see Kac's discussion in his address: Probability methods in some problems of analysis and number theory, *Bull. Amer. Math. Soc.* 55 (1949), 390-408.

PROBLEM 162: H. STEINHAUS

July 3, 1937; Prize: Dinner at "George's"

We assume that $f(x)$ IS measurable (L), periodic, $f(x+1) = f(x)$ and $f(x) = +1$ or -1. Do we have, almost everywhere,

$$\limsup_{n \to \infty} f(nx) = +1, \qquad \liminf_{n \to \infty} f(nx) = -1?$$

More generally: If $f_n(x)$ are measurable, uniformly bounded, and $f_n(x+1/n) \equiv f_n(x)$, do we have then

$$\limsup_{n \to \infty} f_n(x) = \text{constant almost everywhere?}$$

$$\liminf_{n \to \infty} f_n(x) = \text{constant almost everywhere?}$$

Addendum. A more general theorem, formulated by Professor Banach, is true: If $f(x)$ is an arbitrary measurable function with period 1, then one has almost everywhere the relations:

$$\overline{\lim_{n\to\infty}} f(nx) = \underset{0\leq x\leq 1}{\to} \text{ essential upper bound } f(x),$$

$$\underline{\lim_{n\to\infty}} f(nx) = \underset{0\leq x\leq 1}{\to} \text{ essential lower bound } f(x),$$

M. Eidelheit

October 16, 1937

Second Edition Remark

The theorem quoted by Eidelheit was published in a more general form in S. Mazur and W. Orlicz, Sur quelques propriétés de fonctions périodiques et presque-périodiques, *Studia Math.* 9 (1940), 1–16; MR 3-107 (Théorème 1, p. 5). While Eidelheit attributes this theorem to Banach, Mazur and Orlicz do not mention Banach's name.

As for the second half of problem 162, Mazur and Orlicz give an equivalent formulation and a partial answer on pp. 13–14 of their paper.

PROBLEM 163: J. von NEUMANN

July 4, 1937; Prize: A bottle of whiskey of measure > 0

Given is a completely additive and multiplicative Boolean algebra B. That is to say:

(1) B is a partially ordered set with the relation $a \subset b$.
(2) Every set $S \subset B$ has the least upper (greatest lower) bound $\Sigma(S)(\Pi(S))$. [We write: $\Sigma(a,b) = a+b$, $\Pi(a,b) = ab$, $\Sigma(B) = 1$, $\Pi(B) = 0$.]
(3) We have a general "distributive law" $(a+b)c = ac + bc$.
(4) Every element $a \in B$ has an (according to (3), unique) "inverse" in B: $a+(-a) = 1$, $a(-a) = 0$.

A measure in B is a numerical function:

$$1.\ \mu(a) \begin{cases} = 0, \text{ for } a = 0, \\ > 0, \text{ for } a \neq 0. \end{cases}$$

$$2.\ a_i \in B\ (i = 1,2,\ldots),\ a_i a_j = 0 \text{ for } i \neq j$$
$$\text{imply } \mu(\Sigma_i(a_i)) = \Sigma_i \mu(a_i).$$

Obviously, one has to determine:

(5) If $S \subset B$, $(a,b \in S, a \neq b) \implies ab = 0$, then S is at most countable.

Question: When does there exist a "measure" in B? Remarks: As one verifies without difficulty, the following "generalized distributivity" law is necessary!

(6) Let $a_1^i \leq a_2^i \leq \cdots$ for $i = 1,2,3,\ldots$, then we have

$$\prod_i \sum_j (a_j^i) = \sum_{j(i)} \prod_i (a_{j(i)}^i)$$

without the assumption that $a_1^i \leq a_2^i \leq \cdots$ this characterizes, according to Tarski, the "atomic" Boolean algebras.

(1) to (5) do not imply (6). Counterexample: The Boolean algebra of Borel sets, modulo sets of first category. Example for (1) to (6): Measurable sets (or Borel sets) modulo sets of measure 0 when one employs Lebesgue measure. Is (5), (6) sufficient?

Commentary

This is really two problems. The more specific one—given a complete Boolean algebra B, do conditions (5) (the "countable chain condition") and (6) (a distributive law) imply the existence of a (finite, strictly positive, countably additive) measure on B—is still not completely solved. It is known that the answer no is relatively consistent with the usual axioms (ZFC) of set theory. In fact, I have given a counterexample [8, Th. 5, pp. 164–166] assuming the falsity of Souslin's hypothesis; and it is known that there are models of set theory in which Souslin's hypothesis is false [6, 12]. There are also models in which Souslin's hypothesis is true [11]; it is not known whether the answer to Problem 163 would then be affirmative. (But I conjecture that the answer, in ZFC, is always no.)

The more general problem raised here by von Neumann is that of finding (assuming conditions (1)–(6)) necessary and sufficient conditions for the existence of a measure. Such conditions have been given (a) by Maharam [8], simplified by Hodges and Horn [4], (b) by Kelley [7]. A simpler condition and a variant on the conditions of Kelley were given by Ryll-Nardzewski (quoted in [7, p. 1176]). Under stronger hypothesis, simpler conditions have been given by Horn and Tarski [5]. A survey of this question is in Sikorski's book [10, pp. 201–204]. None of these answers is entirely satisfactory, in that they require the existence of a sequence of subsets of B with certain properties, and this is not easy to verify in specific cases. Perhaps this is inevitable from the nature of the question. A different (but also not easily applicable) answer is implicit from the structure theory of [9]; it is necessary and sufficient that B be isomorphic to a countable direct sum $B_1 + !B_2 + \cdots$, where B_n is either an atom or the measure algebra of some product of unit intervals with Lebesgue product measure.

The method of attack in [8] led to a further interesting question. If B does have a measure μ, it is then an abelian group (under symmetric difference) with invariant metric $d(x,y) = \mu(x \Delta y)$. Thus a necessary condition is that B be metrizable. It will then have an invariant metric ρ; and on defining $\lambda(x) = \rho(x,o)$ (where o is the zero element) we obtain a "continuous outer measure" λ on B; that is, a finite nonnegative function, vanishing only at o, satisfying (i) $\lambda(x \vee y) \le \lambda(x) + \lambda(y)$, and (ii) if $x_1 \ge x_2 \ge \cdots$ and $\lim_n x_n = o$ then $\lim_n \lambda(x_n) = 0$. In [8] a further condition was imposed to pass from a continuous outer measure to a measure. But it was asked whether the existence of a continuous outer measure on B implies by itself the existence of a measure. This, the "control measure" question is still open. For some partial results and applications, see [1]! and [2].

The counterexample in [8] (assuming Souslin's hypothesis false) has the following remarkable property: each countably generated complete subalgebra B_1 of B does have a measure, though B does not. This raises two further open questions concerning complete Boolean algebras with the countable chain condition:

(a) Is there such an algebra, with the above remarkable property, even if Souslin's hypothesis is true?
(b) If each countably generated complete subalgebra of B (a complete Boolean algebra satisfying (5)) has a measure, what further conditions ensure that B has a measure? (Of course, the measures on the subalgebras need not be consistent with one another.) Would the existence of a continuous outer measure be sufficient?

Von Neumann's Problem 163 leads naturally to an even more fundamental one: Given a (finitely additive) Boolean algebra A, under what conditions will A admit a *finitely additive* (strictly positive, finite) measure? The solution of this problem was the first step in Kelley's treatment of the countably additive case; he showed that A admits a finitely additive measure if and only if the nonzero elements of A can be partitioned into countably many subsets each having positive "intersection number" [7, pp. 1166, 1167]. That this condition does not hold automatically, even if A satisfies the countable chain condition (5), is shown by an example of Gaifman [3].

As mentioned above, the complete Boolean algebras with countably additive measures have an easily described structure [9]. It would be very interesting to have a structure theory for finitely additive measures, the structure of which can be much more complicated.

1. J.P.R. Christenson, Some results with relation to the control measure problem, *Vector space measures and applications Proc. Conf. Univ. Dublin, Dublin 1977), II*, Lecture Notes in Math. vol. 645, Springer-Verlag, Berlin-New York 1978, 27–34.
2. J.P.R. Christenson and W. Herer, On the existence of pathological submeasures and the construction of exotic topological groups, *Math. Ann.* 213 (1975), 203–210.
3. H. Gaifman, Concerning measures on Boolean algebras, *Pacific J. Math.* 14 (1964), 61–73.
4. J.L. Hodges and A. Horn, On Maharam's conditions for a measure, *Trans. Amer. Math. Soc.* 64 (1948), 594–595.
5. A. Horn and A. Tarski, Measures in Boolean algebras, *Trans. Amer. Math. Soc.* 64 (1948), 467–497.

6. T. Jech, Non-provability of Souslin's hypothesis, *Comment. Math. Univ. Carolinae* 8 (1967), 291–305.
7. J.L. Kelley, Measures on Boolean algebras, *Pacific J. Math.* 9 (1959), 1165–1177.
8. Dorothy Maharam, An algebraic characterization of measure algebras, *Ann. of Math.* (2) 48 (1947), 154–167.
9. _____, On homogeneous measure algebras, *Proc. Nat. Acad. Sci. USA* 28 (1942), 108–111.
10. R. Sikorski, *Boolean Algebras,* 2nd. ed., Ergebnisse der Math. u. ihrer Grenzgebiete, N.S. vol. 25, Springer-Verlag, Berlin 1964.
11. R.M. Solovay and S. Tennenbaum, Iterated Cohen extensions and Souslin's problem, *Ann. of Math.* (2) 94 (1971), 201–245.
12. S. Tennenbaum, Souslin's problem, *Proc. Nat. Acad. Sci. USA* 59 (1968), 60–63.

Dorothy Maharam

PROBLEM 164: ULAM

Let a finite number of points including 0 and 1, be given on the interval $[0,1]$, a number $\varepsilon > 0$, and a transformation of this finite set into itself T, with the following property: For every point p, $|p, T(p)| > \varepsilon$. Let us call a "permissable step" passing from the point p to $T(p)$ or to one of the two neighbors (points nearest from the left or from the right side) of the point $T(p)$.

Question: Does there exist a universal constant k such that there exists a point p_0 from which, in a number of permissable steps $[k/\varepsilon]$ one can reach a point q which is distant from p_0 by at least $1/3$?

PROBLEM 165: ULAM

Prize: Two bottles of wine

Let p_n be a sequence of rational points in the n-dimensional unit sphere. The first N points $p_1,\ldots,p_N$ are transformed on N points (also located in the same sphere) $q_1,\ldots,q_N$, all different. We define a transformation on the points p_n, $n > N$, by induction as follows: Assume that the transformation is defined for all points p_v, $v < n$, and their images are all different. This mapping has a certain Lipschitz constant L_{n-1}. The Lipschitz constant of the inverse mapping we denote by L'_{n-1}. We define the mapping at the point p_n so that the sum of the constants $L_n + L'_n$ should be minimum. (In the case where we have several points satisfying this postulate we select one of them arbitrarily.)

Question: Is the sequence $\{L_n + L'_n\}$ bounded?

PROBLEM 166: ULAM

Let M be a topological manifold, f a real-valued continuous function defined on M. We denote by G_f^M the group of all homeomorphic mappings T of M onto itself such that $f(T(p)) = f(p)$ for all $p \in M$.

Question: If N is a manifold *not* homeomorphic to M, does there exist f_0 such that $G_{f_0}^M$ is not isomorphic to any G_f^N?

PROBLEM 167: ULAM

Let S denote the surface of the unit sphere in Hilbert space. Let $f_1,\ldots,f_n$ be finite system of real-valued, continuous functions defined on S. Let T be a continuous transformation of S into part of itself. Does there exist a point p_0 such that $f_v(T(p_0)) = f_v(p_0)$, $v = 1,\ldots,n$?

Commentary

Klee [2] constructed a homeomorphism h of S onto the entire Hilbert space E, and Bessaga [1] showed h could even be made a diffeomorphism. Now choose $z \in E \sim \{0\}$, let f be a continuous linear functional on E such that $f(z) \neq 0$, and for each $x \in E$ let $t(x) = x + z$. Finally, let $f_1 = fh$ and $T = h^{-1}th$. Then f_1 is a differentiable real-valued function on S and T is a diffeomorphism of S onto S. If $f_1(T(p_0)) = f_1(p_0)$, then with $y = h(p_0)$ it follows that $f(t(y)) = f(y)$. That is impossible, for $f(t(y)) = f(y) + f(z)$.

1. C. Bessaga, Every infinite-dimensional Hilbert space is diffeomorphic with its unit sphere, *Bull. Acad. Polon. Sci. Ser. Sci. Math. Astronom Phys.* 14 (1966), 27–31.
2. V. Klee, Convex bodies and periodic homeomorphisms in Hilbert space, *Trans. Amer. Math. Soc.* 74 (1953), 10–43.

V. Klee

PROBLEM 168: ULAM

Prize: Two bottles of beer

Does there exist a sequence of sets A_n such that the smallest class of sets containing these, and closed with respect to the operation of complementation and countable sums, contains all the analytic sets (on the interval)?

Commentary

If the continuum hypothesis or Martin's axiom holds, then the answer is yes [1]. B.V. Rao [3] and R. Mansfield [2] showed that there is no sequence of sets A_n having the required properties which are Lebesgue measurable.

Finally, Rao [4] showed that if one assumes the axiom of determinacy, then the answer is no.

1. R.H. Bing, W.W. Bledsoe, R.D. Mauldin, Sets generated by rectangles, *Pac. J. Math.* 51 (1974), 27–36.
2. R. Mansfield, The solution to one of Ulam's problems concerning analytic sets II, *Proc. Amer. Math. Soc.* 26 (1970), 539–540.
3. B.V. Rao, Remarks on analytic sets, *Fund. Math.* 66 (1969/70), 237–239.
4. _____, Remarks on generalized analytic sets and the axiom of determinateness, *Fund. Math.* 69 (1970), 125–129.

R. Daniel Mauldin

Second edition Commentary

In Miller [1] Theorem 3 it is shown to be relatively consistent with ZFC that there is no countable family $\mathscr{H}$ of sets of the reals such that every analytic set is in the σ-algebra generated by $\mathscr{H}$.

1. Miller, Arnold W.; Generic Souslin sets. Pacific J. Math. 97 (1981), no. 1, 171–181.

Arnold W. Miller

PROBLEM 169: E. SZPILRAJN

Does there exist an additive function $\mu(E)$, equal for congruent sets, defined for all plane sets, and which is an extension of the linear measure of Caratheodory? $(0 \leq \mu(E) \leq +\infty)$?

Commentary

In this problem, "additive" means "finitely additive" (for "countably additive" the answer would, of course, be negative). The answer is yes, see, e.g., J. Mycielski, Finitely additive invariant measures (I), *Coll. Math.* 42 (1979), 309–318.

J. Mycielski

PROBLEM 170: E. SZPILRAJN

Is every plane set all of whose homeomorphic plane images are Lebesgue measurable (L), measurable absolutely? [That is to say, measurable with respect to every Caratheodory function ("Massfunction").]

This is true for linear sets; for plane sets an analogous theorem is true if one replaces homeomorphisms by generalized homeomorphisms in the sense of Mr. Kuratowski.

Addendum. Affirmative answer follows from an unpublished result of von Neumann.

Commentary

The solution announced at the end of the problem holds for all R^n and rests on the following theorem of von Neumann, later proved by Oxtoby and Ulam [1, 2]: If μ is a Borel probability measure on $I^n = [0, 1]^n$, then μ is homeomorphic to the usual product Lebesgue measure if and only if it is positive for nonempty open sets, zero for points, and $\mu(\partial I^n) = 0$. To deduce the solution of the problem from this theorem suppose that $A \subseteq R^n$ is not measurable relative to some Caratheodory measure μ.

Without loss of generality, we can assume $A \subseteq I^n$ and μ is homeomorphic to Lebesgue measure by a homeomorphism h, but then $h(A)$ is not Lebesgue measurable.

1. J.C. Oxtoby and S.M. Ulam, Measure-preserving homeomorphisms and metrical transitivity, *Ann. of Math.* (2) 42 (1941), 874–920.
2. J.C. Oxtoby and V.S. Prasad, Homeomorphic measures in the Hilbert Cube, *Pac. J. Math.* 77 (1978), 483–497.

Jan Mycielski

PROBLEM 163: J. SCHEIER, S. ULAM

Let $T(A)$ denote the set of all mappings of a set A into itself. An operation is defined for pairs of elements of the set T: $U(f,g) = h$ for all $h \in T(A)$. ($U(f,g) \neq 1$.)
Assumptions:

(1) $U(f,g)$ is associative; that is, $U(f, U(g,h)) = U(U(f,g),h)$.
(2) $U(f,g)$ is invariant with respect to permutations of the underlying set; i.e., if p is a permutation of the set A, then $U(p^{-1}fp, p^{-1}gp) = p^{-1}U(f,g)p$.

Theorem. $U(f,g) = f(g)$ (composition).

Second Edition Commentary

Trivially, if U is any solution to the problem, then so also is $U'(f,g) := U(g,f)$. In particular, there is the trivial solution $U(f,g) = g \circ f$ (which should probably be excluded).

Here is an example of a U of a different form. Let A be a finite set. Define the function $s: A^A \to A^A$ by the rule $s(f) = Id$ if $f \in \text{Sym}(A)$ and $s(f) = f$ otherwise. Set $U(f,g) = s(f) \circ s(g)$. Note that $s(f \circ g) = s(f) \circ s(g)$ if f and g are in the image of s. The proof that U satisfies the desired properties is immediate from the further relations: $s^2(f) = s(f)$ and $s(f^\sigma) = (sf)^\sigma$ for σ in $\text{Sym}(A)$ and $f \in A^A$, where $f^\sigma = \sigma^{-1}f\sigma$.

For an arbitrary set A one can set $t(f) = Id$ if $f: A \to A$ is an onto function and $t(f) = f$ otherwise. Then $W(f,g) = t(f) \circ t(g)$ works with the same proof as above.

However, there is another possible interpretation of Problem 171. The third line "$U(f,g) = h$ for all $h \in T(A)$" may mean that one insists that the product be onto. To this end, define $V(f,g) = f \circ g$ if both f and g are in $\text{Sym}(A)$ and otherwise $V(f,g) = s(f) \circ s(g)$. This V satisfies properties (1) and (2), and moreover is surjective in the sense that any h is of the form $h = V(f,g)$ for some $f, g \in A^A$.

Daniel Goldstein

PROBLEM 172: M. EIDELHEIT

June 4, 1938

A space E of type (B) has the property (a) if the weak closure of an arbitrary set of linear functionals is weakly closed. [A sequence of linear functionals $f_n(x)$ converges weakly to $f(x)$ if $f_n(x) \to f(x)$ for every x.]

The space E of type (B) has the property (b) if every sequence of linear functionals weakly convergent converges weakly as a sequence of elements in the conjugate space $\bar{E}$.

Question: Does every separable space of type (B) which has property (a) also possess property (b)?

PROBLEM 173: M. EIDELHEIT

July 23, 1938

Let A denote the set of all linear operations mapping a given space of type (B) into itself. Is the set of operations in A which have continuous inverses dense in A (under the usual norm)?

Commentary

It is now well known that the set of invertible linear operators in an infinite dimensional Hilbert space is connected and open, but is not dense. See Problem 109 in P. Halmos: *A Hilbert space problem book,* Van Nostrand, Princeton, 1967. For a characterization of the closure of the invertible operators, see R. Bouldin, *Proc. Amer. Math. Soc.* 108 (1990), 721-726.

R. Daniel Mauldin

PROBLEM 174: M. EIDELHEIT

July 23, 1938

Let $U(x)$ be a linear operation defined in a space of type (B_0) mapping this space into itself and such that the operation $x - \lambda U(x)$ has inverses for sufficiently small λ. Can we then have

$$(x - \lambda U)^{-1} = x + \lambda U(x) + \lambda^2 U(U(x)) + \cdots ?$$

PROBLEM 175: BORSUK

August 10, 1938

(a) Is the product (Cartesian) of the Hilbert cube Q with the curve which is shaped like the letter T, homeomorphic with Q?
(b) Is the product space of an infinite sequence of letters T homeomorphic to Q?

Commentary

The answer to this striking problem has been given by R.D. Anderson [1], who showed that (a) the products $T \times \prod_i[-1,1]_i$ and $\prod_i[-1,1]_i$ are strongly homeomorphic, and (b) if $Y_n \times \prod_i X_i$ is for each n strongly homeomorphic to $\prod_i X_i$ and the X_is are compact then $\prod_i Y_i$ is homeomorphic ($\cong$) to $\prod_i X_i$. The details of Anderson's proof have never been published, but the definition of "strongly homeomorphic" and another proof of (b) appeared in [15]. Anderson's proof of S(a) depended on a construction of lattice-isomorphic bases of open sets in $T \times Q$ and Q, respectively. By the same method he showed that (c) countable products of nondegenerate dendra yield Hilbert cubes.

The above results and Anderson's further study of the topological properties of the Hilbert cube initiated intense investigations on infinite products of compact *AR*s. A. Szankowski [12] has given an alternate proof of (c) based on a Hilbert-cube analogue of the lakes of Wada construction. A crucial step has been made by J.E. West, who in a series of papers [15, 16, 17] has developed a technique of establishing homeomorphisms of product spaces which enabled him to show that (d) the class $\mathfrak{X} = \{X : X \times Q \cong Q\}$ is closed with respect to the mapping cylinder construction, and (e) if X_1, X_2 and $X_1 \cap X_2$ are in $\mathfrak{X}$ then so is $X_1 \cup X_2$. He also proved that if $X \times Q \cong Q$ then $X \times \prod_i[-1,1]_i$ is strongly homeomorphic to $\prod_i[-1,1]_i$, thereby establishing that (f) $\prod_i X_i \cong Q$ whenever all the X_is are in $\mathfrak{X}$ and contain more than one point. Both Szankowski's and West's proofs heavily depended on the properties of Z-sets discovered by Anderson in [2].

West's theorems have been re-proved by several authors who, in particular, replaced West's "interior-approximation" technique by the use of a theorem of M. Brown giving a sufficient condition for the inverse limit of a sequence of compacta to be homeomorphic to each of them. See West's expository article [18] for a closer discussion. (Subsequent to this article were [20, 5, 10, 11, 13, 6, 9].)

In 1975, R.D. Edwards gave the definitive characterization of Q-factors by proving that $X \times Q \cong Q$ whenever X is a compact *AR* (see [16]). Combined with the Anderson-West result (f) mentioned above, this shows that infinite cartesian multiplication of nondegenerate compact *AR*s always yields Hilbert cubes. A slightly stronger result is that $X \cong Q$ whenever X is a compact *AR* and for each n there are nondegenerate spaces $X_1, \ldots, X_n$ with $X \cong X_1 \times \cdots \times X_n$ (see [14]).

In fact, West and Edwards established their results more generally for factors of manifolds modeled on Q; i.e., West showed that $\mathscr{Y} = \{Y : Y \times Q$ is a Q-manifold$\}$ is closed with respect to the mapping cylinder construction, and Edwards showed that locally compact *ANR*s are in $\mathscr{Y}$. The implications are crucial to (1) identifying Q-manifolds and specifically copies of the Hilbert cube, and (2) theory of *ANR*s. Not going into details we mention here that West's results were the basis for both Curtis-Schori's solution of Wojdysawski's problem if hyperspaces of Peano continua were Hilbert cubes [7, 8] and West's [19] solution of Borsuk's problem on finiteness of homotopy type of compact *AR*s. Similarly, Edwards' theorem was the basis for a general characterization of Q-manifolds (see [14]).

Results analogous to that of Edwards remain valid also for non-locally compact *ANR*s once Q is replaced by a suitable normed linear space (see [18] and [14] for references).

K. Borsuk's Problem 175 appeared to be of great significance, especially if viewed as an obvious specification of the more general problem of determining which infinite products are homeomorphic to the Hilbert cube and how the singularities of *AR*s behave under infinite multiplication (the latter is in the spirit of other questions of Borsuk; see [4]). It was raised 18 years before any nontrivial factor of a finite- or infinite-dimensional cube was constructed [3] and, with Wojdysawski's question on hyperspaces of Peano continua (posed in *Fundamenta Math.* in 1939), has stimulated the very basic research on the topology of the Hilbert cube.

1. R.D. Anderson, The Hilbert cube as a product of dendrons, *Amer. Math. Soc. Notices* 11 (1964), 572.
2. _____, On topological infinite deficiency, *Michigan Math. J.* 14 (1967), 365–383.
3. R.H. Bing, The cartesian product of a certain nonmanifold and a line is E^4, *Ann. of Math.* 70 (1959), 399–412.
4. K. Borsuk, *Theory of Retracts,* Polish Scientific Publishers, Warsaw, 1967.
5. M. Brown and M. Cohen, A proof that simple-homotopy equivalent polyhedra are stably homeomorphic, *Michigan Math. J.* 21 (1974), 181–191.
6. T.A. Chapman, *Lectures on Hilbert Cube Manifolds,* CBMS Lecture Notes #28, 1976.
7. D.W. Curtis and R.M. Schori, 2^X and $C(X)$ are homeomorphic to the Hilbert cube, *Bull. Amer. Math. Soc.* 80 (1974), 927–931.
8. _____, Hyperspaces of Peano continua are Hilbert cubes, *Fund. Math.* 101 (1978), 19–38.
9. A. Fathi, A. Marin Y.M. Visetti, New proofs of Chapman's *CE* mapping theorem and West's mapping cylinder theorem, *Proc. Amer. Math. Soc.* 67 (1977), 327–334.
10. M. Handel, The Bing staircase construction for Hilbert cube manifolds, Gen. Topology and Appl. 9 (1978), 19-28.
11. _____, On certain sums of Hilbert cubes, preprint.
12. A. Szankowski, On factors of the Hilbert cube, *Bull. Polon. Acad. Sci* 17 (1969), 703–709.
13. H. Torunczyk, On factors of the Hilbert cube and metric properties of retractions, *Bull. Polon. Acad. Sci.* 24 (1976), 757–765.
14. _____, Characterization of infinite-dimensional manifolds, *Proceedings of the International Conference on Geometric Topology (Warsaw, 1978)* , 431–437, PWN, Warsaw, 1980..
15. J.E. West, Infinite products which are Hilbert cubes, *Trans. Amer. Math. Soc.* 150 (1970), 1–25.
16. _____, Mapping cylinders of Hilbert cube factors, *General Topol. and Appl.* 1 (1971), 111–125; Part II, 5 (1975), 35–44.
17. _____, Sums of Hilbert cube factors, *Pacific J. Math.* 54 (1974), 293–303.
18. _____, *Cartesian factors of infinite-dimensional spaces,* Topology Conference, Virginia Polytechnic Inst. and State University, Springer Lecture Notes in Math. #375, 1974, 249-268.
19. _____, Mapping Hilbert cube manifolds to *ANR*s, A solution of a conjecture of Borsuk, *Annals of Math.* 106 (1977), 1–18.
20. R.Y.T. Wong and Nelly Kroonenberg, Unions of Hilbert cubes, *Trans. Amer. Math. Soc.* 211 (1975), 289–297.

H. Torunczyk

PROBLEM 176: M. EIDELHEIT

September 12, 1938

In a ring of type (B) (normed, complete linear ring with the norm satisfying the condition: $|xy| \leq |x| |y|$) containing a unit element there is given an element a possessing an inverse a^{-1}.

Question: Does there exist a sequence of polynomials $c_0^{(n)} I + c_1^{(n)} a + \cdots + c_m^{(n)} a^n$ converging to a^{-1}? (I = the unit element, c are numbers.)

Addendum. Answer is negative. Example: The ring of linear operations $U(x)$ of the space (C) into itself; $U(x) = x(t^2)$, $0 \leq t \leq 1$.

M. Eidelheit

November 11, 1938

PROBLEM 177: M. KAC

September 11, 1938

What are the conditions which a function $\Phi(x, y)$ must satisfy in order that for every pair of Hermitian matrices A and B the matrix $\Phi(A, B)$ is "positive definite"?

Remark

This problem is discussed in Mark Kac's conference lecture, on pages 17–26.

PROBLEM 178: M. KAC

Let

$$\phi(x, y) = \frac{1}{\frac{1}{x} + \frac{1}{y} - 1}.$$

Prove that if

$$\phi\left(\int_{-\infty}^{+\infty} e^{i\xi x} d\sigma_1(x), \int_{-\infty}^{+\infty} e^{i\xi x} d\sigma_2(x)\right) = \frac{1}{1 + \xi^2},$$

then $\sigma_1(x) = \alpha_1 e^{-\beta_1|x|}$ and $\sigma_2(x) = \alpha_2 e^{-\beta_2|x|}$. (This is analogous to Cramer's theorem that if

$$\int_{-\infty}^{+\infty} e^{i\xi x} d\sigma_1(x) \times \int_{-\infty}^{+\infty} e^{i\xi x} d\sigma_2(x) = e^{-\xi^2/4},$$

then $\sigma_1(x)$ and $\sigma_2(x)$ are of the form $e^{-\beta_1 \xi^2}$ and $e^{-\beta_2 \xi^2}$.)

Remark

This problem is discussed in Mark Kac' conference lecture, on pages 17–26.

PROBLEM 179: OFFORD

January 10, 1939; Original manuscript in English
If $a_0, \ldots, a_n$ are any real or complex numbers and if $\varepsilon_v = \pm 1$, $v = 1, 2, \ldots, n$, then the following theorem is true:

$$|a_0 + \varepsilon_1 a_1 + \varepsilon_2 a_2 + \cdots + \varepsilon_n a_n| \geq \min_{0 \leq v \leq n} |a_v|$$

except for a proportion of at most $A/n^{1/4}$ of the 2^n sums.
    Problems:

 (i) To find a short proof of this result.
(ii) When the as are all equal to 1 the size of the exceptional set is $(A/\sqrt{n})2^n$. Is this the right upper bound whatever the numbers a_v?

Commentary

Littlewood and Offord (On the number of real roots of a random algebraic equation (III), *Mat. Sbornik,* 12 (1943), 277–285) showed that the proportion of the 2^n sums which fail is at most $c \log n/\sqrt{n}$ and Erdős (On a lemma of Littlewood and Offord, *Bull. Amer. Math. Soc.,* 51 (1945), 898–902) improved this to $c/\sqrt{n}$. Kleitman (On a lemma of Littlewood and Offord on the distribution of certain sums, *Math. Zeitschr.* 90 (1965), 251–259) improved the bound to

$$\binom{n}{\left[\frac{n}{2}\right]}$$

and showed this is the best possible. Thus the answer to the second question is no. Kleitman (On a lemma of Littlewood and Offord on the distribution of linear combinations of vectors, *Adv. in Math.* 5 (1970), 155–157) generalized the problem and result to vectors in Hilbert space.
    Littlewood and Offord (op. cit.) used their result to consider the number of real roots of an equation of the form

$$\sum_0^n \varepsilon_v a_v x^v = 0, \qquad \varepsilon_v = \pm 1.$$

Let

$$M = \sum_0^n |a_v|.$$

They showed that except for a set of these equations of the proportion

$$O\left(\frac{\log\log n}{\log n}\right),$$

the remainder of the equations have not more than

$$10\log n\left\{\log\left(\frac{M}{\sqrt{|a_0 a_n|}}\right) + 2(\log n)^5\right\}$$

real roots for each equation.

W.A. Beyer

PROBLEM 180: KAMPÉ DE FÉRIET
May 16, 1939

Let $v(t,E)$ be a stationary random function (in the sense of E. Slutsky, A. Khinchine): (E a random event)

$$\overline{v(t,E)} = 0 \qquad \overline{v(t,E)^2} = \text{constant} \left.\vphantom{\begin{array}{c}1\\1\end{array}}\right\} \quad \text{for all } t$$

$$\overline{v(t,E)v(t+h,E)} = \text{function of } h \text{ alone} \qquad\qquad -\infty < t < +\infty.$$

Does there exist a random variable A which, with uniform probability, assumes every value α between 0 and 1,

$$\text{Prob } [A < \alpha] = \alpha, \qquad 0 \le \alpha \le 1,$$

such that

(1) $E = \phi(\alpha)$,
(2) $v[t_1, \phi(\alpha)]$ and $v[t_2, \phi(\alpha)]$ are two independent functions (in the sense of H. Steinhaus) for every couple t_1, t_2 ($t_1 \ne t_2$)?

PROBLEM 181: H. STEINHAUS

Find a continuous function (or perhaps an analytic one) $f(x)$, positive and such that one has

$$\sum_{n=-\infty}^{\infty} f(x+n) \equiv 1$$

(identically in x in the interval $-\infty < x < +\infty$); examine whether $(1/\sqrt{\pi})e^{-x^2}$ is such a function; or else prove the impossibility; or else prove uniqueness.

Addendum. The function $(1/\sqrt{\pi})e^{-x^2}$ does not have the property—this follows from the sign of the second derivative for $x = 0$ of the expression

$$\sum_{-\infty}^{+\infty} \frac{1}{\sqrt{\pi}} e^{-(x+n)^2}.$$

H. Steinhaus

We take a function $g(x)$ positive, continuous, and such that

$$\sum_{n=-\infty}^{+\infty} g(x+n) = f(x) < +\infty$$

in the interval $(-\infty, +\infty)$ for example, $g(x) = e^{-x}$ and the function $g(x)/f(x)$ satisfies the conditions.

S. Mazur

December 1, 1939

PROBLEM 182: B. KNASTER
December 31, 1939; Prize: Small light beer
    The disk cannot be decomposed into chords (not reducing to a single point), but a sphere can be so decomposed (noneffectively). Give an effective decomposition of a sphere into chords. The same for the n-dimensional sphere for "chords" of dimension $k \leq n - 2$.

Commentary

This problem is still open. Concerning the partition of a spherical surface and the plane into arcs, see J.H. Conway and H.T. Croft: Covering a sphere with congruent great circle arcs, *Proc. Cambridge Phil. Soc.* 60 (1964), 787–800.

Jan Mycielski

PROBLEM 183: BOGOLUBOW
February 8, 1940; Prize: A flask of brandy; Original manuscript in French
    Given is a compact, connected, and locally connected group of transformations of the n-dimensional Euclidean space. Prove (or give a counterexample) that one can introduce in this space such coordinates that the transformations of the group will be linear.

Remark

The answer is no. One of the earliest counterexamples (given by Conner and Floyd) is presented by G.E. Bredon, *Introduction to Compact Transformation Groups,* Vol. 46, Pure and Applied Mathematics, Academic Press, New York, 1972, 58–61.

PROBLEM 184: S. SAKS
February 8, 1940; Prize: One kilo of bacon
   A subharmonic function ϕ has everywhere partial derivatives $\partial^2\phi/\partial x^2$, $\partial^2\phi/\partial y^2$. Is it true that everywhere $\Delta\phi \geq 0$?
   Remark: It is obvious immediately that $\Delta\phi \geq 0$ at all points of continuity of $\partial^2\phi/\partial x^2$, $\partial^2\phi/\partial y^2$, therefore on an everywhere dense set.

PROBLEM 185: S. SAKS
   Is it true that for every continuous surface $z = f(x,y)$ ($0 \leq x \leq 1, 0 \leq y \leq 1$) the surface area is equal to

$$\lim_{h\to 0}\int_0^1\int_0^1 \sqrt{\left[\frac{f(x+h,y)-f(x,y)}{h}\right]^2 + \left[\frac{f(x,y+h)-f(x,y)}{h}\right]^2 + 1}\,dxdy.$$

   Remark: The theorem is true for curves [for surfaces it was given by L.C. Young, but the proof (cf. S. Saks, *Theory of the Integral,* 1937) contains an essential error].

Commentary

S. Saks had written, in *Theory of the Integral* (Warsaw 1937, p. 182): "... as proved by L.C. Young ...

$$S(F;I_0) = \lim_{\alpha,\beta\to 0}\int\int_{I_0}\left\{\left[\frac{F(x+\alpha,y)-F(x,y)}{\alpha}\right]^2 \right.$$
$$\left. + \left[\frac{F(x,y+\beta)-F(x,y)}{\beta}\right]^2 + 1\right\}^{1/2} dxdy." \qquad (1)$$

   Here I_0 is any rectangle with sides parallel to the axes, and $S(F;I_0)$ is the area of the continuous surface $z = F(x,y)$ over I_0. Young (An expression connected with the area of a surface $z = F(x,y)$, *Duke Math. J.* 11 (1944), 43–57) states that the proof is based on a false inequality (which appears near the bottom of 183); the existence of an error had been pointed out by V. Jarník and also by T. Rado and P.V. Reichelderfer. Young writes, "The error was not mine, but I am partly to blame for suggesting that the theorem could easily be proved in this sort of way, and I failed to detect the error during proof reading." (According to Saks' preface, Young "greatly exceeded his role of translator in his collaboration with the author.") Young then analyzes the situation in depth.

Denote the integral in (8.4) by $S^*(\alpha, \beta)$. Young observes that the correct part of Saks' proof establishes only that $S(F; I_0) \leq \lim_{\alpha, \beta \to 0} S^*(\alpha, beta)$. He shows that there is strict inequality for $I_0 = [0, \sqrt{2}] \times [0, \sqrt{2}]$ and $F(x, y) = g(x + y)$, where g is the well-known singular continuous monotone function constant on the complementary intervals of Cantor's set, extended similarly to the whole line. By taking the average F and two functions corresponding to rotations of the surface through angles $\pm \pi/3$, he obtains a surface $z = F_0(x, y)$ such that after an arbitrary rotation of the axes there is still strict inequality.

Young also establishes the following definitive criterion. In order that, for a continuous surface $z = F(x, y)$ of finite area, this area be the limit of $S^*(\alpha, \beta)$, it is necessary and sufficient that there exists a decomposition of I_0 into two Borel sets E_1, E_2 such that on E_1 the function F is absolutely continuous in x on the sections determined by almost all constant values of y, while in E_2 it is absolutely continuous in y on the sections determined by almost all constant values of x.

Young's condition is presumably not necessary (as well as sufficient) for the equality $S(F; I_0) = \lim_{h \to 0} S^*(h, h)$ with which Problem 185 is concerned, but it might be regarded as neither sensible nor interesting to seek a criterion for the possession of this rather artificial property, tied as it is to a particular choice of axes. Modern research has moved in the direction of results independent of such a choice.

Roy O. Davies

PROBLEM 186: S. BANACH

March 21, 1940

Does there exist a sequence $\{\phi_i(t)\}$ of functions, orthogonal, normed, and complete in the interval $(0 \leq t \leq 1)$ with the property that for every continuous function $f(t), 0 \leq t \leq 1$ (not identically zero) the development

$$\sum_{i=1}^{\infty} \phi_i(t) \int_0^1 f(t) \phi_i(t) \, dt$$

is at almost every point unbounded?

Remark

The answer is yes. See the commentary to problem 86

PROBLEM 187: P. ALEXANDROFF

April 19, 1940; Original manuscript in French

(1) Let P be a mutilated polyhedron (that is to say, in its simplest decomposition one has deleted a certain number of simplices of arbitrary dimension) contained in R^n. $R^n - P$ is then also a mutilated polyhedron. We understand by the Betti group of this polyhedron the usual Betti group in the sense of Vietoris. The duality law of Alexander is then true for mutilated polyhedra. Prove that if $P \subset R^n$ is a mutilated *topological* polyhedron (that is to say, a topological image of a mutilated polyhedron) the duality theorem of Alexander still holds.

(2) Prove (or refute) the theorem: For every Hausdorff space which is bicompact, the inductive definition of dimension is equivalent to the definition given with the aid of coverings (Uberdeckungen).

(3) Prove (or refute) the impossibility of an interior continuous transformation of a cube with p dimensions into a cube of p dimensions for $p < q$.

Commentary

Concerning part 2, examples of compact spaces whose covering and inductive dimensions are distinct were given by Lunc [5] and Lokuchevsky [3]. A description of Lokuchevsky's example can be found in [2, 178–180].

Concerning part 3, an open (and monotone) mapping of I^m onto I^n with $m < n$ was defined by L. Keldyš [3]. Similar examples were announced earlier by R.D. Anderson [1].

1. R.D. Anderson, Some upper semi-continuous collections of continuous curves filling up R^3, *Bull. Amer. Math. Soc.* 59 (1953), 559.
2. R. Engelking, *Dimension Theory,* North-Holland, New York, 1978.
3. L. Keldyš, Transformation of a monotone irreducible mapping into a monotone-open one, and monotone-open mappings of the cube which raise the dimension. *Mat. Sb.* 43 (1957), 187–226.
4. O.V. Lokuchevsky, On the dimension of bicompacta, *Doklady* 67 (1949), 217–219.
5. A.L. Lunc, A bicompactum whose inductive dimension is greater than its dimension defined by means of coverings, *Doklady* 66 (1949), 801–803, *MR* 11–46.

Ryszard Engelking

PROBLEM 188: S. SOBOLEW

April 20, 1940; Prize: For a solution of the problem, a bottle of wine; Original in both Polish and Russian

One has proved the existence of a Cauchy problem

$$u\Big|_{x_n=0} = \phi_0(x_1,\ldots,x_{n-1})$$

$$\frac{\partial u}{\partial x_n}\Big|_{x_n=0} = \phi_1(x_1,\ldots,x_{n-1}),$$

for the quasilinear partial differential equation of the form

$$\sum_{j=1}^{n}\sum_{i=1}^{n} A_{ij}\frac{\partial^2 u}{\partial x_i \partial x_j} = F$$

of the hyperbolic type (where A_{ij} and F depend on $x_1,\ldots,x_n$, u, $\partial u/\partial x_1,\ldots,$ $\partial u/\partial x_n$), if the function ϕ_0 possesses square integrable derivatives up to the order $[n/2]+3$ and function ϕ_1 partial derivatives up to the order $[n/2]+2$ (also square integrable). We assume in addition that the derivatives of functions A_{ij} and F with respect to $\partial u/\partial x_i$ and to u are continuous.

For the nonlinear equation of the general form

$$\phi\left(x_1,\ldots,x_n,u,\frac{\partial u}{\partial x_1},\ldots,\frac{\partial u}{\partial x_n}\frac{\partial^2 u}{\partial x_1^2},\ldots,\frac{\partial^u}{\partial x_n^2}\right) = 0$$

one can easily show the existence of a solution if only ϕ_0 has derivatives up to the order $[n/2]+4$, and ϕ_1 up to the order $[n/2]+3$, square integrable. One should construct an example of such an equation and such boundary conditions having derivatives of the order less by 1, square integrable, such that the solution would not exist, or else lower the number of derivatives necessary for the existence of a solution to the number necessary in the case of quasilinear equations. (This latter number cannot be lowered any more as shown by known examples.)

Problem 188.1 M. EIDELHEIT
November 27, 1940

Let $z(x,y)$ be a function absolutely continuous on every straight line parallel to the axes of the coordinate system In the square $0 \le x,y \le 1$. Let $f(t)$ and $g(t)$ be two absolutely continuous functions in $0 \le t \le 1$ with values also in $(0,1)$. Is the function $h(t) = z(f(t),g(t))$ also absolutely continuous? If not, then perhaps under the additional assumptions that

$$\int_0^1\int_0^1\left|\frac{\partial z}{\partial x}\right|^p dxdy < \infty, \qquad \int_0^1\int_0^1\left|\frac{\partial z}{\partial y}\right|^p dxdy < \infty,$$

where $p > 1$.

As this edition went to press, a negative solution to Eidelheit's problem has been given by L. Maligranda, V. Mykhaylyuk, and A. Plichko.

PROBLEM 189: A. F. FERMANT
Original manuscript in Russian

Let $w = f(z)$ be a regular function in the circle $|z| < 1$, $f(0) = 0$, $f'(0) = 1$. We shall call the "principal star" of this function the following one-leafed star-like domain: On the leaf of the Riemann surface corresponding to the function $w = f(z)$ to which the point $w = f(z) = 0$ belongs, we take the biggest one-leafed region belonging to the surface.

Prove the theorem: The principal star of the function $w = f(z)$ contains a circle of a radius not less than an absolute constant B (generalization of a theorem of A. Bloch).

Commentary

The answer does not appear to be known. The idea of the principal star goes back to W. Gross who used it for entire functions. E. Landau showed that the biggest disk on the Riemann surface centered at 0 can be arbitrarily small. Otherwise, nothing seems to be known about the star as defined in the problem.

1. W. Gross, Über die Singularitäten analytischer Funktionen, *Monatshefte f. Math. u. Phys.* 29 (1918), 3-47.
2. E. Landau, Der Picard-Schottkysche Satz und die Blochsche Konstante, *Sitz. Der. d. Preuss. Ak. d. Wiss. phys.-math. Kl.* 1926.

Lars V. Ahlfors

PROBLEM 190: L. LUSTERNIK

February 4, 1941; Prize: For the solution, a bottle of champagne for the solver; Original manuscript in Russian

Let there be given in the Hilbert space L_2 an additive functional $f(x)$ defined on a part of L_2, and a self-adjoint operator A. If f is linear, then it is an element of L_2 and $Af = f(Ax)$. Let us extend the operation A over all additive functionals f by the formula: $Af = f(Ax)$. If there is a point λ of the continuous spectrum of A, then we can find an infinite set of additive functionals f, not identically equal to zero, for which $(A - \lambda E)(f) \equiv 0$.

These $f(x)$ can be considered as, so to say, ideal associated elements for the points λ of the continuous spectrum since the properties of the continuous spectrum are reflected on the structure of the sets of the ideal associated elements.

Commentary

This problem does not seem to be precisely formulated since there is no yes or no answer requested or conjecture stated. By the term "linear" in the second sentence is meant additive and continuous, as was common in those days. See Banach, *Theorie des operations Linearires,* Monografje Matematyczne, 1932, page VI. In the third sentence, f presumably should have domain containing the range of A. The proposer's method of associating a functional f with A and λ,

$$f(Ax) = \lambda f(x) \qquad \forall x \in L^2,$$

anticipates, in a rather vague way, the construction of generalized eigenfunctions as invented by Gelfand and Kostyuchenko in 1955 (see Gelfand and Shilov, *Generalized Functions,* Academic Press, Volume 3, 1967, Chapter 4, and Volume 4, 1964,Chapter 1, Section 4). In the Gelfand and Kostyuchenko theory, let A be an operator in a linear topological space Φ. A linear functional $f \in \Phi'$ such that

$$f(Ax) = \lambda f(x)$$

for every $x \in \Phi$ is called a generalized eigenvector of A corresponding to the eigenvalue λ. It is then shown that if we have a rigged Hilbert space $\Phi \subset H \subset \Phi'$ and if the operator A can be extended to a unitary or self-adjoint operator in H, then the system of generalized eigenfunctions of A is complete.

W.A. Beyer

R. Daniel Mauldin

PROBLEM 191: E. SZPILRAJN

April , 1941

   Auxiliary definitions: I call a *measure* every nonnegative, completely additive set function defined on a certain completely additive class of sets K, subsets of a fixed set χ and such that $\mu(\chi) = 1$. The measure μ is *convex* (according to M. Fréchet, "sans singularités"), if for every set A such that $\mu(A) > 0$ there exists a set $B \subset A$, such that $\mu(A) > \mu(B) > 0$. The measure μ is *separable* if there exists a countable class $D \subset K$, such that for every $\eta > 0$ and every $M \in K$ there exists $L \in D$, such that $\mu[(M - L) + (L - M)] < \eta$. The class K is a class of sets stochastically independent with respect to μ if $\mu(A_1 A_2 \cdots A_n) = \mu(A_1)\mu(A_2)\cdots\mu(A_n)$ for every disjoint sequence $\{A_k\}$ of sets belonging to K.

   Definitions of a base. The class $B \subset K$ is called a base of a measure μ if

(1) B is a class of sets stochastically independent with respect to μ; and
(2) All sets of the class K can be approximated, up to sets of measure 0, by sets of the smallest countably additive class of sets containing B.

   Remarks: Let B_n denote the set of numbers from the interval $\langle 0, 1 \rangle$, whose nth binary digit $= 1$. The sequence $\{B_n\}$ is a base for the Lebesgue measure in the interval $\langle 0, 1 \rangle$. It follows easily that every convex, separable measure has a base. In the known examples of nonseparable measures, there also exists a base.

   Problem: Does every convex measure possess a base?

Solution

The answer is no. One counterexample would be provided by the direct sum of the Lebesgue measure spaces $2^{\aleph_0}$ and 2^c (scaled to make the total measure 1). That this space is a counterexample follows from the following theorem.

Theorem A non-atomic (= convex, in Szpilrajn's terminology) measure has a base if and only if the corresponding measure algebra is homogeneous.

Recall that a measure μ defined on a σ-algebra K of subsets of a set X is homogeneous provided that $h(A) = h(X)$, for every measurable subset A of positive measure, where $h(A) =$ least cardinal of a family F of measurable subsets of A such that every measurable subset of A differs by at most a null set (with respect to μ restricted to A) from a member of the Borel field of subsets of A generated by F.

Sketch of proof: If the measure algebra E of (X,μ) is homogeneous, then according to the results in [1], E is isomorphic to the measure algebra of 2^m (with Lebesgue product measure) for some infinite cardinal m, and this clearly provides a "base" for μ, in the sense of Problem 191. Conversely, if μ has a base B in this sense, then E is isomorphic to the measure algebra of 2^m where $m =$ cardinality of B: and this is well known (and easily seen) to be homogeneous.

The space $2^{\aleph_0} \oplus 2^c$ is not homogeneous, since $h(2^m) = m$, for every infinite cardinal m.

1. D. Maharam, On homogeneous measure algebras, *Proc. Nat. Acad. Sci. USA* 28 (1942), 108–111.

Dorothy Maharam

PROBLEM 192: B. KNASTER, E. SZPILRAJN
May, 1941

Definition. A topological T has the property (S) (of Suslin) if every family of disjoint sets, open in T, is at most countable.

Definition. A space T has property (K) (of Knaster) if every noncountable family of sets, open in T, contains a noncountable subfamily of sets which have elements common to each other. *Remarks:*

(1) One sees at once that the condition (K) implies (S) and, in the domain of metric spaces, each is equivalent to separability.
(2) B. Knaster proved in April 1941 that, in the domain of continuous, ordered sets, the property (K) is equivalent to separability. The problem of Suslin is therefore equivalent to the question whether, for ordered continuous sets, the property (S) implies the property (K).

Problem (B. Knaster and E. Szpilrajn).   Does there exist a topological space (in the sense of Hausdorff, or, in a weaker sense, e.g., spaces of Kolmogoroff) with the property (S) and not satisfying the property (K)?

Remark:

(3) According to Remark (2), a negative answer would give a solution of the problem of Suslin.

Problem (E. Szpilrajn). Is the property (S) an invariant of the operation of Cartesian product of two factors?

Remarks:

(4) One can show that if this is so, then this property is also invariant of the Cartesian product of any number of (even noncountably many) factors.
(5) E. Szpilrajn proved in May 1941 that the property (K) is an invariant of the Cartesian product for any number of factors and B. Lance and M. Wiszik verified that if one space possesses property (S), and another space has property (K), then their Cartesian product also has property (S).

Commentary

B. Knaster proved in April 1941 that the existence of a Souslin line was equivalent to the existence of a connected, linearly ordered space without property (K). The square of a Souslin line does not have property (S). But E. Szpilrajn proved in May 1941 that a product of any family of spaces with property (K) has property (K). B. Lance and M. Wiszik verified that the product of a space with property (S) and a space with property (K) has property (S). Thus these two problems came to be posed.

We now know that the answer to both questions is independent of Zermelo Frankel set theory.

Definitions: We say that a topological space has the associated property if every uncountable family of open sets has:

(c) (for caliber $\aleph_1$) an uncountable subfamily with nonempty intersection.
(p) (for precaliber $\aleph_1$) an uncountable subfamily with the finite intersection property.
(K) (for Knaster) an uncountable subfamily each two of whose members have nonempty intersection.
(S) (for Souslin; now called *ccc* for countable chain condition) at least two members with nonempty intersection.

Clearly $(c) \to (p) \to (K) \to (S)$.

Property (c) is actually a topological property. There is a simple subset of 2^{ω_1} with property (p) but not (c); for compact T_2 spaces (p) and (c) are clearly equivalent.

However (p), (K), and (S) depend only on the associated Boolean algebra. If one assumes Martin's axiom together with the negation of the continuum hypothesis $(MA + \neg CH)$, then (p), (K), and (S) are all equivalent (independently proved by K. Kunen, F. Rowbottom, R. Solovay; see I. Juhasz, Martin's axiom solves Ponomarev's problem, *Bull. Acad. Polon. Sci.* Ser. Sci. Math. Ast. Phys. 18 (1970), 71–74. Thus it is consistent with *ZFC* that the answers to both questions be yes.

The existence of a Souslin line, long known to be independent of *ZFC* (with or without *CH*: see Souslin's conjecture, *Amer. Math. Monthly* 76 (1969), 1113–1119) clearly implies that the answer to both questions is no.

More recently (p), (K), and (S) have all been shown to be different if the continuum hypothesis holds. See K. Kunen and F. Tall, Between Martin's axiom and Souslin's hypothesis, *Fund. Math.* 102 (1979), 173–181) for a proof that (K) does not imply (p) under *CH* (proved independently by P. Erdős, R. Laver and independently and more simply, F. Galvin, Chain conditions and products, *Fund. Math.* 108 (1980), 33–48) have shown that if *CH* holds there are spaces with (S) whose square does not have (K). Thus, (S) does not imply (K) and the product question also has a negative answer if one assumes *CH*. For other proofs that *ccc* being productive does not imply (K) when *CH* holds, see (M.L. Wage, Almost disjoint sets and Martin's axiom, *J. Symbolic Logic*, 44 (1979), 313–318) and (E. van Douwen and K. Kunen, *L*-spaces and *S*-spaces in $\wp(\omega)$, *Topology Appl.* 14 (1982), 143–149).

M.E. Rudin

PROBLEM 193: H. STEINHAUS

May 31, 1941
The "expected" number of matches: 7
Probability that $x \leq 9 \longrightarrow 0.68$
Probability that $x \leq 18 \longrightarrow 0.95$
Probability that $x \leq 27 \longrightarrow 0.997$
"The probable" number of matches: 6
The probability that $x \leq 6$ is 0.5
(Two boxes with fifty matches)
(The exact solution requires lengthy computations.)

Commentary

These calculations seem to be the result of a problem which Steinhaus called the Banach match box problem. Since nowadays most mathematicians do not smoke, it should be explained that a certain mathematician always carries two boxes of matches, one in his right pocket and one in his left pocket. He picks a box at random to light his pipe. Initially, the boxes each have N matches. When he first finds a box empty, what is the distribution of the number of matches in the other box? The distribution is given in Feller, *An Introduction to Probability Theory and its Applications,* Vol. 1, 2nd edition, John Wiley & Sons, 1957, 157.

We observe that these calculations make some sense if "median" is replaced by "mean" and the value ".45" in the fourth line of the Los Alamos edition is replaced by ".95". It may be that there were transcription errors. Perhaps there is another interpretation whereby these calculations make sense. [Editor note. To see several papers devoted to the Banach match box problem, check MathSciNet.]

W.A. Beyer

Chapter 7
Appendices to the Problems

7.1 Problem Subject Index

absolutely continuous functions: 188.1
analytic functions: 17.1, 158, 181, 189
analytic sets: 28, 52, 168
approximation property: 153
approximation theory: 14, 120, 159
associative mappings: 171
automatic continuity: 45
automorphism: 112
Baire function: 132
Baire property: 45, 50
basis: 108
Banach match box problem: 193
Banach Spaces (spaces of type (B)): 1, 7, 24, 27, 41, 46, 47, 64, 69, 87, 106, 108, 122, 144, 153, 155, 172, 173, 174, 176, 190
billiards: 147
Boolean algebras: 37, 163
Borsuk-Ulam theorem: 123
Cauchy problem: 188
compact groups: 32
constrained minima: 104, 105
convergence of rearrangements: 106
control measure problem: 163
Convexity: 7, 10.1, 19, 23, 36, 41, 42, 54, 64, 65, 68, 72, 76, 84, 89, 91, 138, 149, 191
curves: 93, 117
Denjoy integral: 50
descriptive set theory: 22, 33, 40, 43, 50, 52, 80, 168

© Springer International Publishing Switzerland 2015
R.D. Mauldin, *The Scottish Book*, DOI 10.1007/978-3-319-22897-6_7

dimension raising maps: 136, 137

diophantine approximation: 152

distribution of sums: 179

dynamical systems: 102, 110, 115, 141, 147, 164

Eidelheit separation theorem: 64

ellipsoid: 84, 91, 128, 129

essential map: 114

everywhere divergent trigonometric series: 121

fixed points: 42, 54, 107, 115, 167

floating bodies: 19

Fréchet spaces (spaces of type (F): A space of type (F) is the terminology from Banach's monograph, Théorie des Opérations Linéaires, for a Fréchet space, a completely metrizable topological vector space.) 15.1

Functional analysis: 10, 24, 26, 72

functional equations: 39

functions of several variables: 23, 66, 70

general topology 58, 77, 127, 133, 134 135, 136, 192

geometric topology 12, 21, 29, 35, 36, 46, 68, 107, 110, 114, 115, 123, 167, 175, 187

geometry: 44, 59, 61, 76, 78, 81, 182

graph theory: 38, 140, 164

group theory: 29, 30, 62, 63, 77, 142, 160, 183

Hadamard's determinant problem: 118

Ham sandwich theorem: 123

harmonic functions: 151

Hilbert cube: 7, 12, 77, 110, 170, 175

Hilbert's fifth problem: 70

homeomorphisms: 20, 97, 98, 100, 170

homeomorphism group: 29

homogeneous polynomials: 74

homothetic: 129

indecomposable continua: 134 Integral Equations: 25, 124

interpolation: 87

lacunary: 130

Lebesgue density: 146

Lie groups: 90

Lipschitz functions: 23, 87, 165

local isometry: 155

magnetic fields: 18, 19

manifolds 21, 68, 166

matrix theory: 6, 47, 177

measure theory: 2, 11, 16, 40, 139, 144, 145, 146, 163, 169, 170, 191

multilinear mappings: 73

orthogonal functions: 51, 85, 86, 119, 130, 154, 186

partial differential equations: 15, 125, 188

permutations: 46, 71, 95, 96, 101
physics: 18, 19, 125
Poisson distribution: 17
Polish groups: 31, 45
polyhedron: 187 polynomials: 20.1, 70, 75, 79, 128, 148
polynomials in Banach spaces: 55, 56
probability theory: 17, 82, 94, 126, 161, 178, 180, 193
quasi-homeomorphisms: 97
Real analysis: 1, 2, 3, 4, 5, 6, 8, 50, 57, 83, 109, 113, 131, 157, 162, 165, 181
real roots of random algebraic equations: 179
recursion theory: 143
set theory: 9, 13, 22, 34, 37, 40, 43, 99, 168
Schauder basis: 119, 153
Schauder fixed point theorem: 54
S_∞: 71, 95, 96, 101, 111
spaces of type (F_0) (Defined in Problem 26): 26, 72
spaces of type (B_0): 138 strategy: 43, 67
subharmonic functions: 184
summability: 4,5, 8, 28, 88, 92, 154
surface area: 185
surface integrals: 149, 150, 185
Suslin line: 192
theory of games: 43
theory of surfaces: 44, 23, 53, 61, 78, 156
topological dimension: 127
transformation groups: 183
unconditional convergence: 122
uniform distribution in groups: 116
universal Banach spaces: 49
universal groups: 103 unicoherent: 46, 128, 135

7.2 Authors of the problems

Alexandroff: 187
Auerbach: 39, 83, 84, 148
Auerbach, Banach, Mazur, Ulam: 10.1
Auerbach, Mazur: 147
Auerbach, Ulam: 90, 114
Banach: 1, 12, 43, 45, 46, 47, 50, 52, 53, 67, 85, 86, 87, 106, 186
Banach, Mazur: 7, 10, 48, 49
Banach, Mazur, Ulam: 108
Banach, Ulam 2, 3, 11, 40, 100
Bogolubow: 183 Borsuk: 175

Eidelheit: 172, 173, 174, 176, 188.1
Eilenberg: 133, 134, 135, 136, 137, 138
Fermant: 189
Fréchet: 117, 118
Infeld: 125
Kac: 126, 161, 177, 178
Kaczmarz: 130
Kampé de Fériet: 180
Knaster: 182
Knaster, Szpilrajn: 192
Kuratowski: 127
Kuratowski, Ulam: 97
Lomnicki, Ulam: 94
Lusternik: 190
Marcinkiewicz: 124
Mazur: 5, 8, 24, 28, 41, 43, 51, 55, 64, 65 66, 72, 75, 76, 80, 88, 89, 91, 92, 93, 143, 153, 154
Mazur, Orlicz: 6, 9, 15.1, 20.1, 26, 27, 56, 73, 74, 79, 122
Mazur, Schauder: 14
Mazur, Sternbach: 155
Mazur, Ulam: 62, 63, 69, 109, 144
Nikliborc: 128, 129, 149, 150
Offord: 179
Orlicz: 119, 120 121
Ruziewicz: 57, 58, 59, 60, 159
Saks: 184, 185
Schauder: 15, 23, 25, 54, 104, 105
Scheier: 4, 111, 112 113
Scheier, Ulam: 22, 95, 98, 103, 116, 171
Sierpinski: 132
Sobolew: 188
Steinhaus: 44, 61, 78, 81, 82, 123, 152, 162, 181, 193
Sternbach: 107
Stoilow: 158
Szpilrajn: 169, 170, 191
Ulam: 13, 16, 17, 17.1, 18, 19, 20, 21, 29, 30, 31, 32, 33, 34, 35, 36, 37, 38, 42, 43, 68, 70, 71, 77, 96, 99, 101, 102, 110, 139, 140, 141, 142, 145, 146, 164, 165, 166, 167 168
von Neumann: 163
Ward: 156, 157
Wavre: 151
Zygmund: 131

7.3   Problems of Note

We end the part of this book dedicated to the original *Scottish Book* with three lists of particularly notable problems. They are highlighted not necessarily for their mathematical content (which is, of course, fascinating and rich), but for their sundry extrinsic values.

The numbers below refer to the problem numbers in Chapter 6.

- Prize problems: 6, 15.1, 24, 27, 28, 43, 77, 106, 110, 115, 151, 152, 153, 157, 162, 163, 165, 168, 182, 183, 184, 188, 190
- Partially solved problems: 1, 2, 18, 37, 55, 75, 80, 92, 97, 98, 99, 111, 115, 168, 187
- Unsolved Problems, Problems with No Commentary, and Problems of Unknown Status: 14, 20.1, 22, 23, 25, 30, 31, 45, 56, 58, 60, 61, 70, 71, 76, 78, 81, 84, 102, 104, 105, 107, 118, 125, 128, 129, 147, 150, 151, 155, 156, 158, 164, 165, 166, 172, 174, 180, 182, 184, 188, 189, 190

Part III
A Brief History of Wrocław's New Scottish Book

Chapter 8
Lwów of the West, *P. Biler, P. Krupski, G. Plebanek* **and** *W. A. Woyczyński*

At the end of World War 2, the Yalta/Potsdam Agreements made Lwów part of the Soviet Union's Ukraine Republic, and the Silesian capital, Wrocław, part of Poland. Most of the faculty of the Lwów's *Jan Casimir University* moved 320 miles west to the newly Polish city of Wrocław. That included several major figures of the Lwów (and Warsaw) School of Mathematics, settled at the new University and Polytechnic of Wrocław led by the former Rector (President) of the Lwów University, Stanisław Kulczyński, a biologist. Figure 1 shows the dramatic shift West of the political boundaries between Germany, Poland and the Soviet Union executed in 1945, and the related move of the academic community from Lwów University to Wrocław.

At the very end of WW 2 Wrocław was surrounded by the Soviet army, and surrendered on May 6, 1945, a couple of days before Germany formally capitulated. But the result of the four-month siege of Festung Breslau was devastating, and some 90 percent of the city, including the university buildings was in ruin. Today the city is completely rebuilt and returned to its past glory. The main university building (constructed during the period 1728-1740 after the university was established in 1702 as a continuation of a Jesuit College) is again dominating the view of the Oder River downtown, and the new Mathematical Institute building where the New Scottish Book is currently stored is crowning the Grunwaldzki Square axis (see, Fig. 2). And, for 2016, Wrocław was designated by the European Union as the Europe's Culture Capital of the Year, the fact which is advertised across the continent with the moniker **WrocLOVE**.

The New Scottish Book (NSB) was conceived in 1946 as a continuation of the original Lwów Scottish Book (SB). The transition may seem seamless as the last entry in SB was contributed by Hugo Steinhaus on May 31, 1941, and he also contributed the first entry in the New Scottish Book on July 2, 1946.

But in-between those dates the four mathematicians (see, Fig. 3), who are recognized as the founding fathers of the Wrocław School, had very different traumatic experiences during the German occupation. Hugo Steinhaus, Edward Marczewski, and Bronisław Knaster were of Jewish background, and had to hide

© Springer International Publishing Switzerland 2015
R.D. Mauldin, *The Scottish Book*, DOI 10.1007/978-3-319-22897-6_8

Figure 1 Poland in 1939, and in 1945. The Lwów community moves to Wrocław.

to avoid concentration camps. Steinhaus, aged 58 in 1945, who spent the war years as a forester's assistant in the Carpathian woods, came with an established reputation as a world class mathematician, a student of David Hilbert in Göttingen, the founder of the Lwów school, and the "discoverer" of Stefan Banach. Marczewski (the name he adopted during the war to avoid persecution, his original name was Szpilrajn), a measure theorist, only 38 in 1945, and the youngest of the group, was arrested by the Germans towards the end of the war and, in an improbable coincidence, was sent to Festung Breslau as a forced laborer to help clear the rubble from Grunwadzki Square at the exact location where the new Institute of Mathematics was to be built under his leadership in the 1960s (see, Fig. 2). Knaster, a Warsaw topologist

Figure 2 Wrocław from 1945 to 2015. From left to right: Destruction of the 1945 siege; the restored main building of the University; Institute of Mathematics of the University.

Figure 3 The Founding Fathers of the Wrocław School of Mathematics: From left to right: Hugo Steinhaus, Władysław Slebodziński, Bronisław Knaster, and Edward Marczewski.

(see, Fig. 4) who spend the war years hiding in Lwów, earning subsistence by serving (with Stefan Banach and other Lwów scientists) as a human "guinea pig" in the German-run Rudolf Weigl Institute developing typhus vaccines for the war effort.

Władysław Ślebodziński, a distinguished differential geometer, although not Jewish, did not avoid Auschwitz and spent the war years in the concentration camp lucky to survive, but with his health damaged.

The significance of Steinhaus' entries to the SB and NSB was more than symbolic. His series of paper "Sur les fonctions indépendentes," based on the ideas from his pioneering 1923 *Fundamenta Mathematica* paper, "Les probabilités dénombrables et leur rapport à la théorie de la mesure" that was the precursor of Kolmogorov's *Grundbegriffe*, was the inspiration for the probability group. The first six papers in the series, written jointly with his student Mark Kac (later at Cornell) were published in Lwów in *Studia Mathematica* between 1936 and 1940, resumed with the paper number 7 published in the same journal in 1948, with the last, tenth, appearing in 1953. Immediately, more problems were contributed by Marczewski, and Knaster, as well as many distinguished Polish mathematicians who survived the war: A. Alexiewicz, S. Gołąb, A. Mostowski, W. Orlicz, W. Sierpiński, R. Sikorski, J. Szarski, T. Ważewski, Z. Zahorski. At the initiative of Marczewski, most of the problems entered in NSB have been reprinted in the journal *Colloquium Mathematicum* established in Wrocław by the Founding Fathers.

The combination of the mathematical interests of the Founding Fathers was a strike of luck for Wrocław as they provided diverse background for the future

Figure 4 New Scottish Book notebooks in their new elegant covers.

developments. Steinhaus was a leader in functional analysis, probability theory, and orthogonal expansions, Marczewski's work in measure theory (and later in universal algebras) was already significant before the war, Knaster had been the leader of the topology seminar at Warsaw University in the 1930s (together with Mazurkiewicz) where the attendees included Samuel Eilenberg (later at Columbia), Szpilrajn-Marczewski, Nachman Aronszajn (later at University of Kansas), and Kazimierz Kuratowski, and which attracted up to thirty participants on a regular basis. And Ślebodziński was already acknowledged in the differential geometry community as the inventor of the concept of Lie derivative. So, from the very beginning, there was a solid and broad basis for creation of the mathematical research center in Wrocław.

The New Scottish Book (in Polish, Nowa Księga Szkocka) was active from 1946 through 1979[1] after which date there were no entries. with many entries by some of the most prominent researchers of the second half of the 20th century. Mathematicians from all over the world started visiting Wrocław and contributing to NSB. Poland, although governed by the communist party, was not like the Soviet Union. The communications with the West, while controlled, remained open, and the Catholic Church never stopped playing an important role in the society (unlike the Orthodox Church in Russia which was completely dismantled by the communists). So already in July of 1946 we find NSB contributions by Gustave Choquet from Paris, and Vaclav Jarnik from Prague.

The international reputation of the Wrocław mathematics community made it also very influential in running the local academic establishment. By 1953 Marczewski was elected Rector of Wrocław University, and his position facilitated

---

[1] Just seven problems were entered between 1979 and 1987, the time of a historical Solidarity movement in which many Wrocław mathematicians played major roles.

more visits from well-respected foreign mathematicians. Alfred Renyi visits from Budapest, and Szolem Mandelbrojt from Paris, both contributing problems to NSB. The Mandelbrojt problem led directly to Louis de Brange's proof of the Bieberbach conjecture. We also see the rise of Czesław Ryll-Nardzewski, a young Wrocław mathematician who was to make the name for himself by seminal contributions in ergodic theory, harmonic and complex analysis, probability theory and foundations of mathematics (joined later in the latter field by Jan Mycielski), and had a paramount influence on the life of the mathematical community in Wrocław (and, more broadly, in Poland).

By the end of the 1950s, we also see the emergence of another young superstar, Kazimierz Urbanik (later, also Rector of the University of Wrocław) who started out with a series of papers offering an alternative development of the Gelfand-Itô theory of generalized stochastic processes. In the 1960s Jean-Pierre Kahane visits from Paris, contributes several problems to NSB, and develops a close and lasting interaction with the Wrocław harmonic analysis group led by Ryll-Nardzewski and Stanisław Hartman (and later, by later Andrzej Hulanicki), and we see a steady stream of distinguished visitors (and contributors to NSB) such as Pavel Sergeevich Aleksandrov from Moscow, R.H. Bing from Madison, Wisconsin, and Aryeh Dvoretzky from Jerusalem.

One of the last problems (No. 961) was entered in October 1979 by Fulvio Ricci, from Turin, the intellectual exchange with the international mathematical community was very much alive. Just in October of 1978, Paul Erdös visited from Budapest, and contributed a problem, and Hubert Delange from Orsay, suggested a problem of finding extremal points of the set of entire functions satisfying an exponential inequality.

Initially, the NSB, a collection of soft cover notebooks, was stored in the joint library of the Institute of Mathematics of the University of Wrocław and the Wrocław Polytechnic, with access guarded by the inimitable Ms Marietta Wilanowska, the chief librarian, an approach quite different from the flamboyant Scottish Café environment of the original SB. After some twenty years of its existence the NSB was moved to the library located on the top floor of the new Mathematical Institute building with a spectacular view on the Oder River and the Cathedral Island with its several romanesque and gothic churches dating back to the 12th Century. In 2014, on the initiative of Dean Biler, the booklets were sent to the bookbinder and given an elegant hard leather-bound covers pictured in Fig 5. But they are still stored by the chief librarian under lock-and-key with an on-demand access. In 1959, young Wrocław mathematicians, Henryk Fast and Stanisław Świerczkowski, prodded by Edward Marczewski, prepared a privately typed little volume of 70 pages, where they collected NSB problems for the years 1946 -1958.

P.B., P.K., and G.P,
Institute of Mathematics,
University of Wrocław,
Wrocław, Poland

W.A.W.,
Department of Mathematics, Applied Mathematics and Statistics, and
Center for Stochastic and Chaotic Processes
in Science and Technology,
Case Western Reserve University,
Cleveland, OH, USA

Part IV
The New Scottish Book

At least 280 problems can be classified as topological ones. The vast majority of them emerged from research activities of the famous Topology Seminar conducted by Bronisław Knaster in Wrocław in the years 1946–1977. 'CM P. . . ' refers to problems published in Problem Section of *Colloquium Mathematicum*.

Instytut Matematyczny, Uniwersytet Wrocławski, pl. Grunwaldzki 2/4, 50-384 Wrocław, Poland Piotr.Biler@math.uni.wroc.pl; Pawel.Krupski@math.uni.wroc.pl; Grzegorz.Plebanek@math.uni.wroc.pl

Chapter 9
Selected Problems

Instytut Matematyczny, Uniwersytet Wrocławski, pl. Grunwaldzki 2/4, 50-384 Wrocław, Poland
Piotr.Biler@math.uni.wroc.pl; Pawel.Krupski@math.uni.wroc.pl;
Grzegorz.Plebanek@math.uni.wroc.pl

1. July 2, 1946;   Hugo Steinhaus (Wrocław) (in Polish)

Let $0.a_1a_2a_3\dots$ and $0.b_1b_2b_3\dots$ be decimal representations of transcendental numbers. Is $0.a_1b_1a_2b_2a_3b_3\dots$ also a transcendental number?

4. July 6, 1946;   Gustave Choquet (Grenoble) (in French)

Let E be a compact metric space covered by a finite number of open sets C_i of diameter less than $\varepsilon > 0$. Let δ_i denote the diameter of the boundary of C_i, and $A(\varepsilon) = \inf \sum \delta_i^2$ over all such coverings. Finally, define $A = \lim_{\varepsilon \to 0} A(\varepsilon)$. Show that $A = 0$ for every space E of dimension 1, in the sense of Menger.

Prize: a bottle of champagne (to be drunk in Paris) for a positive answer; a bottle of bordeaux for a negative answer.

5. July 7, 1946;   Edward Marczewski (Wrocław) (in Polish)

Let μ be a probability measure on a σ-algebra $\mathscr{A}$ of subset of a space X. Two subfamilies $\mathscr{P}_1$ and $\mathscr{P}_2$ of $\mathscr{A}$ are *stochastically independent* if $\mu(P_1 \cap P_2) = \mu(P_1) \cdot \mu(P_2)$ for every $P_i \in \mathscr{P}_i$, $i = 1, 2$.

Two real-valued measurable functions $f_1, f_2 : X \to \mathbb{R}$ are *stochastically independent in the sense of Kolmogorov* if $\mathscr{P}_1$ and $\mathscr{P}_2$ are stochastically independent, where $\mathscr{P}_i = \{f_i^{-1}(Z) \in \mathscr{A} : Z \subseteq \mathbb{R}\}$. Functions f_1 and f_2 are *stochastically independent in the sense of Steinhaus* if the corresponding condition is satisfied by the families $\mathscr{P}_i = \{f_i^{-1}(a, b) : a < b\}$.

Functions stochastically independent in the sense of Kolmogorov are stochastically independent in the sense of Steinhaus. Does the converse implication hold? CM 1, P3

Answer: It is noted that the answer is 'yes' if the measure in question is the Lebesgue measure, which was proved by S. Hartman. In general, the answer is 'no,' which follows from an unpublished note by S. Banach.

Remark: First counterexample was published a few years later. Apparently, the answer is positive whenever the measure μ is perfect.

7. November 29, 1946;   Bronisław Knaster (Wrocław) (in Polish)
The Borsuk-Ulam Antipodal Theorem (K. Borsuk, *Drei Sätze über die n-dimensionale euklidische Sphäre*, Fund. Math. 20 (1933), 178, Satz II, known as the "Antipodensatz") was generalized by Hopf (H. Hopf, *Eine Verallgemeinerung bekannter Abbildungs- und Überdeckungssätze*, Portugaliae Math. 4 (1944), 131) as follows: For each continuous map $f(p)$ of the n-dimensional sphere S_n of diameter d onto the n-dimensional Euclidean space E_n and for $\delta \leq d$, there exists at least one pair of points p_1, p_2 in S_n such that the distance between them is δ and $f(p_1) = f(p_2)$.

I noticed that from a certain "tripod" problem (which provokes me to be called "affaire Dreifuss" ("Dreifuß problem"), [with an allusion to "l'affaire Dreyfus," of course; note of the editor]) of Steinhaus one can derive, among others, the following topological content:

Given any continuous $f(p)$ from S_2 onto E_1, does there exist at least one triple p_1, p_2, p_3 in S_2 which is isometric (i.e., congruent) to a given triple of points in S_2 and such that $f(p_1) = f(p_2) = f(p_3)$?

The problem is open. Is the following more general problem true: Given any continuous $f(p)$ from S_n onto E_k, $k = 1, 2, \ldots n$, does there exist at least one set $P = \{p_1, p_2, \ldots, p_{n-k+2}\}$ which is isometric to a given $(n - k + 2)$-point subset of S_n and such that $f(p_1) = f(p_2) = \cdots = f(p_{n-k+2})$?

How many and when do such sets P exist?

Prize: A dinner with vodka in the restaurant "Monopol" (in Wrocław) for a proof of the theorem; a coffee with a cake in the University café for disproving.

CM 1, P4

Answer: Yes. for S_2, see E.E. Floyd, Real-valued mappings of spheres, Proc. Amer. Math. Soc. 6 (1955), 957-959, CM 1, P4, R1; the positive answer in full generality was given by R. P. Jerrard, On Knaster's conjecture, Trans. Amer. math. Soc. 170 (1972), 385–402, see CM 30, P4, R3.

14. December 13, 1946;   Stanisław Gołąb (Kraków) (in Polish)
Prove that in the two-dimensional Minkowski geometry the length of the indicatrice ("Eichkurve"), measured in the induced metric, is at least 6. J. Kawaler – a mathematician from Wadowice – was supposed to prove this. However, I could not contact him from the beginning of the war.

16. December 13, 1946;   Zygmunt Zahorski (Kraków) (in Polish)
Let us call a function f *smooth* iff

$$\lim_{h \to 0} \frac{f(x-h) + f(x+h) - 2f(x)}{h} = 0.$$

[NB continuous functions with this property form the Zygmund class in modern terminology.]
Question 1: A *smooth* function must it be measurable?
(If it were measurable this is of Baire's first class.)
Question 2: Determine the category and the measure of the set of points where a continuous *smooth* function has the derivative.
It is easy to prove that this set is of power continuum, and each function φ with $\varphi(x) \in [\underline{f'}(x), \overline{f'}(x)]$ has the Darboux property.

Answer: Q1; No.
CM 1, P8, P9, p. 32; R1, R2, p. 148. CM 5, P220, P221, p. 234.

41. March 9, 1947;   Stanisław Hartman (Wrocław) (in Polish)
It is easy to see that $\liminf |\cos n|^n = \liminf |\sin n|^n = 0$. A bit harder is to prove $\limsup (\cos n)^n = 1$ and $\limsup |\sin n|^n = 1$. Find $\liminf (\cos n)^n$, $\liminf (\sin n)^n$, $\liminf (\sin n)^n$.

Prize: A coffee and cakes in café "Bagatela," Wrocław.

51. May 11, 1947;   Mieczysław Biernacki (Lublin) (in Polish)
Is it true that the Cauchy product of two convergent series with monotone decreasing terms has eventually monotone terms?

Answer: No. An example (in French) is given by Gustave Choquet, May 13, 1947.

52. May 11, 1947;   Jan Mikusiński (in Polish)
The following theorem is true: If a function f satisfies the differential equation $f^{(n)}(x) + f(x) = 0$ on an interval $[a,b], f(a) = f(b) = 0$, then $b - a \leq N(n)$ for some quantity which depends on n only. Determine the least $N(n)$. The solution is known for $n = 2, \ldots, 6$.

70. June 20, 1948;   Karol Borsuk (Warsaw) (in Polish)
Let A be a compact 0-dimensional subset of the Euclidean space E_n, $n \geq 4$.
Can the fundamental group of $E_n \setminus A$ be non-trivial?

CM 1, P45

71. June 20, 1948;   Karol Borsuk (Warsaw) (in Polish)
A set A is said to be a topological divisor of a space E if there is a set B such that E is homeomorphic to the Cartesian product $A \times B$.
Must any divisor of a polyhedron be a polyhedron?

CM 1, P46

76. September 29, 1948;   Eduard Čech (Prague) (in French)
If $\dim E \leq n$ is defined by the existence of arbitrarily small finite open covers of order $\leq n+1$ (i.e., each point of E belongs to at most $n+1$ elements of the cover) and if E is completely normal, is then the dimension of a subspace less than or equal to the dimension of entire space E? The answer is affirmative if each closed subset of E is G_δ, in particular if E is metrizable.

CM 1, P53

102. May 03, 1950;   Bronisław Knaster (Wrocław) (in Polish)
A point p of a set Z is of order n if there exist arbitrarily small neighborhoods of p whose boundaries intersect Z in exactly n points; in the case when n grows ad infinity if diameters of the neighborhoods approach 0, p is said to be of order ω. A curve is regular if it consists of points of order $\leq \omega$. A cycle is a continuous image of the interval such that each pair of points lies on a simple closed curve. The cardinality of the preimage of a point is called the multiplicity of the map at this point.
Question: Does there exist, for each cycle which is a regular curve, a continuous map from a circle onto the cycle of multiplicity at each point equal to the order of the point?

Prize: A coffee with cakes for "yes", a cod (a meat-ball for Ms. Nosarzewska [who solved many problems in topology but didn't like fishes; note of the editor]) for "no."

107. May 26, 1950;   M. Katětov (Prague) (in Czech)
Let P be a normal topological space. Is it true that for every decreasing sequence $(F_n)_n$ of closed subsets of P, if $\bigcap_{n=1}^{\infty} F_n = \emptyset$, then there is a sequence $(G_n)_n$ of open subsets of P such that $F_n \subseteq G_n$ for every n and $\bigcap_{n=1}^{\infty} G_n = \emptyset$?
CM, P96

Answer: No. In the modern terminology this amounts to asking if every normal space is countably paracompact. The class of countably paracompact spaces was introduced by Dowker and Katetov in 1951; see section 5.2 in Engelking's book. The answer is 'no' and every counterexample is now called a *Dowker space*. The first Dowker space (of large size) was constructed by M.E. Rudin in 1971. The question whether there are small Dowker spaces became quite popular in set-theoretic topology (see M.E. Rudin's survey in *Handbook of Set Theoretic topology* (North Holland, 1984)); cf. *Borel measures* by R.J. Gardner and W.F. Pfeffer, ibidem.

108. May 31, 1950;   M. Katětov (Prague) (in Czech)
Is every perfectly normal space fully normal?
CM, P97

Answer: No. A space is fully normal if every open cover has a open star refinement. By Theorem 5.1.12 in Engelking's book a space is fully normal if (and only if) it is paracompact. See historical remarks in Engelking, p. 389 in the 1977 edition; by 5.5.3 the answer is negative.

110. 1950;   M. Katětov (Prague) (in Czech)
Which Banach spaces are paracompact in the weak topology?
CM, P98

Remark: H.H. Corson showed in 1961 that for the weak topology of a Banach space paracompactness is equivalent to the Lindelöf property (Trans. Amer. Math. Soc. 101 (1961), 1-15). There is a well-studied natural class of Banach space which are weakly Lindelöf — weakly compactly generated spaces (WCG). R. Pol showed that a Banach space may be weakly Lindelöf but not WCG.

120. October 28, 1950;   Edward Marczewski (Wrocław) (in Polish)
Is every indirect product of two countably additive set functions μ and ν defined on algebras $\mathscr{A} \subseteq P(X)$ and $\mathscr{B} \subseteq P(Y)$ is countably additive on the product algebra in $X \times Y$? What if $\mathscr{A}$ and $\mathscr{B}$ are σ-algebras?

   Here an additive set function λ defined on $\mathscr{A} \otimes \mathscr{B}$ is called an indirect product of μ and ν if $\lambda(A \times Y) = \mu(A)$ for every $A \in \mathscr{A}$ and $\lambda(X \times B) = \nu(B)$ for every $B \in \mathscr{B}$.

Answer: No, see E. Marczewski; Cz. Ryll-Nardzewski, *Remarks on the compactness and non direct products of measures*, Fund. Math. 40 (1953), 165–170. Indirect product of measures became essential in investigating compact measures in the sense of Marczewski and perfect measures, see, e.g., J. Pachl, *Disintegration and compact measures*, Math. Scan. 43 (1978), 157–168 and W. Adamski, *Factorization of measures and perfection*, Proc. Amer. Math. Soc. 97 (1986), 30–32.

121. November 13, 1950;   Roman Sikorski (Warsaw) (in Polish)
Is every Hausdorff space with a countable open basis a continuous interior image of a separable metric space? (A mapping is interior if it sends open sets onto open sets).
CM 2, P78

Answer: Yes. A.S. Schwarz, *On a problem of Sikorski*, Uspekhi Mat. Nauk, 12:4(76) (1957), 215.

129. December 1, 1950;   Stanisław Hartman (Wrocław) (in Polish)
Is it true that the limit of a uniformly convergent sequence of functions on the interval $[0, 1]$ satisfying the Darboux property has also this property?

Answer: Henryk Fast gave a negative answer on January 4, 1951 (not recorded in NSB).

131. January 4, 1957;   Edward Marczewski (Wrocław) (in Polish)
Suppose that μ_1, μ_2 are measures such that some (indirect) product of them is compact. Do μ_i have to be compact themselves?

Answer: A measure μ is *compact in the sense of Marczewski* if there is a family of μ measurable sets $\mathscr{C}$ such that $\mu(A) = \sup\{\mu(C) : C \subseteq A, C \in \mathscr{C}\}$ and whenever $(C_n)_n$ is a sequence in $\mathscr{C}$ with the finite intersection property then $\bigcap_n C_n \neq \emptyset$. The answer is 'yes' by a highly nontrivial result due to Pachl that a restriction of a compact measure to any sub-σ-algebra is compact.

Remark: The Lebesgue measure on $[0,1]$ is trivially compact in the sense of Marczewski. However the problem, if any probability measure on some σ-algebra $\mathscr{A} \subseteq Bor[0,1]$ is compact has not been fully answered. The positive solution under the continuum hypothesis was given by D.H. Fremlin (*Weakly α-favourable measure spaces*, Fund. Math. 165 (2000), 67–94).

132. February 24, 1951;   Stanisław Hartman (Wrocław) (in Polish)
It is known that for almost every real x the sequence $\{n!x\}_{\text{mod}1}$ has the equipartition property. Find such an x explicitly.

Answer: Hugo Steinhaus, February 26, 1951. E.g., the number $x = \sum_{k=1}^{\infty} \frac{1}{k!}[\xi k \sqrt{k}]$ has this property for every real ξ.

150. May 12, 1951;   Kazimierz Kuratowski (Warsaw) (in Polish)
Can a generalized (i.e., non-compact) Janiszewski space be of dimension > 2?

172. November 30, 1951;   Bronisław Knaster (Wrocław) (in Polish)
Czesław Ryll-Nardzewski has proved that the Cantor set C, so every perfect 0-dimensional separable space, is homogeneous in the strongest (known) sense: if two compact subsets F_1 and F_2 of C are homeomorphic, then each homeomorphism between their boundaries can be extended to a homeomorphism of C onto itself that maps F_1 onto F_2 (and conversely—the condition is necessary). This has been presented at today's meeting of the Polish Mathematical Society, Wrocław Branch. Question: Are perfect 0-dimensional spaces the only ones with this type of homogeneity?

Prize: Two isometric (leather) gloves.

See B. Knaster and M. Reichbach, *Notion d'homogénéité et prolongements des homéomorphies*, Fund. Math. 40, (1953), 180193.

189. May 14, 1952;   Karol Borsuk (Warsaw) (in Polish)
Is this true that a curve whose Cartesian square is homeomorphic to a subset of the Euclidean 3-space is either an interval or a circle?

194. July 2, 1952;   Czesław Ryll-Nardzewski (Wrocław) (in Polish)
If a complex function $f(x) = \sum_{n=-\infty}^{\infty} a_n e^{int}$ satisfies $\sum_{n=-\infty}^{\infty} |a_n| < \infty$, is it true that $|f(x)|$ has absolutely convergent Fourier series?

Answer: No; Jean-Pierre Kahane (Montpellier, Paris), *Sur les fonctions sommes de séries trigonométriques absolument convergentes.* C. R. Acad. Sci., Paris **240** (1955), 36–37.

197. April 12, 1951;   Adam Rybarski, Abraham Götz (Wrocław) (in Polish)
Consider a convex surface and the shortest paths between two given distinct points on that surface. The shortest path could be nonunique, as is in the case of antipodal points on a sphere. Is it true that the sphere is the only surface such that if the shortest paths are nonunique, there are infinitely many such paths?

199. September 25, 1952;   Alfréd Rényi (Budapest) (in English)
What are necessary and sufficient conditions regarding the system $\{f_n(x)\}, n = 1, 2, \ldots$, of independent measurable functions, defined in the interval $[a, b]$, which ensure the completeness of the system $\{f_1^{m_1}(x) f_2^{m_2}(x) \ldots f_n^{m_n}(x)\}$, where $m_1, m_2, \ldots$ run independently over all nonnegative integers and $n = 1, 2, 3, \ldots$?

200. September 25, 1952;   Alfréd Rényi (Budapest) (in English)
Let us suppose that we know the projections of a probability measure (or mass distribution) of the plane on an enumerable infinite set of straight lines. Is the distribution uniquely determined or not?

Remark. I have proved that the distribution is uniquely determined in case the distribution is such that its characteristic function is an analytic function of both variables; this is true, e.g., in the case when the distribution is bounded.

Answer: No; A. Heppes, *On the determination of probability distributions of more dimensions by their projections.* Acta Math. Acad. Sci. Hung. **7** (1956), 403–410.

215. December 16, 1952;   J. Novák (Prague) (in Czech)
Is it true that in the power set $P(X)$ of the set X of cardinality $\aleph_1$ every sequence has a converging subsequence? (A sequence $(A_n)_n$ of subsets of X converges to A if $A = \bigcap_n \bigcup_{k \geq n} A_k = \bigcup_n \bigcap_{k \geq n} A_k$.)
CM, P135

Answer: Equivalently, this is the question if the Cantor cube 2^{ω_1} is sequentially compact in its usual product topology. The question was related to the fact that the answer is clearly negative under the continuum hypothesis (CH). However, the statement is independent of the usual axioms of set theory — it holds true under Martin's axiom and the negation of CH, see D.H. Fremlin, *Consequences of Martin's axiom* (CUP, 1984).

220. March 25, 1953;   Bronisław Knaster (Wrocław) (in Polish)
How many topological types are there of 0-dimensional dense-in-itself G_δ's?
In particular, how many topological types are there among them having a homeomorphic image in the Cantor set C that differs from C on a non-compact subset of left ends of contiguous intervals?

CM 3, P123

240. October 23, 1953;   Bronisław Knaster (Wrocław) (in Polish)
Is it true that for each continuous mapping of the projective plane onto a (necessarily) bounded continuum in the Euclidean plane there exist three points of the first plane which are mapped onto a single point of the second one?

252. May 31, 1954;   Stanisław Mrówka (in Polish)
Does the Axiom of Choice follow from the theorem that each T_1 compact topology can be enriched to a T_2 compact one?
Remark: The theorem above together with the fact that the Tikhonov product of two-point spaces is compact imply the Axiom of Choice.

272. September 7, 1955;   Jan Mycielski (Wrocław) (in Polish)
Does there exist a set M of $2^{\aleph_0}$ continuous functions on the interval $[0,1]$ such that for every $f,g \in M, f \neq g$, the limit

$$\frac{df}{dg}(x) = \lim_{h\to 0} \frac{f(x+h)-f(x)}{g(x+h)-g(x)}$$

does not exist for any $x \in [0,1]$?

Remark. Using Baire's category method one can show the existence of such a set of power $\aleph_1$, so that the problem is of interest without assuming the Continuum Hypothesis.

Answer: Yes; J. de Groot, *A system of continuous, mutually non-differentiable functions.* Math. Z. **64** (1956), 192–194.

322. September 27, 1956;   Paul Erdös (Budapest) (in English)
Prove that for every integer $n > 0$

$$\frac{4}{n} = \frac{1}{x} + \frac{1}{y} + \frac{1}{z}$$

is solvable in positive integers x, y, z.

328. December 23, 1956; Hugo Steinhaus (Wrocław) (in Polish)
Let $p \geq 3$ denote a prime number and consider regular p-polygon. Prove (or disprove) that any three diagonals of this polygon do not intersect at a point.

Prize: a picture of the regular 23-gon.

Answer: Yes; proof in: H. T. Croft, M. Fowler, *On a problem of Steinhaus about polygons.* Proc. Cambridge Phil. Soc. **57** (1961), 686–688.

334. January 25, 1957; Hugo Steinhaus (Wrocław) (in Polish)
Suppose that $\{n_k\}$ is an increasing sequence of integers. Conjecture: either the circle $|z| = 1$ is a natural boundary (singular, "coupure") for the function $f(z) = \sum_{k=1}^{\infty} z^{n_k}$, or f is a rational function.

338. February 10, 1957; Edward Marczewski (Wrocław) (in Polish)
Two metric spaces are called quasi-homeomorphic if they admit continuous mappings from the first one onto the second one and from the second one onto the first one, both with arbitrarily small fibers.
What is an analogue of the Menger-Nöbeling theorem for quasi-homeomorphisms? In other words, for each natural n, find the least number $m = \phi(n)$ such that every separable (or compact) space of dimension $\leq n$ is quasi-homeomorphic to a subset of the Euclidean m-space.
Clearly, $n \leq \phi(n) \leq 2n + 1$. [It is known that for compact spaces (1) the dimension is an invariant of quasi-homeomorphisms (K. Kuratowski and S. Ulam, Fund. Math. 20 (1933), 252) while (2) topological embeddability, e.g., in the plane—is not (a Borsuk's note submitted to CM)].

Answer: T. Ganea proved in Bull. Math. Polon. Sci. 7 (1959), 27–32, that $\phi(n) = 2n + 1$.

344. May 14, 1957; Szolem Mandelbrojt (Paris) (in French)
Let $f(z) = \sum_{n=1}^{\infty} a_n z^n$ and $\varphi(z) = \sum_{n=1}^{\infty} b_n z^n$ be univalent functions for $|z| < 1$. Prove that $\sum \frac{a_n b_n}{n} z^n$ is also a univalent function for $|z| < 1$. If it is true, then this will settle the Bieberbach conjecture (i.e., if $a_1 = 1$, then $|a_n| \leq n$).

Answer: No. Counterexamples are in: B. Epstein, I.J. Schoenberg, *On a conjecture concerning schlicht functions.* Bull. Amer. Math. Soc. **65** (1959), 273–275.
C. Loewner, E. Netanyahu, *On some compositions of Hadamard type in classes of analytic functions.* Bull. Amer. Math. Soc. **65** (1959), 284–286.

A much more general result was proved by L. de Branges in *A proof of the Bieberbach conjecture*, Acta Math. 154 (1985), 137–152.

349. May 14, 1957; Marshall H. Stone (Chicago) (in English)
Let $\sum_{k=0}^{n} p_k f^{(n-k)} = 0$ be a differential equation with sufficiently smooth coefficients p_k on the interval $a < x < b$; finite or infinite, with $p_0 > 0$ there. Let n_{ab} be the number of solutions in $L^2(a, b)$, n_a — the number of solutions in $L^2(a, c)$ but not in

$L^2(a,b)$ (where $a < c < b$), and n_b — the number of solutions in $L^2(c,b)$ but not in $L^2(a,b)$. Find the relation, if any, which must hold among the integers n_{ab}, n_a, n_b; and discover properties of the coefficients $p_0, \ldots, p_n$ which condition these integers. In the case $n = 2$, the works of A. Wintner and Ph. Hartman should be considered.

350. May 15, 1957;   Bronisław Knaster (Wrocław) (in Polish)
There exists a continuous complex-valued function on the circle that admits each of its values at two points (e.g., antipodal ones) but there is no such, i.e. of constant multiplicity 2, real-valued function either on a circle or on an interval.
More generally, prove that none of the continuous functions on an interval with values in an arbitrary topological space has multiplicity 2.
Characterize topologically the class of continua such that no continuous function defined on them has constant multiplicity $n \geq 2$ (n finite).

CM 5, P223

Comments: (Knaster, April 4, 1958): O. Hanner proved it for an interval but the paper in Fund. Math. 45 (1958) cannot appear. It turned out that this and many other theorems on 2-to-1 functions were done by Americans: Harold in Duke Math. J. 5 (1939) and Amer. J. Math. 62 (1940), Roberts in Duke Math. J. 6 (1940), Martin and Roberts in Trans. Amer. Math. Soc. 49 (1941), Gilbert in Duke Math. J. 9 (1942) and Civin in Duke Math. J. 10 (1943). It has been possible to discover these results because the war-time Zentralblatts, etc. have arrived only now!

376. March 14, 1958;   Boris Gnedenko (Kiev) (in Russian)
Let $\xi_1, \xi_2, \ldots, \xi_n, \ldots$ be mutually independent random variables such that for some positive constants $b_1, b_2, \ldots, b_n, \ldots$, the law of the random variable $b_n \xi_n$ is independent of n. Find the class of limit laws of the sums

$$s_n = \frac{\xi_1 + \cdots + \xi_n}{B_n} - A_n,$$

with some constants A_n and $B_n \to \infty$ as $n \to \infty$.

383. April 22, 1958;   Tudor Ganea (Bucarest) (in French)
Does there exist a finite-dimensional compact absolute retract X which is homogeneous, i.e., for $a, b \in X$ there exists a homeomorphism $h : X \to X$ such that $h(a) = b$?

CM 6, P275

393. May 28, 1958;   Johannes de Groot (Amsterdam)
Does there exist a (plane) continuum which does not admit any continuous mapping to itself, except trivial ones?

Comments: (Knaster, January 6, 1959): I posed the same question at the Topological Seminar in Warsaw University on October 22, 1930. It is recorded in the seminar book. I have found it out due to Professor K. Borsuk's letter in which he reminded me of the fact.

Answer: Similar questions were also asked by J. de Groot [Fund. Math. 42 (1955), 203–206] (about the existence of a connected set with the property) and by R. D. Anderson in 1955 (about continua with this kind of rigidity). H. Cook in [Fund. Math. 60 (1967), 241–249] presented two famous examples. The first one is an indecomposable, 1-dimensional continuum admitting only the identity and constant self-maps. His second example is a hereditarily indecomposable 1-dimensional continuum such that any continuous map between its nondegenerate subcontinua must be the identity map. Since both examples are non-planar, the Knaster - de Groot question was open until 1985 when T. Maćkowiak [Houston J. Math. 11 (1985), no. 4, 535–558] described a hereditarily decomposable arc-like (hence planar) continuum for which the identity is the only mapping onto a nondegenerate subcontinuum. A little bit later, in [Dissertationes Math. (Rozprawy Mat.) 257 (1986)], he constructed an arc-like hereditarily decomposable continuum with the same rigidity as the second Cook's example, i.e. with no nonconstant, nonidentity maps between subcontinua.

Cook's and Maćkowiak's examples have found many interesting applications in continuum theory, fixed-point theory, topological groups, functional analysis (see citations in MR).

394. May 28, 1958;   Johannes de Groot (problem raised by H. de Vries) (Amsterdam)
Does there exist a rigid topological group T? (rigid means: the automorphism group of T equals identity).

Comments: The problem reduces to groups whose all elements are of order 2.
CM 7, P286

Answer: The problem should be meant as to whether there is a topological group without automorphisms different from translations. A solution was given by J. van Mill in 1983 [Trans. Amer. Math. Soc. 280 (1983), 491-498] who constructed a separable metric, connected and locally connected topological group H, the only autohomeomorphisms of which are group translations. Moreover, for each $x \in H$, each autohomeomorphism of $H \setminus \{x\}$ is the identity map.

Under the Continuum Hypothesis, he obtained a connected and locally connected topological group H with a stronger rigidity property: each continuous self-map of H is either constant or a translation.

399. May 30, 1958;   Paul Turán (Budapest) (in English)
Let $F(n,A,c)$ be the set of those α's in $0 < \alpha < 1$ for which the inequality $\left| \alpha - \frac{\mu}{q} \right| < \frac{A}{q^2}$ is soluble with an $n \leq q \leq cn$. Denoting by $|F(n,A,c)|$ the measure of this set, it is to be proved that $\lim_{n\to\infty} |F(n,A,c)|$ exists (and as it is easy to see, positive and less than 1).

460. March 12, 1959;   Bronisław Knaster (Wrocław) (written in Polish by A. Lelek)
It follows from the Hahn-Mazurkiewicz theorem that for each continuous function that maps an interval onto a square the set of points which belong to at least 3-point preimages is dense.
Does every continuous function mapping an n-dimensional cube onto an $(n+k)$-dimensional cube have a dense set of at least $(k+2)$-point preimages?

462. September 9, 1959;   Mark Kac (Ithaca, NY) (in English)
[Problem from Bell Telephone Laboratories, Murray Hill, NJ, US] If n points are placed on the surface of a three-dimensional sphere and if they repel each other according to the inverse square law, can one characterize the equilibrium position(s) of the points?

466. October 10, 1959;   Zbigniew Semadeni (Warsaw) (in Polish)
A Banach space X has property P_s (is said to be s-injective) if for every Banach superspace $Y \supseteq X$ there is a projection from Y onto X of norm $\leq s$. Is it true that a space with property P_s is isomorphic to a P_1-space?
CM, P308

Answer: This is still open (as far as I know, some form of the question appears on a recent list of problems composed by H.G. Dales). Though 1-injective Banach spaces were characterized in early fifties, there has been a little progress in understanding s-injective spaces for $s > 1$.

467. October 10, 1959;   Aleksander Pełczyński (Warsaw) (in Polish)
Is there a projection from $\ell^\infty[0,1]$ onto its subspace of elements with countable support.
CM, P309?

Answer: No, see A. Pełczyński, V. Sudakov, *Remark on non-complemented subspaces of the space $m(S)$*, Colloq. Math. 9 (1962) 85–88.

Remark: The question was related to the classical Phillips theorem, saying that c_0 is not complemented in ℓ^∞.

471. November 03, 1959;   Bronisław Knaster (Wrocław) (in Polish)
I call a dendroid any arcwise connected continuum such that between each pair of its points it contains only one irreducible continuum.
Characterize topologically planable dendroids (i.e., having a homeomorphic image in the plane).
CM 8, P323

485. March 7, 1960;   Kazimierz Urbanik (Wrocław) (in Polish)

Let E be a perfect subset of the positive half-line containing the origin. Show that if for each $\varepsilon > 0$ the Hausdorff dimension of $E \cap [0, \varepsilon]$ is positive, then the additive semigroup generated by E contains the entire half-line.
CM 8, P322, p. 139.

526. November 22, 1960;   Bronisław Knaster (Wrocław) (in Polish)
I call a dendroid any arcwise connected and hereditarily unicoherent continuum (this definition is equivalent to that given in Problem 471). If, in the definition, one replaces the arcwise connectedness with the λ-connectedness (i.e., every two points can be joined by an irreducible subcontinuum of type λ in the sense of Kuratowski, Topologie II, p. 137, reference), then I call such continuum a λ-dendroid.
Question: Does the Borsuk's fixed-point theorem for dendroids (Bull. Polon. Math.Sci. 2 (1954), 17–20) extend to λ-dendroids?

Answer: This was a famous problem in fixed-point theory of continua. J. J. Charatonik commented (NSB, October 7, 1969) he has a positive answer but his argument was wrong. The problem was solved by Roman Mańka, a Ph.D. student of Knaster, in 1974 (Fund. Math. 91 (1976), no. 2, 105–121; see also an extensive review MR0413062 (54 #1183)). He proved that λ-dendroids have the fixed-point property even for upper semi-continuous continuum-valued mappings.

533. April 6, 1961;   Jean-Pierre Kahane (Montpellier, Paris) (in French)
Determine a condition on the sequence of integers $\{\lambda_n\}$ such that continuous functions with the Fourier series $\sum_{n=-\infty}^{\infty} a_n \exp(i\lambda_n x)$ are either everywhere smooth (C^∞) or nowhere (on any interval) smooth.

545. April 14, 1961;   Aleksander Pełczyński (Warsaw) (in Polish)
Is the space S of all measurable functions on $[0, 1]$ (with the asymptotic convergence) homeomorphic to the Hilbert space ℓ^2?
CM 9, P365

546. April 14, 1961;   Aleksander Pełczyński (Warsaw) (in Polish)
Let R be a metric compact absolute retract.
Is it true that for each two metric compact infinite spaces Q_1 and Q_2 the spaces $C_R(Q_1)$ and $C_R(Q_2)$ are homeomorphic?
(Symbol $C_R(Q)$ denotes the space of continuous functions from Q to R with the uniform convergence).

Remark: If R is convex, the answer is positive.
CM 9, P366

547. April 14, 1961;   Aleksander Pełczyński (Warsaw) (in Polish)
Let R be a set homeomorphic to letter T.
Is the space $C_R(Q)$ homeomorphic to the Hilbert space for an arbitrary metric compact infinite space Q?
CM 9, P367

554. April 24, 1961;   Bronisław Knaster (Wrocław) (in Polish)
Let $\mathscr{F}$ be a family of dendroids.
Does there exist a continuum C such that all $D \in \mathscr{F}$ are continuous images of C?
For curves which are not dendroids the answer is negative (Z. Waraszkiewicz, Fund. Math. 22 (1934), 180–205).

555. June 2, 1961;   Karl Menger (Chicago) (in English)
Let k be a positive integer. What is the dimension of the set of all points of the Hilbert space ℓ^2 that have exactly k irrational coordinates?

559. June 26, 1961;   Alexander Doniphan Wallace (New Orleans)
Does the closed 2-cell admit a continuous associative multiplication such that $\{x : x^2 = x\}$ coincides with the boundary?

CM 9, P381

562. September 16, 1961;   Clifford Hugh Dowker (London)
It can be shown by direct computation that the set R of points with all coordinates rational in a non-separable real Hilbert space has the same dimension in terms of coverings as in terms of neighbourhoods of points; indeed, $\dim R = R = 1$.
Problem: To find a sufficient condition on a space X for equality of dimensions, $\dim X = X$; moreover, a condition satisfied by the above space R.

CM 9, P383

563. November 07, 1961;   Jerzy Mioduszewski (Wrocław) (in Polish)
Let $f, g : Y \to Y$ be continuous functions onto a triod (letter Y) which commute, i.e., $fg = gf$. Does there exist a point x such that $f(x) = g(x)$?

CM 9, P383

565. November 12, 1961;   E. Sparre Andersen (Århus) (in English)
Let $X_1, X_2, \ldots$ be independent random variables, all with distribution function $F(x)$, and let $a_n = \mathrm{Prob}\,(X_1 + \cdots + X_n > 0)$. Is it possible to choose $F(x)$ in such a way that the sequence $a_1, a_2, \ldots$ is divergent and Cesàro summable of order 1? It is known that there exists $F(x)$ such that $a_1, a_2, \ldots$ is not Cesàro summable.

569. February 2, 1962;   Yuriĭ Mikhaĭlovich Smirnov (Moscow) (in Russian)
Is every metric space R countable-dimensional (i.e., $R = \bigcup_{i=1}^{\infty} N_i$, $\dim N_i = 0$) if it has transfinite dimension R?

CM 10, P412

570. February 2, 1962;   Yuriĭ Mikhaĭlovich Smirnov (Moscow) (in Russian)
Is any weakly infinite-dimensional space (compact) with a countable base countable-dimensional?
A space R is weakly infinite-dimensional if for every sequence of pairs of closed subsets $A_n, B_n, A_n \cap B_n = \emptyset$, there are closed sets C_n separating R between A_n and B_n with $\bigcap_{n=1}^{\infty} C_n = \emptyset$.

CM 10, P413

582. May 26, 1962; Pavel Sergeevich Aleksandrov (Moscow) (in Russian)
Let n be a natural number and τ an uncountable cardinal number.
Does there exist a compactum $A^{n\tau}$ of dimension dim $= n$ and weight τ which is universal for all at most n-dimensional compacta of weight $\leq \tau$ (i.e., containing topologically every such compactum)?
CM 10, P422

583. May 26, 1962; Pavel Sergeevich Aleksandrov (Moscow) (in Russian)
What are compacta which are images of an n-dimensional cube under 0-dimensional open continuous mappings?
In particular, can a p-dimensional cube, $p \geq 3$, be mapped by such mapping onto a q-dimensional cube for $q > p$?
CM 10, P423

586. May 26, 1962; Pavel Sergeevich Aleksandrov (Moscow) (in Russian)
Is it true for each compact space X that $X = X$?

587. May 26, 1962; Pavel Sergeevich Aleksandrov (Moscow) (in Russian)
I proved the inequality $\dim X \leq X$ for compact spaces (1940), Yu. M. Smirnov proved it for strongly paracompact spaces.
Does there exist a paracompact space for which $X < \dim X$?

592. May 26, 1962; Aleksander Pełczyński (Warsaw) (in Polish)
Can every infinite extremally disconnected compact Hausdorff space Q be continuously mapped onto the generalized Cantor set D^τ, where τ is the weight of Q?
CM 10, P425

601. August 30, 1962; R. H. Bing (Madison, WI)
Suppose P is a pseudoarc, U is an open subset of P and h is a homeomorphism of P onto itself that leaves each point of U fixed. Must h be the identity homeomorphism?

CM 10, P431

603. August 30, 1962; R. H. Bing (Madison, WI)
Is a 2-sphere S tame if it is homogeneous under a space homeomorphism? That is—for each pair of points p, q of S there is a homeomorphism $h : E^3 \to E^3$ such that $h(S) = S, h(p) = q$.
(This problem was proposed by a student named Becker in one of my seminars).

604. August 30, 1962; R. H. Bing (Madison, Wisc.)
Is a 2-sphere S in E^3 tame if it can be approximated from either side by a singular 2-sphere? That is—for each component U of $E^3 \setminus S$ and each $\varepsilon > 0$ there is a map of S into U such that f moves no point more than ε.
CM 10, P429

608. 1962; Arieh Dvoretzky (Jerusalem) (in English)
Let C be a convex symmetric body in $\mathbb{R}^4$. If all orthogonal three-dimensional projections of C can be mapped on one another in C, is C an ellipsoid?

The answer is yes if $C \subset \mathbb{R}^3$ and two-dimensional projections are considered. H. Auerbach, S. Mazur, S. Ulam, *Sur une propriété caractéristique de l'ellipsoïde.* Monatshefte Math. **42** (1935), 45–48.

647. May 24, 1963;   Richard D. Anderson (Baton Rouge, LA)
Does there exist any homeomorphism α of the Cantor set C onto itself such that for any homeomorphism β of C onto itself there exists a mapping η of C onto C such that $\eta\alpha = \beta\eta$?

CM 10, P460

Answer: No. E. Nunnally, There is no universal-projecting homeomorphism of the Cantor set. Colloq. Math. 17 (1967), 51–52.

CM 17, P460, R2

648. May 24, 1963;   Richard D. Anderson (Baton Rouge, LA)
Let X be a compact metric continuum which is locally the product of the Cantor set and the open interval. Suppose one arc component (or, alternatively, all arc components) is dense in X.
Does X admit a continuous flow (one-parameter group of homeomorphisms) fully transitive on each arc component?

661. June 21, 1963;   Jun-iti Nagata (Osaka)
Is every n-dimensional metric space topologically imbedded in a topological (Cartesian) product of n 1-dimensional metric spaces? (Covering dimension).
Every n-dimensional metric space can be imbedded in a product of $(n + 1)$ 1-dimensional metric spaces. The prediction for the problem of Prof. Borsuk is negative.

CM 12, P463

674. November 21, 1963;   Jan Jaworowski (Warsaw) (in Polish)
A space X is called strongly contractible to a point $x \in X$ if there is a homotopy contracting X to x which keeps x fixed. A compact metric space, strongly contractible to each of its points, of finite dimension is an absolute retract, since it is locally contractible.
Do there exist compact, strongly contractible to each of its points spaces which are not absolute retracts?

CM 12, P476

677. January 11, 1964;   Sibe Mardešić (Zagreb)
Is there a simple closed curve C in the plane such that every straight line intersects C exactly in $0, 1$ or $\aleph_0$ points?

679. January 11, 1964;   Sibe Mardešić (Zagreb)
Let X be a metric disc-like continuum, i.e., such that it can be ε-mapped onto a 2-disc $I \times I$ for each $\varepsilon > 0$.
Does X possess the fixed point property?

683. January 11, 1964;   Sibe Mardešić (Zagreb)
Is dim $X = X = X$ for all locally connected compact spaces?

684. January 11, 1964;   Sibe Mardešić and P. Papić (Zagreb)
Let X be a locally connected continuum which is at the same time the image of a totally ordered compact space K under some continuous mapping.
Is X also obtainable as the continuous image of some totally ordered continuum C?

728. May 1, 1965;   Nachman Aronszajn (Lawrence, KS) (in Polish)
Let $f(t) = \sum_{k=1}^{N} a_k r_k(t)$ be a linear combination of Rademacher functions r_k, $E \subset (0,1)$ a set of measure $|E| \in (0,1]$. Consider the inequality

$$c \sum_{k=1}^{N} |a_k|^2 \le \int_E |f(t)|^2 \, dt$$

with a constant c which depends on $|E|$ only. Such an inequality is false for $|E| \le \frac{1}{2}$, with $f = r_1 - r_2$. This is true for $|E| \in \left(\frac{2}{3}, 1\right]$, and I do not know what happens if $|E| \in \left(\frac{1}{2}, \frac{2}{3}\right]$.

774. February 12, 1967;   Bronisław Knaster (Wrocław) (in Polish)
Do there exist n-dimensional continua, $1 < n < \infty$, which contain no nontrivial (i.e., proper, non-singleton) continuous self-images?

777. February 21, 1967;   Roman Duda (Wrocław) (in Polish)
Is it true that if a polyhedron can be decomposed as the Cartesian product of two spaces which are different from it, then it can also be decomposed as the product of two different from it polyhedra?

CM 19, P634

822. October 28, 1968;   Władysław Narkiewicz (Wrocław) (in Polish)
Show that if $f(n) \ge 0$ is an integer valued multiplicative function such that for all j and $N \ge 2$

$$\lim_{x \to \infty} \frac{1}{x} |\{n \le x : f(n) \equiv j \,(\mathrm{mod}\, N)\}| = \frac{1}{N},$$

then $f(n) = n$.

H. Delange, *Sur les fonctions multiplicatives à valeurs entières.* C. R. Acad. Sci., Paris, Sér. A **284** (1977), 1325–1327.

823. March 3, 1969;   Wojbor A. Woyczyński (Wrocław, Cleveland) (in Polish)
Let $\langle \Omega, \mathscr{F}, m \rangle$ be a measure space, B – a Banach space, and $\mathscr{B}$ – the algebra of Borel sets on $[0,1]$, and

$$M : \Omega \times \mathscr{B} \ni (\omega, A) \mapsto M_\omega(A) \in B$$

such that for each fixed $A \in \mathscr{B}$, M is $\mathscr{F}$-measurable, and for each fixed $\omega \in \Omega$, M is a B-valued measure on $\mathscr{B}$. It is known that for each fixed $\omega \in \Omega$ there exists a finite positive measure m_ω on $\mathscr{B}$ such that $m_\omega(A) \leq \|M_\omega\|(A)$, $A \in \mathscr{B}$, and $\lim_{m_\omega(A) \to 0} \|M_\omega\|(A) = 0$, where $\|M\|(A) = \sup|\sum_{i=1}^n a_i M(A_i)|$ with the upper bound taken over all $|a_i| \leq 1$ and finite partitions of A into A_i, cf. N. Dunford, J. Schwartz, Linear Operators I, (IV.10.5). Can we choose m_ω in a measurable way, i.e. so that for each $A \in \mathscr{B}$ the map $\Omega \ni \omega \mapsto m_\omega(A) \in \mathbb{R}^+$ is $\mathscr{F}$-measurable?

Answer. Zbigniew Lipecki (Wrocław), January 14, 1974.
Yes, B^* has a denumerable set of functionals separating points in B (with a sketch of proof).

827. March 11, 1969;   Roman Duda (Wrocław) (in Polish)
Does every metric separable connected space have dimension ≤ 1 if it is hereditarily locally connected (i.e., any connected subset is locally connected)?
If yes, can such space be compactified with a hereditarily locally connected continuum?

CM 21, P682

832. June 14, 1969;   Zbigniew Zieleźny (Wrocław, Buffalo) (in Polish)
Let $K_{x,y}$ be a (Schwartz) distribution on $\mathbb{R}^n \times \mathbb{R}^n$ which is C^∞ with respect to x, and for each x_0 $K_{x_0,y}$ is in $\mathscr{E}'_y$ (a compactly supported distribution). Define the convolution (symbolically) $Tu = \int K_{x,x-y} u_y \, dy$, where $u \in \mathscr{D}'(\mathbb{R}^n)$. Question: When $Tu \in \mathscr{E}$ implies $u \in \mathscr{E}$, for every $u \in \mathscr{D}'$, i.e. when the convolution operator T is hypoelliptic?

850. November 3, 1970;   Jean-François Méla (Paris) (in French)
Does there exist a set $E = (n_k)$ of positive integers, $nk < n_{k+1}$, such that both the properties below a), b) are satisfied?

a) There is a constant $C > 0$ such that for each $p \geq 1$ there exists a bounded measure μ on the torus whose Fourier transform satisfies $\hat{\mu}(n_k) = 1$ for $1 \leq k \leq p$, $\hat{\mu}(n_k) = 0$ for $k > p$, with $\|\mu\| \leq C$;
b) There exists a function bounded on E which cannot be continued onto $\mathbb{Z}$ as the Fourier transform of a measure on the torus.

854. November 6, 1970;   Dame Mary Lucy Cartwright (Cambridge) (in English)
By a theorem of Nemytskii and Stepanov the necessary and sufficient condition for a set to be a minimal set of a uniformly almost periodic flow is that it is a connected separable Abelian group. In 1940 Kodaira and Abe showed that if such a group is of $(n-1)$-dimensions and embedded in $\mathbb{R}^n$, then it is an $(n-1)$-dimensional torus $\mathbb{T}^{n-1}$. Does this remain true without the hypothesis of almost periodicity? By Denjoy's irregular case on the torus it seems necessary to assume that the minimal set contains an $(n-1)$-dimensional ball. In particular, does it remain true for solutions of autonomous systems of differential equations in n dimensions if the minimal set of a solution contains an $(n-1)$-dimensional cylinder formed by the product of a small $(n-2)$-dimensional ball in the plane normal to the solution through the center of the ball and an interval of length 1 of all solutions through

points of the ball? This is true for $n = 2$ when the ball reduces to a point. Is it true for $n > 2$, if it is true for $\mathbb{R}^2$ is it true for other n-dimensional manifolds?

867. December 8, 1971; Kazimierz Urbanik (Wrocław) (in Polish)
Let $\mathscr{D}$ denote the space (*càdlàg*) of right continuous functions on $[0, 1]$ with limits on the left, with the Skorokhod topology. Moreover, let μ be a Borel measure induced on $\mathscr{D}$ by a homogeneous process with independent increments. Show that μ does not have structure of Hilbertian measure, i.e. there is no linear set $\mathscr{H} \subset \mathscr{D}$ which is μ-measurable and has the structure of a Hilbert space with a μ-measurable inner product with the topology stronger than the Skorochod topology, and $\mu(\mathscr{H}) = 1$. For the Wiener process the answer is yes (Stanisław Kwapień (Warszawa) and Małgorzata Guerquin (Wrocław)).

868. March 01, 1972; Bronisław Knaster (Wrocław) (in Polish)
Is every closed subset of an absolute retract X the fixed-point set of a continuous map $f : X \to X$?

CM 27, P846

871. April 25, 1972; Karol Borsuk (Warsaw) (in Polish)
Can any locally connected continuum in the Euclidean 3-space E^3, disconnecting E^3, be continuously mapped into itself without a fixed point?

CM 27, P847

872. April 25, 1972; Karol Borsuk (Warsaw) (in Polish)
Does every dendroid have a trivial shape?
Does any finite union of dendroids have the planar shape (i.e., the shape of some plane compactum)?

CM 27, P848

873. April 28, 1972; Wojbor A. Woyczyński (Wrocław, Cleveland) (in English)
Let X be a Banach space. The gradient $g : S_X \to S_X$ of the norm $\| \, . \, \|$ in X is defined by the equality $g(x)\xi = \lim_{t \to 0} \frac{1}{t} (\|x + t\xi\| - \|x\|)$. Does the gradient in L^p-spaces, $1 < p < 2$, satisfy the Hölder condition with exponent $p - 1$?

Answer: Yes; W. Woyczynski, *Random series and laws of large numbers in some Banach spaces*. Teor. Veroyatn. Primen. **18** (1973), 361–367. reprinted: Theory Probab. Appl. **18** (1973), 350–355.

893. May 30, 1974; Richard M. Schori (Baton Rouge, LA)
If X is a compact metric space, by 2^X we mean the collection of all nonempty closed subsets of X metrized with the Hausdorff metric. If $\varepsilon > 0$, the ε-hyperspace of X, $2^X_\varepsilon = \{A \in 2^X : \text{diameter} A \leq \varepsilon\}$.
If X is a compact ANR, does there exist a metric on X (perhaps any convex metric will suffice) such that for sufficiently small $\varepsilon > 0$, 2^X_ε is a) locally contractible, b) an ANR, c) of the same homotopy type as X, d) a Hilbert cube manifold, e) a Hilbert cube manifold homeomorphic to $X \times Q$, where Q is the Hilbert cube?

Furthermore, if $X \in AR$, is 2_ε^X homeomorphic to Q for each $\varepsilon > 0$?
If e) and this last question are both true, then $X \in AR$ implies that $X \times Q \cong Q$. If this
is the case, then, by Chapman, each compact ANR has finite homotopy type.

CM 35, P981

898. December 13, 1974;   Sam B. Nadler, Jr.
Let H_n denote the Hausdorff metric for the space $\mathscr{C}_n$ of all nonempty subcontinua of Euclidean space $\mathbb{R}^n$. Let $\mathscr{A}_n = \{A \subset \mathbb{R}^n : A \text{ is an arc}\}$, $\Sigma_n = \{A \subset \mathbb{R}^n : A \text{ is a simple closed curve}\}$, $\mathscr{T}_n = \{A \subset \mathbb{R}^n : A \text{ is a simple triod}\}$, $\mathscr{P}_n = \{A \subset \mathbb{R}^n : A \text{ is a pseudoarc}\}$.
Question 1: Is $\mathscr{A}_3$ homogeneous?
By the Schönflies theorem, $\mathscr{A}_2$ is homogeneous; note: homeomorphisms $h : \mathbb{R}^2 \to \mathbb{R}^2$ lift up to $\mathscr{A}_2$. But in $\mathbb{R}^3$ there are wild arcs. Are these arcs really wild in $\mathscr{A}_3$?

Prize: 1 bottle of wine.
Question 2: Same as Question 1 for the spaces Σ_n, $\mathscr{T}_n$ and $\mathscr{A}_{n \geq 3}$.

Prize: 1 bottle of wine.
Bing has shown $\mathscr{P}_n$ is a residual G_δ in the space $\mathscr{C}_n$.
Question 3: Is $\mathscr{P}_n$ homogeneous and, if so, is it a topological group?
If $\mathscr{P}_n$ is homogeneous, then I would conjecture that $\mathscr{P}_n$ is homeomorphic to ℓ^2!
This would be interesting because then every functional analysist would know what a pseudoarc is — namely, a point in the Hilbert space!

Prize: 1 bottle of wine.

909–910. January 2, 1975;   Bronisław Knaster (Wrocław) (in Polish)
Let *GFPP* denote the existence of a point $x \in F(x)$ for any upper semi-continuous function defined on points x of a continuum X whose values are subcontinua $F(x) \subset X$. R. Mańka proved (Fund. Math. 1976) property *GFPP* for all hereditarily decomposable and hereditarily unicoherent X (so called λ-dendroids).
Question 1: For what continua X does property *GFPP* of X induce property *GFPP* of the hyperspace $C(X)$ of all subcontinua of X?
Question 2: For what X and F does property *GFPP* of X induce property *GFPP* of $F(X)$?

911. January 2, 1975;   Bronisław Knaster (Wrocław) (in Polish)
Is (and when) a deformation retract of a contractible continuum (in particular, of a contractible fan) contractible?

912. January 2, 1975;   Bronisław Knaster (Wrocław) (in Polish)
For what continua X is the uniform limit of a sequence of confluent maps $X \to Y$ (in the sense of J. J. Charatonik, Fund. Math. 56 (1964)) a confluent map?

Prize: For solving any of Problems 908–913: an opportunity to tell two jokes of moderate length.

915. April 8, 1975;   Nigel J. Kalton (Swansea) (in English)
Let (E, τ) be an F-space with a separating dual. Prove or disprove that E contains an infinite dimensional weakly closed subspace G and a closed infinite dimensional subspace H such that $G \cap H = \{0\}$ and $G + H$ is closed.

918. June 8, 1975;   Aline Bonami (Orsay) (in French)
Consider the set Λ of all the integers of the form $2^k 3^\ell$, $k, \ell \in \mathbb{N}$. Using the Paley-Littlewood theory one can prove that there exist constants A_p, $p > 2$, such that for each trigonometric polynomial P with its spectrum in Λ, $P(x) = \sum_{\lambda \in \Lambda} a_\lambda \exp(i\lambda x)$, one has $\|P\|_p \leq A_p \|P\|_2$. Does there exist a direct proof?
Moreover, can one prove more precisely that $\sum_{\lambda \in \Lambda} |a_\lambda|^2 < \infty$ implies $\exp(\mu|\sum a_\lambda \exp(i\lambda x)|) \in L^1$ for each $\mu > 0$?

921. July 30, 1975;   F. Burton Jones (Riverside, CA)
After 40 years, does there exist a normal Moore space (i.e., a space satisfying R. L. Moore's axiom 0 and parts 1,2,3 of axiom 1) which is not metric?
One would hope for a "honest" example of such a space without some weird set theory assumption)

922. October 10, 1975;   Fleming Topsøe (Copenhagen) (in English)
Is it possible to find in $\mathbb{Q}$ – the space of rationals, a sequence $(P_n)_{n \geq 1}$ of probability measures such that P_n converges in topology of weak convergence to some measure P, and such that the only compact subset $K \subset \mathbb{Q}$ for which $P_n K \to P K$ holds is the empty set? With $\mathbb{Q}$ replaced by a Polish space such an example does not exist.

Prize: A week in Copenhagen for an example.

920. July 30, 1975;   F. Burton Jones (Riverside, CA)
After 50 years, Professor Knaster's question: "Is the circle the only homogeneous plane continuum?" has been reduced to:
Is every hereditarily indecomposable homogeneous plane continuum a pseudo-arc? i.e., is it chainable or circularly chainable?

923. December 11, 1975;   David P. Bellamy (Warsaw)
Can every finite-dimensional metric hereditarily indecomposable continuum be embedded into a finite product of pseudo-arcs?
As a trivial application of my results in: *Mapping hereditarily indecomposable continua onto a pseudo-arc*, Lecture Notes in Math. 375 (1974), 6–14, it can be established that (1) every Hausdorff hereditarily indecomposable continuum can be embedded into some product of pseudo-arcs; (2) every metric hereditarily indecomposable continuum can be embedded into a countable product of pseudo-arcs.

CM 37, P1006

Remark: The question was repeated by the author in [Proceedings of the International Conference on Geometric Topology, PWN Warszawa, 1980, p. 459] and [Open Problems in Topology II, Edited by E. Pearl, Elsevier, Amsterdam,

2007, p. 259]. R. Pol proved in [Topology Proc. 16 (1991), 133–135] that each n-dimensional hereditarily indecomposable metric continuum can be embedded into the product of n one-dimensional hereditarily indecomposable metric continua.

924. December 11, 1975;   David P. Bellamy (Warsaw)
If M is an open n-manifold for $n \geq 2$ and $\beta M \setminus M$ is connected, is $\beta M \setminus M$ an aposyndetic continuum?
This is true if $M = \mathbb{R}^n$.

CM 37, P1007

925. December 11, 1975;   David P. Bellamy (Warsaw)
What are necessary and sufficient conditions to ensure that a metric continuum X cannot be mapped onto the cone over itself?

926. December 11, 1975;   David P. Bellamy (Warsaw)
If X is an indecomposable Hausdorff continuum with infinitely many composants, is the cardinality of the set of composants of X always equal to $2^{\mathfrak{m}}$ for some infinite cardinal number $\mathfrak{m}$?

929. March 23, 1977;   Tadeusz Maćkowiak (Wrocław)
Let a continuum X be such that $\dim X \geq 2$. Does it follow that X contains an hereditarily indecomposable continuum?

Remark: By a well-known Mazurkiewicz theorem, any metric continuum of dimension at least 2 contains a nondegenerate indecomposable subcontinuum.

932. August 3, 1977;   Ryszard Frankiewicz (Wrocław)
Can the spaces $\beta D(\omega_1) \setminus D(\omega_1)$ and $\beta \mathbb{N} \setminus \mathbb{N}$ be homeomorphic?
$D(\omega_1) = $ the discrete space of cardinality ω_1.

941. April 05, 1978;   James E. West (Ithaca, NY)
Let $E \xrightarrow{p} B$ be a locally trivial fiber bundle with fiber F and suppose E, B, and F to be compact metric.
Under what conditions (if any) is $2^E \xrightarrow{2^p} 2^B$ a locally trivial fiber bundle with fiber 2^F? In particular, if E, B, F are ANR's? What about Hurewicz fibrations? Serre fibrations?

CM 43, P1187

942. April 05, 1978;   James E. West (Ithaca, NY)
Let $E \xrightarrow{p} B$ be a Hurewicz fibration with each fiber homeomorphic to a given Hilbert cube manifold F.
If E and B are also Hilbert cube manifolds, under what conditions is p a locally trivial bundle?

CM 43, P1188

943. April 05, 1978;   James E. West (Ithaca, NY)

Let $E \xrightarrow{p} B$ be a Hurewicz fibration with $B =$ the Hilbert cube, E a compact ANR, and each fiber of p a nondegenerate absolute retract.

Must there exist two cross-sections, σ and τ of p, with disjoint images? (Problem of West and Toruńczyk).

If not, give a characterization of those cross-sections σ for which there exists a cross-section τ with image disjoint from that of σ.

CM 43, P1189

945. September 15, 1978;   Charles L. Hagopian (Sacramento, CA)

Does every arcwise connected disc-like continuum have the fixed-point property?

CM 44, P1201

946. October 21, 1978;   Paul Erdös and Endre Szemerédi (Budapest) (in English)

Let $1 \leq a_1 < \cdots < a_n$ be n integers. Let $f(n)$ be the smallest integer so that the set $a_i + a_j$, $a_i a_j$, $1 \leq i \leq j \leq n$, contains at least $f(n)$ distinct integers. Szemerédi and I proved ($\ell > 0$ is an absolute constant)

$$(*)\quad n^{1+\ell} < f(n) < n^2 \exp(\ell \log n / \log \log n).$$

The upper bound may give the right order of magnitude; $f(n) > n^{2-\varepsilon}$ seems certain.

Denote by $F(n)$ the smallest integer so that there are at least $F(n)$ distinct integers of the form $\sum \varepsilon_i a_i$, $\prod a_i^{\varepsilon_i}$, $\varepsilon_i = 0$ or 1. We conjecture $F(n) > n^k$ for every k if $n > n_0(k)$. We proved $F(n) < n^{\ell \log n / \log \log n}$ which perhaps gives the right order of magnitude for $F(n)$.

953. May 15, 1979;   Hubert Delange (Orsay) (in English)

Let $\mathscr{H}(\mathbb{C})$ be the vector space of entire functions of one complex variable, with the topology of convergence on compact sets. In $\mathscr{H}(\mathbb{C})$ consider the set S consisting of those entire functions F which satisfy $|f(x+iy)| \leq \exp(|y|)$ for every real x and y (which is equivalent to $|f^{(n)}(x)| \leq 1$ for every $n \geq 0$ and every real x). S is a compact convex set. Find its extremal points.

The functions $z \mapsto C \exp(iaz)$, where $|C| = 1$ and $a \in [-1, 1]$, are extremal points. If there were no other extremal points, every function of S could be put into the form $f(z) = \int_{-1}^{1} \exp(itz)\, d\mu(t)$, μ – a complex measure on $[-1, 1]$. But this is not the case. Example: $f(z) = \int_0^z \frac{\sin u}{u}\, du = \text{p.v.} \int_{-1}^{1} \exp(itz) \frac{dt}{t}$.

961. October 27, 1979;   Fulvio Ricci (Torino, Pisa) (in English)

Given two pseudomeasures T and U on the torus $\mathbb{T}$ such that $\hat{T}(0) = \hat{U}(0) = 0$, consider the distributional derivative $(P_T P_U)'$ of the pointwise product $P_T P_U$, where P_T and P_U are primitives of T and U, respectively. In general, $(P_T P_U)'$ is not a pseudomeasure. This product has been introduced by J. Benedetto.

Are there closed subsets E of the torus $\mathbb{T}$ which are not sets of strong spectral resolution, and such that if T and U above have support in E, then $(P_T P_U)'$ is a pseudomeasure?

Let $E = \{0\} \cup \{\frac{1}{2^m} + \frac{1}{2^n} : m, n \geq 0\}$. What is the answer to the first question in this case?

Remark. If $U \in \mathcal{M}(E)$, then $(P_T P_U)' \in \mathcal{A}'(E)$, see F. Ricci, *A multiplicative structure on some spaces of pseudomeasures on the circle and related properties*, Bull. Sci. Math., II. Ser. 103 (1979), 423–434.